Pathophysiology and Imaging Diagnosis of Demyelinating Disorders

Special Issue Editor
Evanthia Bernitsas

MDPI • Basel • Beijing • Wuhan • Barcelona • Belgrade

MDPI

Special Issue Editor
Evanthia Bernitsas
Wayne State University School of Medicine
USA

Editorial Office
MDPI
St. Alban-Anlage 66
Basel, Switzerland

This edition is a reprint of the Special Issue published online in the open access journal *Brain Sciences* (ISSN 2076-3425) in 2017 (available at: http://www.mdpi.com/journal/brainsci/special_issues/ multiple_sclerosis_demyelinating).

For citation purposes, cite each article independently as indicated on the article page online and as indicated below:

Lastname, F.M.; Lastname, F.M. Article title. *Journal Name* **Year**, *Article number*, page range.

First Edition 2018

ISBN 978-3-03842-943-2 (Pbk)
ISBN 978-3-03842-944-9 (PDF)

Table of Contents

About the Special Issue Editor

Evanthia Bernitsas, Dr., is Director of the Multiple Sclerosis Center and an Associate Professor at Wayne State University School of Medicine, Detroit, Michigan. She is affiliated with multiple hospitals in the area, including DMC Harper University Hospital and Karmanos Cancer Center. She received her medical degree from Aristotle University of Thessaloniki School of Medicine and has been in practice for more than 20 years. She is involved as Principal Investigator in clinical trials and is the author of numerous articles on MS.

Preface to "Pathophysiology and Imaging Diagnosis of Demyelinating Disorders"

The spectrum of "demyelinating disorders" is broad and it includes various disorders with central nervous system (CNS) demyelination, such as multiple sclerosis (MS), Neuromyelitis optica spectrum disorders (NMOSD), transverse myelitis, optic neuritis, acute disseminated encephalomyelitis, overlap and unclassified disorders, with MS being the most common. MS is a complex, multifaceted autoimmune disorder and the most common cause of non-traumatic disability in young adults [1,2]. Considerable research over the recent years has improved our knowledge and led to earlier diagnosis, novel therapeutic strategies, and an overall longer time in the workforce and improved quality of life for MS patients. However, diagnosis and management remain challenging. The disease burden on patients and caregivers is immense. Up until now, there are no FDA-approved remyelinating therapies. MS is still an incurable disease and many questions regarding pathogenesis, diagnosis and treatment remain unanswered.

In this special issue, Zabad et al. review extensively the wide spectrum of demyelinating syndrome, classification, rare and atypical presentations, differential diagnosis and evolution from the first demyelinating episode to the full- blown disease. Serum biomarkers, key imaging findings and management strategies are discussed. Specifically, the presence and significance of aquaporin 4 (AQP-4) and myelin oligodendrocyte glycoprotein (MOG) antibody, myelin basic protein (MBP), glial fibrillary acidic protein (GFAP), S100, MOG, specific cytokines, such as interleukin 6 (IL-6) in the diagnostic evaluation and management is highlighted. This review emphasizes practical points in the "real world" practice that are of valuable assistance to the clinician [3].

Misdiagnosis of MS may occur, especially early in the disease process, as there is a significant number of diseases with similar presentation [4,5]. Over the last decade, the diagnostic accuracy of demyelinating disorders has improved, as advanced diagnostics, especially magnetic resonance imaging (MRI) techniques appear. The development of biomarkers is a necessity, as they have diagnostic, prognostic, and therapeutic value [6]. Using a calibrated functional MRI, Hubbard et al. investigate a new imaging biomarker, the visual-evoked cerebral metabolic rate of oxygen (veCMRO$_2$), its contribution in improving diagnostic accuracy and the possibility of being used as a prognostic biomarker in the future in the context of a "gold standard" model of MS diagnostics that combine many relevant factors [7].

The symptoms of MS are non- specific, not always obvious and a number of them cannot be measured objectively. As fatigue is one of the most common, multifactorial, disabling and difficult to treat symptoms, with a severity that can only be evaluated by self-reporting scales, more insight into its pathophysiology and imaging characteristics is needed [8]. The article by Bernitsas et al. sheds light on the pathophysiology of MS-related fatigue and specifically focuses on its volumetric and neural integrity measures in patients with different degrees of pure MS- fatigue and low disability, using advanced MRI technology [9].

Comorbidities in MS patients have been extensively studied, as they have a negative impact on the quality of life, management and overall prognosis on MS patients. Comorbidities may delay initiation of disease-modifying treatment, limit therapeutic options and complicate treatment decisions. There is growing evidence that comorbidities may increase relapse rate and disability progression [10–14]. Painful paresthesias are part of the MS symptomatology; however painful sensations can be seen in other conditions co-existing with MS and may lead to diagnostic confusion. The review article by Purvis et al. focuses on the concurrent presence of cervical spondylotic myelopathy in MS patients that is commonly seen in everyday clinical practice and evaluates the results of decompressive surgery on pain management and quality of life in this population. The need for a comprehensive approach and multidisciplinary collaboration is emphasized [15].

Pathophysiology of demyelinating disorders is complex and not very well understood. The contribution of B-lymphocytes has been increasingly acknowledged, in addition to the traditional view regarding the role of T-lymphocytes in demyelinating pathophysiology. There are various B and T subsets, as well as different cell populations that are key players in the immune response and their involvement has been further investigated.

In this special issue, three review articles discuss MS pathogenesis and address old and new knowledge. In a very comprehensive review by Dargahi et al., the pathophysiology of MS is explained and the role of specific cells, including T and B-lymphocytes and their subsets, macrophages, microglia, natural killer and dendritic cells, in the pathogenesis of demyelination is further analyzed [16]. Kinzel et al. review the role of humoral immunity in demyelinating disorders and further explore the role of peripheral CNS-specific antibodies in initiating a cascade of events that lead to CNS demyelination [17]. As MS encompasses both an inflammatory and a neurodegenerative component, with neurodegeneration being more prominent later in the disease course and especially during the progressive stage and associated with disability, Salapa et al. discuss the role of neuronal and axonal damage in MS, emphasize the multifactorial nature of neurodegeneration and summarize potential mechanisms that contribute to neuro-axonal injury. [18].

A new, deep insight into MS pathogenesis may promote novel neuroprotective and remyelinating therapeutic strategies. The review by Bose focuses on a very specific population of cells in MS pathophysiology, the T, B and resident memory cells, their role in MS pathophysiology, the effect of the disease modifying agents on this cell population and their potential of being a therapeutic target [19]. Lisak and Benjamins review melanocortins and their receptors (MCR), and analyze the direct effect of melanocortins on the CNS (neurons and glia) as well as their effect on the immune cells in the periphery. The role of adrenocorticotropic hormone (ACTH) in treating MS relapses is discussed and comparative efficacy results between ACTH and intravenous steroids from clinical trials are presented. In this review article, future research targets are explored and the potential for developing innovative neuroprotective therapies involving MCR agonists is highlighted [20]. As there is growing interest in cell-based therapeutic strategies for MS [21], more research is needed. Emerging immunotherapeutic approaches, such as stem cells, nanoparticles, mannan, DNA vaccines, altered peptide ligands and cyclic peptides, are presented by Dargahi et al. [16], after reviewing current and approved disease-modifying agents.

Conflicts of Interest: The author declares no conflict of interest.

References

1. Noseworthy, J.H.; Lucchinetti, C.; Rodriguez, M.; Weinshenker, B.G. Multiple Sclerosis. *NEJM* **2000**, *343*, 938–952. [CrossRef] [PubMed]
2. Confavreux, C.; Vukusic, S. Natural history of MS: A unifying concept. *Brain* **2006**, *129*, 606–616. [CrossRef] [PubMed]
3. Zabad, R.K.; Stewart, R.; Healey, K.M. Pattern Recognition of the Multiple Sclerosis Syndrome. *Brain Sci.* **2017**, *7*, 138. [CrossRef] [PubMed]
4. Brownlee, W.J.; Hardy, T.A.; Fazekas, F.; Miller, D.H. Diagnosis of Multiple Sclerosis: Progress and challenges. *Lancet* **2017**, *389*, 1336–1346. [CrossRef]
5. Rudick, R.A.; Miller, A.E. Multiple Sclerosis or multiple possibilities? *Neurology* **2012**, *78*. [CrossRef] [PubMed]
6. Matute-Blanch, C.; Montalban, X.; Comabella, M. Multiple Sclerosis and other demyelinating and autoimmune inflammatory diseases of the central nervous system. *Handb. Clin. Neurol.* **2017**, *146*, 67–84. [PubMed]
7. Hubbard, N.A.; Sanchez Araujo, Y.; Caballero, C.; Ouyang, M.; Turner, M.P.; Himes, L.; Faghihahmadabadi, S.; Thomas, B.P.; Hart, J., Jr.; Huang, H.; et al. Evaluation of Visual-Evoked Cerebral Metabolic Rate of Oxygen as a Diagnostic Marker in Multiple Sclerosis. *Brain Sci.* **2017**, *7*, 64. [CrossRef] [PubMed]
8. Bakshi, R. Fatigue associated with MS: Diagnosis, impact and management. *Mult. Scler.* **2003**, *9*, 219–227. [CrossRef] [PubMed]

9. Bernitsas, E.; Yarraguntla, K.; Bao, F.; Sood, R.; Santiago-Martinez, C.; Govindan, R.; Khan, O.; Seraji-Bozorgzad, N. Structural and Neuronal Integrity Measures of Fatigue Severity in Multiple Sclerosis. *Brain Sci.* **2017**, *7*, 102. [CrossRef] [PubMed]

10. Marrie, R.A. Comorbility in multiple sclerosis: Implications for patient care. *Nat. Rev. Neurol.* **2017**, *13*, 375–382. [CrossRef] [PubMed]

11. McDonnell, G.; Cohen, J.A. Comorbidities in MS are associated with treatment intolerance and disability. *Neurology* **2017**, *89*. [CrossRef] [PubMed]

12. Thormann, A.; Sorensen, P.S.; Koch-Henriksen, N.; Laursen, B.; Magyari, M. Comorbidity in multiple sclerosis is associated with diagnostic delays and increased mortality. *Neurology* **2017**, *89*, 1668–1675. [CrossRef] [PubMed]

13. Kowalec, K.; McKay, K.A.; Patten, S.B.; Fisk, J.D.; Evans, C.; Tremlett, H.; Marrie, R.A. Comorbidity increases the risk of relapse in multiple sclerosis: A prospective study. *Neurology* **2017**, *89*, 2455–2461. [CrossRef] [PubMed]

14. Zhang, T.; Tremlett, H.; Zhu, F.; Kingwell, E.; Fish, J.D.; Bhan, V.; Campbell, T.; Stadnyk, K.; Carruthers, R.; Wolfson, C.; et al. Effects of physical comorbidities on disability progression in multiple sclerosis. *Neurology* **2018**, *90*, e419–e427. [CrossRef] [PubMed]

15. Purvis, T.E.; Lubelski, D.; Mroz, T.E. Is Decompressive Surgery for Cervical Spondylotic Myelopathy Effective in Patients Suffering from Concomitant Multiple Sclerosis or Parkinson's Disease? *Brain Sci.* **2017**, *7*, 39. [CrossRef] [PubMed]

16. Dargahi, N.; Katsara, M.; Tselios, T.; Androutsou, M.E.; de Courten, M.; Matsoukas, J.; Apostolopoulos, V. Multiple Sclerosis: Immunopathology and Treatment Update. *Brain Sci.* **2017**, *7*, 78. [CrossRef] [PubMed]

17. Kinzel, S.; Weber, M.S. The Role of Peripheral CNS-Directed Antibodies in Promoting Inflammatory CNS Demyelination. *Brain Sci.* **2017**, *7*, 70. [CrossRef] [PubMed]

18. Salapa, H.E.; Lee, S.; Shin, Y.; Levin, M.C. Contribution of the Degeneration of the Neuro-Axonal Unit to the Pathogenesis of Multiple Sclerosis. *Brain Sci.* **2017**, *7*, 69. [CrossRef] [PubMed]

19. Bose, T. Role of Immunological Memory Cells as a Therapeutic Target in Multiple Sclerosis. *Brain Sci.* **2017**, *7*, 148. [CrossRef] [PubMed]

20. Lisak, R.P.; Benjamins, J.A. Melanocortins, Melanocortin Receptors and Multiple Sclerosis. *Brain Sci.* **2017**, *7*, 104. [CrossRef]

21. Scolding, N.J.; Pasquini, M.; Reingold, S.C.; Cohen, J.A. Cell-based therapeutic strategies for multiple sclerosis. International Conference on Cell-Based Therapies for MS. *Brain* **2017**, *140*, 2776–2796. [CrossRef] [PubMed]

Evanthia Bernitsas
Special Issue Editor

brain
sciences

MDPI

Review

Is Decompressive Surgery for Cervical Spondylotic Myelopathy Effective in Patients Suffering from Concomitant Multiple Sclerosis or Parkinson's Disease?

Taylor E. Purvis [1], Daniel Lubelski [1] and Thomas E. Mroz [2,3,*]

[1] Department of Neurosurgery, Johns Hopkins University School of Medicine, Baltimore, MD 21287, USA;
    tpurvis2@jhmi.edu (T.E.P.); dlubelski@jhmi.edu (D.L.)
[2] Cleveland Clinic Center for Spine Health, Cleveland Clinic, 9500 Euclid Ave, S-80,
    Cleveland, OH 44195, USA
[3] Department of Neurological Surgery, Cleveland Clinic, 9500 Euclid Ave, S-40, Cleveland, OH 44195, USA
*   Correspondence: mrozt@ccf.org; Tel.: +1-216-445-9232

Academic Editor: Evanthia Bernitsas
Received: 24 February 2017; Accepted: 6 April 2017; Published: 10 April 2017

Abstract: A subset of patients with a demyelinating disease suffer from concurrent cervical spondylotic myelopathy, both of which evince similar symptomatology. Differentiating the cause of these symptoms is challenging, and little research has been done on patients with coexisting diseases. This review explores the current literature on the appropriate surgical management of patients with concurrent multiple sclerosis (MS) and cervical spondylotic myelopathy (CSM), and those with both Parkinson's disease (PD) and CSM. MS and CSM patients may benefit from surgery to reduce pain and radiculopathy. Surgical management in PD and CSM patients has shown minimal quality-of-life improvement. Future studies are needed to better characterize demyelinating disease patients with concurrent disease and to determine ideal medical or surgical treatment.

Keywords: demyelinating disease; multiple sclerosis (MS); cervical spondylotic myelopathy (CSM); Parkinson's disease (PD); demyelination; myelopathy; outcomes

1. Introduction

Demyelinating diseases commonly present symptoms such as muscle weakness, stiffness and spasms, gait disorders, pain, changes in sensation, and disruptions in bowel and bladder function [1,2]. While the pathophysiology of multiple sclerosis (MS) and cervical spondylotic myelopathy (CSM) differs—MS via an autoimmune process and CSM by a mechanical compressive process—both are characterized by damage to myelin and have overlapping presentations [3,4]. Coexisting disorders such as Parkinson's disease (PD) and CSM can also cause similar symptoms that create difficulty when attempting to differentiate the diseases for treatment or monitoring purposes [2,5–7]. The primary objective of decompression and fusion in treatment of CSM is to prevent progression of neurological decline. In many patients, however, there may be improvement in patients' symptoms and functional status [8]. Little is known about the clinical and quality-of-life (QOL) outcomes following spine surgery for cervical myelopathy in patients with a coexistent demyelinating disease with similar symptoms. This review article seeks to describe such surgical outcomes reported in the literature for patients with concurrent MS and CSM and concurrent PD and CSM.

2. Materials and Methods

2.1. Search Strategy

A review of the literature was performed using the US National Library of Medicine PubMed database and a hand-search strategy to identify references from the selected articles. The search query included the following terms: demyelinating disease, multiple sclerosis (MS), cervical spondylotic myelopathy (CSM), Parkinson's disease (PD), amyotrophic lateral sclerosis (ALS), demyelination, and myelopathy.

2.2. Eligibility Criteria

Studies were included if they were written in English or had an English translation, and the patient population was comprised of those with a demyelinating disease and coexisting CSM.

3. Results

A total of nine studies were identified that met the inclusion criteria, including eight with concurrent MS and CSM and one with PD and CSM. The identified studies were case reports or case series (Table 1). No prospective studies were identified.

Table 1. Reviewed literature on demyelinating disease and coexisting disease with similar symptoms.

Authors	Year	Number of Patients	Surgical Intervention	Mean Follow-Up Time (Months)	Main Study Findings
Concurrent Multiple Sclerosis and Cervical Spondylotic Myelopathy					
Surgical Outcomes in Patients with Concurrent MS and CSM					
Brain and Wilkinson [9]	1957	17 with MS and CSM	Laminectomy	——	Patients reported poor outcomes following laminectomy, particularly for those with disseminated sclerosis.
Young et al. [10]	1999	7 with MS and CSM	Decompression	14 (range 6–24)	5 patients showed postoperative improvement in spondylosis symptoms. 1 patient developed acute MS symptoms a day after surgery.
Arnold et al. [11]	2011	15 with MS and cervical myeloradiculopathy	Decompression, fusion, and fixation	47	13 patients demonstrated objective improvement in upper and lower extremity strength and neck and/or upper extremity pain or paresthesias.
Burgerman et al. [12]	1992	6 with MS and CSM	Anterior cervical discectomy or cervical laminectomy	30 (12–72)	Long-term improvement in 2/3 patients with anterior cervical discectomy. 1 patient treated with cervical laminectomy showed only transient clinical improvement. 3 patients (2 laminectomies, 1 anterior cervical discectomy) showed no change in symptoms.
Lubelski et al. [13]	2014	77 with MS and CSM; 77 with CSM	Cervical decompression	57.7 ± 43.3 (MS and CSM); 49.4 ± 42.5 (CSM)	39% in the MS group did not have myelopathy improvement in the short-term vs. 23% in the control group ($p = 0.04$) and, in the long-term, 44% in the MS group did not improve vs. 19% in the control group ($p = 0.004$).

Table 1. *Cont.*

Authors	Year	Number of Patients	Surgical Intervention	Mean Follow-Up Time (Months)	Main Study Findings
Bashir et al. [14]	2000	14 with MS and spinal cord compression	Cervical decompression	45.6 (range, 12.0–117.6)	All patients with neck pain reported improvement in or elimination of their pain (*n* = 11). 6/10 patients with cervical radiculopathy reported complete resolution of their radicular symptoms, and 4 reported a reduction. 7/13 patients with progressive myelopathy experienced no improvement in symptoms.
Tan et al. [15]	2014	18 with MS and CSM	Cervical decompression and fusion	18 (range, 3–45)	4 reported improvement (28.6%), 9 (64.3%) reported stabilization, and 1 (7.1%) described a worsening of myelopathy. All 7 patients with neck pain described elimination of or significant improvement in symptoms.
Quality-of-Life Outcomes in Patients with Concurrent MS and CSM					
Lubelski et al. [16]	2014	13 with MS and CSM; 52 controls with CSM	Cervical decompression	22.3 ± 10.6 (MS and CSM); 18.2 ± 10.8 (CSM)	QALY in the MS and CSM group did not change significantly from pre- to post-operation (*p* = 0.96) vs. a significant change in the control CSM group from a QALY of 0.50 to 0.64 (*p* < 0.0001).
Concurrent Parkinson's Disease and Cervical Spondylotic Myelopathy					
Xiao et al. [17]	2016	11 with PD and CSM; 44 controls with CSM	Cervical decompression	12.4 ± 16.2 (PD and CSM); 13.4 ± 11.3 (CSM)	Patients with PD and CSM reported worse quality-of-life at last follow-up than controls (0.526 vs. 0.707, *p* = 0.01). PD and CSM patients did have improvement in pain-related disability.

PD: Parkinson's Disease; CSM: Cervical Spondylotic Myelopathy; MS: Multiple Sclerosis; QALY: Quality-Adjusted Life-Year.

3.1. Concurrent Multiple Sclerosis and Cervical Spondylotic Myelopathy

MS is a progressive autoimmune demyelinating disease that affects approximately 0.1% of the United States population [2,18–21]. MS can occur together with CSM and, although the incidence of concurrent disease has not been reported, is understood to occur. The symptoms are similar for both diseases, including bowel and bladder dysfunction, spasticity, gait ataxia, and sensory deficits [2]. Treatment for the two conditions differs greatly, as the pathophysiology of the myelopathy is very different. Typically, progressive or advanced CSM is treated with surgical decompression [2,18,19] whereas MS is managed medically with corticosteroids or interferon beta [21,22]. Little is known about the surgical or QOL outcomes in concurrent MS and CSM patients treated with spine surgery.

3.1.1. Surgical Outcomes in Patients with Concurrent MS and CSM

In a 1957 report on patients with coexisting MS and cervical spondylosis, Brain and Wilkinson [9] described 17 patients and the challenges that arose in diagnosis and treatment for both diseases. The authors described poor outcomes following laminectomy, particularly for patients with disseminated sclerosis. Given the progressive nature of MS, the authors recommended against any operation that would provide only transitory relief and instead suggested neck immobilization in a collar as a treatment alternative. The authors recognized that for patients who do not have MS, however, a collar may provide suboptimal relief of the spondylosis.

More recent studies have demonstrated conflicting information that instead shows the potential benefits of surgery in patients with MS and CSM. In a study of seven patients with concurrent disease, Young and colleagues [10] found that five patients treated with decompressive surgery showed postoperative improvement in spondylosis symptoms (mean follow-up, 14 months; range, 6–24 months). One patient developed acute MS symptoms a day after surgery. The authors concluded that surgical treatment of spondylosis in patients with coexisting MS and CSM improves symptoms and that MS flare following surgery is rare.

Arnold et al. [11] came to similar conclusions in a case series of 15 patients with MS and cervical myeloradiculopathy who were treated with surgical decompression and fusion (mean follow-up, 47 months). Thirteen patients demonstrated improvement in upper and lower extremity strength and neck and/or upper extremity pain or paresthesias. In the remaining two patients, symptoms did not improve but did not worsen either. No surgical complications were reported. The authors concluded that surgical intervention for cervical myeloradiculopathy should be considered a safe and effective option in patients with concurrent MS.

One study by Burgerman and colleagues [12] suggested that not all forms of surgical treatment may be effective in patients with coexistent MS and CSM. In a series of six patients, surgery resulted in lasting improvement of symptoms in two of three patients who underwent anterior cervical discectomy (mean follow-up, 30 months; range, 12 months–6 years). One patient treated with cervical laminectomy showed only transient clinical improvement, while three patients (two laminectomies, one anterior cervical discectomy) showed no change in symptoms. The authors suggested that patients who develop progressively worse anatomic compression should be evaluated for surgical treatment.

In a larger retrospective review of 77 patients with concurrent MS and CSM that were matched with 77 patients with only CSM, all of whom underwent cervical decompression surgery, Lubelski et al. [13] reported that both populations had postoperative improvement. MS and control patients were followed for an average of 58 months and 49 months, respectively. Patients with concurrent MS and CSM had improvements that were less dramatic than those in the control group. A significantly greater proportion of patients in the MS group had myelopathic symptoms that did not improve with surgery in both the short-term (39% in the MS group did not improve vs. 23% in the control group; $p = 0.04$) and long-term (44% in the MS group did not improve vs. 19% in the control group; $p = 0.004$). Patients with primary and secondary progressive MS did show poorer outcomes compared to patients with relapsing remitting MS. Both controls and patients with coexisting MS and CSM had similar postoperative improvement in neck pain and radicular symptoms. The authors concluded that surgery can be recommended to MS and CSM patients, although they should be advised of the potential for less relief of myelopathic symptoms than if they had CSM alone.

Bashir et al. [14] published a case series that found similar outcomes in patients with MS and coexisting spinal cord compression due to cervical spondylosis or cervical disc disease. Fourteen patients underwent cervical decompression surgery to address presenting symptoms of neck pain ($n = 11$), cervical radiculopathy ($n = 10$), and progressive myelopathy ($n = 13$) (mean follow-up, 3.8 years; range, 1.0–9.8 years). All patients with neck pain reported improvement in or elimination of their pain ($n = 11$). Six of the 10 patients with cervical radiculopathy reported complete resolution of their radicular symptoms, and four reported a reduction. Seven of the 13 patients with progressive myelopathy experienced no improvement in symptoms, although this group uniformly had improvement in or elimination of radicular complaints and neck pain. These results are consistent with those of Lubelski et al. [13], that demonstrated improvement in neck and radicular pain in MS and CSM patients.

One study by Tan and colleagues [15] did show a reduction in myelopathy in addition to an improvement in radicular symptoms and neck pain. Eighteen patients with concurrent MS and CSM were identified after undergoing cervical spine decompression and fusion (mean follow-up, 18 months; range, 3–45 months). The severity of MS symptoms was assessed using the Expanded Disability Status Scale (EDSS). Of the 14 patients with preoperative myelopathy, four reported improvement

(28.6%), nine (64.3%) reported stabilization, and one (7.1%) described a worsening of myelopathy postoperatively. All seven patients with neck pain described elimination of or significant improvement in symptoms. Improvement of radiculopathy occurred in four of five patients (80%) who had preoperative symptoms. No patients with preoperative bladder dysfunction ($n = 8$) experienced relief following surgery. EDSS scores in 16 patients decreased or stabilized (94.4%), while scores increased in two patients (5.6%). The authors explained that their findings were consistent with those of Lubelski et al. [13] in that most patients with myelopathy achieved only stability in symptoms (62%) rather than improvement (30%). These results, together with those of Young et al. [10], Arnold et al. [11], Burgerman et al. [12], Lubelski et al. [13], and Bashir et al. [14] reported above, suggest that surgical treatment may be indicated for relief of neck pain and radicular symptoms rather than the myelopathic symptoms that will progress with MS. Moreover, the collective evidence suggests that surgery does not result in exacerbations of MS. Finally, although MS would likely demonstrate periods of remission in the most common relapsing/remitting variant [23], CSM would otherwise have continuous and progressive myelopathic symptoms.

3.1.2. Quality-of-Life Outcomes in Patients with Concurrent MS and CSM

While surgical outcomes such as neurological status and complications have been investigated in patients with coexisting MS and CSM, only one study has examined the QOL outcomes in these patients with concurrent disease. Lubelski et al. [16] identified 13 patients with MS and CSM and 52 control patients with CSM alone who were treated with cervical decompression (mean follow-up was 22 and 18 months, respectively). QOL was assessed using the EuroQol 5-Dimensions (EQ-5D) metric that includes the domains of anxiety/depression, usual activities, self-care, mobility, and pain/discomfort. Patients in the control group had significantly improved QOL scores in three domains (mobility, $p = 0.04$; self-care, 0.003; anxiety/depression, $p = 0.03$), measured from pre- to post-operative status, in contrast to patients with concurrent disease. Quality-Adjusted Life-Year (QALY) measurements, or the years of life added as a result of the surgery, in the concurrent MS and CSM group did not change significantly from pre- to post-operation ($p = 0.96$), while those in the control CSM group had a significant change from a QALY of 0.50 to 0.64 ($p < 0.0001$). Only the CSM controls showed a change in QALY that was greater than the minimal clinically important difference (MCID) of 0.1. A majority of patients with CSM and MS did, however, experience improvement in QALY (54%). These results suggest that while surgery may still be indicated for patients with concurrent disease, patients may not experience QOL benefits following the intervention despite an improvement in pain, radicular symptoms, and potentially myelopathy.

These studies demonstrate that MS and CSM have symptoms that are overlapping, making it difficult to correctly attribute any one symptom to the appropriate causative disease entity. The progressive myelopathic symptoms of CSM, as well as the potential benefit of surgery in relieving pain and radicular symptoms, may warrant surgical intervention in patients with concurrent disease. However, outcomes may be suboptimal in these patients compared to those with CSM alone. Patients should be appropriately educated about the potential impact of MS on their surgical outcomes.

3.2. Concurrent Parkinson's Disease and Cervical Spondylotic Myelopathy

PD affects approximately 1% of individuals over the age of 60 [24,25]. Symptoms of PD are many and diverse, and include tremor, weakness, a variety of movement disorders (e.g., ataxia, shuffling gait, involuntary movements, motor retardation), and bladder or bowel dysfunction [5–7,17]. CSM is characterized by similar symptoms [26], and distinguishing between the two pathologies in patients with coexistent diseases can be challenging. Treatment of CSM is most commonly surgical decompression and fusion, which leads to improvement in QOL [27–34]. Among patients with PD, however, spine surgery can be associated with poor post-operative QOL and may lead to high complication and reoperation rates [35–40]. Treatment of PD is typically pharmacologic or, if necessary, deep brain stimulation [41–46]. Untreated CSM, however, is also associated with worsening symptoms and QOL, and accordingly the question arises as to how best treat patients with concurrent PD and CSM.

Research on patient populations with concurrent PD and CSM is scant. The first study in this population examined QOL outcomes following cervical decompression [17]. Xiao et al. [17] performed a retrospective matched cohort analysis that included 11 patients with PD and CSM matched to 44 controls with CSM alone who underwent cervical decompression (mean follow-up was 12.4 and 13.4 months, respectively). QOL was assessed using several patient-reported health status measurements, including the EQ-5D, Pain Disability Questionnaire (PDQ), and Patient Health Questionnaire-9 (PHQ-9). Patients with concurrent PD and CSM demonstrated a statistically significant reduction in postoperative pain-related disability. However, these changes were less substantial than in control patients. Although PD patients and controls had similar preoperative QOL scores, a smaller proportion of PD patients obtained an MCID in EQ-5D (18% vs. 57%, $p = 0.04$). Upon the last follow-up visit, PD patients also reported worse QOL as measured by EQ-5D (0.526 vs. 0.707, $p = 0.01$) and PDQ (80.7 vs. 51.4, $p = 0.03$). PD was an independent risk factor for a smaller improvement in EQ-5D scores ($\beta = -0.09$, $p < 0.01$) and an inability to obtain an MCID in EQ-5D scores (odds ratio: 0.08, $p < 0.01$). The proportion of patients achieving an MCID in PHQ-9 or PDQ scores was not significantly different between groups.

These results suggest that cervical decompression has minimal benefit in a patient population with coexisting PD and CSM. While spine surgery may provide some reduction in pain-related disability, QOL outcomes were poor compared to controls. In this patient population, preoperative counseling of risks and benefits is integral. And while surgery will provide some benefit, it will certainly not be as great as it could be for those with only CSM. Ultimately, the natural history of PD will lead to progressive worsening in symptoms over time. Of note, the small sample size of this study may not achieve adequate power to detect an effect. Future studies with larger numbers of PD and CSM patients are necessary to confirm the findings of Xiao et al. [17].

4. Limitations

This review is limited by the small sample sizes and retrospective nature of the studies included. Surgical outcome measures were not standardized among studies, which reduces their comparability. Selection of inappropriate surgical candidates or differing surgical skill may also have affected success rates. Moreover, the method of diagnosis of CSM was not standardized among the included studies, and this may have led to conflicting findings. Lastly, radiological interpretation by radiologists may result in reporting of non-essential or incidental findings that suggest surgical intervention in patients who may not otherwise have been identified by surgeons' radiological interpretations. Surgical approach during decompression also differed among studies, further limiting comparability.

5. Conclusions

While the primary goal of surgical intervention for CSM may remain prevention of progressive neurological decline, surgery also has the potential for symptomatic and quality-of-life improvement. There exists conflicting information about the success of spine surgery in reducing symptoms in MS and CSM patients, but most recent research suggests that surgery reduces preoperative pain, radicular symptoms, and possibly myelopathy. The improvement, however, is less than in those without MS. In patients with coexisting PD and CSM, surgical management may reduce some axial and radicular pain symptoms but results in QOL outcomes that may not be clinically significant. These findings suggest that surgery reduces clinical symptoms in these populations with concurrent diseases but that the outcomes will not be as good as in those patients with CSM alone. While studies indicate surgical intervention in patients with coexistent diseases (CSM/PD, CSM/MS) results in less favorable outcomes when compared to CSM alone, the authors believe that the former patient population perhaps has more to lose if compressive myelopathy is left untreated given the smaller functional margin at baseline. It is important that a rational and multispecialty approach (spine surgeons, internists and neurologists, and patient) be taken when constructing a treatment plan for this delicate patient population. Future research is needed in these unique patient populations to determine optimal treatment and to better predict for which patients surgery may provide symptomatic relief. Moreover, appropriately counseling patients with concurrent diseases, especially with regards to the natural course of the disease, is crucial.

Author Contributions: D.L. conceived and designed the review; T.E.P. performed the review; T.E.P. analyzed the data; T.E.P., D.L. and T.E.M. wrote the paper.

Conflicts of Interest: T.E.P.: Nothing to disclose. D.L.: Nothing to disclose. T.E.M.: Stock Ownership: PearlDiver Inc. (no monies received), outside the submitted work; Consulting: Globus Medical, outside the submitted work; Speaking and/or Teaching Arrangements: AOSpine, outside the submitted work. No funding was obtained for this manuscript.

References

1. Mehndiratta, M.M.; Gulati, N.S. Central and peripheral demyelination. *J. Neurosci. Rural Pract.* **2014**, *5*, 84–86. [CrossRef] [PubMed]
2. Tracy, J.A.; Bartleson, J.D. Cervical spondylotic myelopathy. *Neurologist* **2010**, *16*, 176–187. [CrossRef] [PubMed]
3. Hurwitz, B.J. The diagnosis of multiple sclerosis and the clinical subtypes. *Ann. Indian Acad. Neurol.* **2009**, *12*, 226–230. [CrossRef] [PubMed]
4. Young, W.F. Cervical spondylotic myelopathy: A common cause of spinal cord dysfunction in older persons. *Am. Fam. Physician* **2000**, *62*, 1064–1073. [PubMed]
5. Amano, S.; Roemmich, R.T.; Skinner, J.W.; Hass, C.J. Ambulation and parkinson disease. *Phys. Med. Rehabil. Clin. N. Am.* **2013**, *24*, 371–392. [CrossRef] [PubMed]
6. Chen, J.J. Parkinson's disease: Health-related quality of life, economic cost, and implications of early treatment. *Am. J. Manag. Care* **2010**, *16*, S87–S93. [PubMed]
7. Jankovic, J. Parkinson's disease: Clinical features and diagnosis. *J. Neurol. Neurosurg. Psychiatry* **2008**, *79*, 368–376. [CrossRef] [PubMed]
8. Lebl, D.R.; Hughes, A.; Cammisa, F.P., Jr.; O'Leary, P.F. Cervical spondylotic myelopathy: Pathophysiology, clinical presentation, and treatment. *HSS J.* **2011**, *7*, 170–178. [CrossRef] [PubMed]
9. Brain, R.; Wilkinson, M. The association of cervical spondylosis and disseminated sclerosis. *Brain* **1957**, *80*, 456–478. [CrossRef] [PubMed]
10. Young, W.F.; Weaver, M.; Mishra, B. Surgical outcome in patients with coexisting multiple sclerosis and spondylosis. *Acta Neurol. Scand.* **1999**, *100*, 84–87. [CrossRef] [PubMed]
11. Arnold, P.M.; Warren, R.K.; Anderson, K.K.; Vaccaro, A.R. Surgical treatment of patients with cervical myeloradiculopathy and coexistent multiple sclerosis: Report of 15 patients with long-term follow-up. *J. Spinal Disord. Tech.* **2011**, *24*, 177–182. [CrossRef] [PubMed]
12. Burgerman, R.; Rigamonti, D.; Randle, J.M.; Fishman, P.; Panitch, H.S.; Johnson, K.P. The association of cervical spondylosis and multiple sclerosis. *Surg. Neurol.* **1992**, *38*, 265–270. [CrossRef]
13. Lubelski, D.; Abdullah, K.G.; Alvin, M.D.; Wang, T.Y.; Nowacki, A.S.; Steinmetz, M.P.; Ransohoff, R.M.; Benzel, E.C.; Mroz, T.E. Clinical outcomes following surgical management of coexistent cervical stenosis and multiple sclerosis: A cohort-controlled analysis. *Spine J.* **2014**, *14*, 331–337. [CrossRef] [PubMed]
14. Bashir, K.; Cai, C.Y.; Moore, T.A., 2nd; Whitaker, J.N.; Hadley, M.N. Surgery for cervical spinal cord compression in patients with multiple sclerosis. *Neurosurgery* **2000**, *47*, 637–642. [CrossRef] [PubMed]
15. Tan, L.A.; Kasliwal, M.K.; Muth, C.C.; Stefoski, D.; Traynelis, V.C. Is cervical decompression beneficial in patients with coexistent cervical stenosis and multiple sclerosis? *J. Clin. Neurosci.* **2014**, *21*, 2189–2193. [CrossRef] [PubMed]
16. Lubelski, D.; Alvin, M.D.; Silverstein, M.; Senol, N.; Abdullah, K.G.; Benzel, E.C.; Mroz, T.E. Quality of life outcomes following surgery for patients with coexistent cervical stenosis and multiple sclerosis. *Eur. Spine J.* **2014**, *23*, 1699–1704. [CrossRef] [PubMed]
17. Xiao, R.; Miller, J.A.; Lubelski, D.; Alberts, J.L.; Mroz, T.E.; Benzel, E.C.; Krishnaney, A.A.; Machado, A.G. Quality of life outcomes following cervical decompression for coexisting parkinson's disease and cervical spondylotic myelopathy. *Spine J.* **2016**, *16*, 1358–1366. [CrossRef] [PubMed]

18. Edwards, C.C., 2nd; Riew, K.D.; Anderson, P.A.; Hilibrand, A.S.; Vaccaro, A.F. Cervical myelopathy. Current diagnostic and treatment strategies. *Spine J.* **2003**, *3*, 68–81. [CrossRef]
19. Klineberg, E. Cervical spondylotic myelopathy: A review of the evidence. *Orthop. Clin. N. Am.* **2010**, *41*, 193–202. [CrossRef] [PubMed]
20. Ransohoff, R.M. Natalizumab for multiple sclerosis. *N. Engl. J. Med.* **2007**, *356*, 2622–2629. [CrossRef] [PubMed]
21. Rudick, R.A.; Cohen, J.A.; Weinstock-Guttman, B.; Kinkel, R.P.; Ransohoff, R.M. Management of multiple sclerosis. *N. Engl. J. Med.* **1997**, *337*, 1604–1611. [PubMed]
22. Noseworthy, J.H.; Lucchinetti, C.; Rodriguez, M.; Weinshenker, B.G. Multiple sclerosis. *N. Engl. J. Med.* **2000**, *343*, 938–952. [CrossRef] [PubMed]
23. National Clinical Guideline, C. National Clinical Guideline, C. National institute for health and care excellence: Clinical guidelines. In *Multiple sclerosis: Management of Multiple Sclerosis in Primary and Secondary Care*; National Institute for Health and Care Excellence: London, UK, 2014.
24. De Lau, L.M.; Breteler, M.M. Epidemiology of Parkinson's disease. *Lancet Neurol.* **2006**, *5*, 525–535. [CrossRef]
25. Nussbaum, R.L.; Ellis, C.E. Alzheimer's disease and parkinson's disease. *N. Engl. J. Med.* **2003**, *348*, 1356–1364. [PubMed]
26. Toledano, M.; Bartleson, J.D. Cervical spondylotic myelopathy. *Neurol. Clin.* **2013**, *31*, 287–305. [CrossRef] [PubMed]
27. Chagas, H.; Domingues, F.; Aversa, A.; Vidal Fonseca, A.L.; de Souza, J.M. Cervical spondylotic myelopathy: 10 years of prospective outcome analysis of anterior decompression and fusion. *Surg. Neurol.* **2005**, *64*, S30–S35. [CrossRef] [PubMed]
28. Fehlings, M.G.; Wilson, J.R.; Kopjar, B.; Yoon, S.T.; Arnold, P.M.; Massicotte, E.M.; Vaccaro, A.R.; Brodke, D.S.; Shaffrey, C.I.; Smith, J.S.; et al. Efficacy and safety of surgical decompression in patients with cervical spondylotic myelopathy: Results of the aospine north america prospective multi-center study. *J. Bone Jt. Surg. Am.* **2013**, *95*, 1651–1658. [CrossRef] [PubMed]
29. Lee, T.T.; Manzano, G.R.; Green, B.A. Modified open-door cervical expansive laminoplasty for spondylotic myelopathy: Operative technique, outcome, and predictors for gait improvement. *J. Neurosurg.* **1997**, *86*, 64–68. [CrossRef] [PubMed]
30. Matz, P.G.; Anderson, P.A.; Holly, L.T.; Groff, M.W.; Heary, R.F.; Kaiser, M.G.; Mummaneni, P.V.; Ryken, T.C.; Choudhri, T.F.; Vresilovic, E.J.; et al. The natural history of cervical spondylotic myelopathy. *J. Neurosurg. Spine* **2009**, *11*, 104–111. [CrossRef] [PubMed]
31. Mummaneni, P.V.; Kaiser, M.G.; Matz, P.G.; Anderson, P.A.; Groff, M.W.; Heary, R.F.; Holly, L.T.; Ryken, T.C.; Choudhri, T.F.; Vresilovic, E.J.; et al. Cervical surgical techniques for the treatment of cervical spondylotic myelopathy. *J. Neurosurg. Spine* **2009**, *11*, 130–141. [CrossRef] [PubMed]
32. Rao, R.D.; Gourab, K.; David, K.S. Operative treatment of cervical spondylotic myelopathy. *J. Bone Jt. Surg. Am.* **2006**, *88*, 1619–1640. [CrossRef]
33. Singh, A.; Gnanalingham, K.; Casey, A.; Crockard, A. Quality of life assessment using the short form-12 (sf-12) questionnaire in patients with cervical spondylotic myelopathy: Comparison with sf-36. *Spine* **2006**, *31*, 639–643. [CrossRef] [PubMed]
34. Tanaka, J.; Seki, N.; Tokimura, F.; Doi, K.; Inoue, S. Operative results of canal-expansive laminoplasty for cervical spondylotic myelopathy in elderly patients. *Spine* **1999**, *24*, 2308–2312. [CrossRef] [PubMed]
35. Babat, L.B.; McLain, R.F.; Bingaman, W.; Kalfas, I.; Young, P.; Rufo-Smith, C. Spinal surgery in patients with Parkinson's disease: Construct failure and progressive deformity. *Spine* **2004**, *29*, 2006–2012. [CrossRef] [PubMed]
36. Chapuis, S.; Ouchchane, L.; Metz, O.; Gerbaud, L.; Durif, F. Impact of the motor complications of Parkinson's disease on the quality of life. *Mov. Disord.* **2005**, *20*, 224–230. [CrossRef] [PubMed]
37. Koller, H.; Acosta, F.; Zenner, J.; Ferraris, L.; Hitzl, W.; Meier, O.; Ondra, S.; Koski, T.; Schmidt, R. Spinal surgery in patients with Parkinson's disease: Experiences with the challenges posed by sagittal imbalance and the Parkinson's spine. *Eur. Spine J.* **2010**, *19*, 1785–1794. [CrossRef] [PubMed]
38. Moon, S.H.; Lee, H.M.; Chun, H.J.; Kang, K.T.; Kim, H.S.; Park, J.O.; Moon, E.S.; Chong, H.S.; Sohn, J.S.; Kim, H.J. Surgical outcome of lumbar fusion surgery in patients with parkinson disease. *J. Spinal Disord. Tech.* **2012**, *25*, 351–355. [CrossRef] [PubMed]

Brain Sci. **2017**, *7*, 39

39. Schrag, A.; Jahanshahi, M.; Quinn, N. What contributes to quality of life in patients with parkinson's disease? *J. Neurol. Neurosurg. Psychiatry* **2000**, *69*, 308–312. [CrossRef] [PubMed]

40. Upadhyaya, C.D.; Starr, P.A.; Mummaneni, P.V. Spinal deformity and parkinson disease: A treatment algorithm. *Neurosurg. Focus* **2010**, *28*, E5. [CrossRef] [PubMed]

41. Abboud, H.; Floden, D.; Thompson, N.R.; Genc, G.; Oravivattanakul, S.; Alsallom, F.; Swa, B.; Kubu, C.; Pandya, M.; Gostkowski, M.; et al. Impact of mild cognitive impairment on outcome following deep brain stimulation surgery for parkinson's disease. *Park. Relat. Disord.* **2015**, *21*, 249–253. [CrossRef] [PubMed]

42. Connolly, B.S.; Lang, A.E. Pharmacological treatment of parkinson disease: A review. *JAMA* **2014**, *311*, 1670–1683. [CrossRef] [PubMed]

43. Floden, D.; Busch, R.M.; Cooper, S.E.; Kubu, C.S.; Machado, A.G. Global cognitive scores do not predict outcome after subthalamic nucleus deep brain stimulation. *Mov. Disord.* **2015**, *30*, 1279–1283. [CrossRef] [PubMed]

44. Floden, D.; Cooper, S.E.; Griffith, S.D.; Machado, A.G. Predicting quality of life outcomes after subthalamic nucleus deep brain stimulation. *Neurology* **2014**, *83*, 1627–1633. [CrossRef] [PubMed]

45. Genc, G.; Abboud, H.; Oravivattanakul, S.; Alsallom, F.; Thompson, N.R.; Cooper, S.; Gostkowski, M.; Machado, A.; Fernandez, H.H. Socioeconomic status may impact functional outcome of deep brain stimulation surgery in Parkinson's disease. *Neuromodulation* **2016**, *19*, 25–30. [CrossRef] [PubMed]

46. Machado, A.; Rezai, A.R.; Kopell, B.H.; Gross, R.E.; Sharan, A.D.; Benabid, A.L. Deep brain stimulation for Parkinson's disease: Surgical technique and perioperative management. *Mov. Disord.* **2006**, *21*, S247–S258. [CrossRef] [PubMed]

brain sciences

MDPI

Article

Evaluation of Visual-Evoked Cerebral Metabolic Rate of Oxygen as a Diagnostic Marker in Multiple Sclerosis

Nicholas A. Hubbard [1,*], Yoel Sanchez Araujo [1], Camila Caballero [1], Minhui Ouyang [2], Monroe P. Turner [3], Lyndahl Himes [3], Shawheen Faghihahmadabadi [3], Binu P. Thomas [3,4], John Hart Jr. [3,5,6], Hao Huang [2], Darin T. Okuda [3,5] and Bart Rypma [3,6]

[1] Massachusetts Institute of Technology, Cambridge, MA 02139, USA; ysa@mit.edu (Y.S.A.); csquared@mit.edu (C.C.)
[2] Children's Hospital of Philadelphia, University of Pennsylvania School of Medicine, Philadelphia, PA 19104, USA; ouyangm@email.chop.edu (M.O.); huangh6@email.chop.edu (H.H.)
[3] University of Texas at Dallas, Dallas, TX 75080, USA; monroe.p.turner@utdallas.edu (M.P.T.); lyndahl.himes@utdallas.edu (L.H.); shawheen.fahih@utdallas.edu (S.F.); binu.thomas@utsouthwestern.edu (B.P.T.); jhart@utdallas.edu (J.H.); darin.okuda@utsouthwestern.edu (D.T.O.); bart.rypma@utdallas.edu (B.R.)
[4] University of Texas Southwestern Medical Center, Advanced Imaging Research Center, Dallas, TX 75235, USA
[5] Department of Neurology and Neurothreapeutics, University of Texas Southwestern Medical Center, Dallas, TX 75235, USA
[6] Department of Psychiatry, University of Texas Southwestern Medical Center, Dallas, TX 75235, USA
* Correspondence: nhubbard@mit.edu; Tel.: +1-617-324-371

Academic Editor: Evanthia Bernitsas
Received: 31 March 2017; Accepted: 5 June 2017; Published: 11 June 2017

Abstract: A multiple sclerosis (MS) diagnosis often relies upon clinical presentation and qualitative analysis of standard, magnetic resonance brain images. However, the accuracy of MS diagnoses can be improved by utilizing advanced brain imaging methods. We assessed the accuracy of a new neuroimaging marker, visual-evoked cerebral metabolic rate of oxygen (veCMRO$_2$), in classifying MS patients and closely age- and sex-matched healthy control (HC) participants. MS patients and HCs underwent calibrated functional magnetic resonance imaging (cfMRI) during a visual stimulation task, diffusion tensor imaging, T$_1$- and T$_2$-weighted imaging, neuropsychological testing, and completed self-report questionnaires. Using resampling techniques to avoid bias and increase the generalizability of the results, we assessed the accuracy of veCMRO$_2$ in classifying MS patients and HCs. veCMRO$_2$ classification accuracy was also examined in the context of other evoked visuofunctional measures, white matter microstructural integrity, lesion-based measures from T$_2$-weighted imaging, atrophy measures from T$_1$-weighted imaging, neuropsychological tests, and self-report assays of clinical symptomology. veCMRO$_2$ was significant and within the top 16% of measures (43 total) in classifying MS status using both within-sample (82% accuracy) and out-of-sample (77% accuracy) observations. High accuracy of veCMRO$_2$ in classifying MS demonstrated an encouraging first step toward establishing veCMRO$_2$ as a neurodiagnostic marker of MS.

Keywords: calibrated functional magnetic resonance imaging; multiple sclerosis; diagnosis; visual system; metabolism

1. Introduction

Current procedures for diagnosing multiple sclerosis (MS) rely primarily upon clinical presentation and qualitative analysis of standard, medical-grade (e.g., lower resolution) magnetic resonance structural, brain images, e.g., [1]. It has been demonstrated that the diagnostic accuracy of MS can be improved when providers implement advanced neuroimaging techniques and analyses that are not presently common in clinical practice, e.g., [2], see also [3]. Further, research using advanced neuroimaging techniques has demonstrated that these techniques can be more sensitive than their traditional counterparts in detecting subtle changes associated with very early manifestations of MS, e.g., [4,5]. Here, we investigated the accuracy of an advanced neuroimaging technique never before used in MS, calibrated functional magnetic resonance imaging (cfMRI), to classify MS patients and closely age- and sex-matched healthy controls (HCs). Specifically, we focused our analyses upon the ability of a new neuroimaging marker, visual-evoked cerebral metabolic rate of oxygen (veCMRO$_2$), to accurately discriminate between MS patients and HCs.

cfMRI is a relatively new neuroimaging technique that capitalizes upon established relationships between blood-oxygen-level dependent (BOLD) signal and cerebral blood flow (CBF) in order to estimate steady-state, oxygen metabolism [6,7] see [8]. The technique gets its name from the use of a BOLD-calibration parameter, often acquired during a gas-inhalation challenge. The CMRO$_2$ metric permitted by cfMRI offers several advantages over the more commonly used BOLD signal. First, CMRO$_2$ offers physiological specificity. CMRO$_2$ represents a true physiological process, oxygen metabolism, whereas BOLD reflects a confluence of processes and as such, is physiologically non-specific. Second, calibration-derived CMRO$_2$ is strongly tied to electrical and chemical neural activity, e.g., [9–15], whereas an appreciable component of BOLD signal is unexplained by neural activity, e.g., [16–20], see [21], but see [9]. Finally, CMRO$_2$ measures are not dependent upon the hemodynamic assumptions of BOLD, making them optimal measures of brain function in populations with atypical hemodynamics, like MS, e.g., [22,23], see [24].

Evaluating CMRO$_2$ as a diagnostic marker of MS is particularly relevant for these patients because MS is associated with changes to neurometabolism. Neuroimaging research has produced considerable evidence of altered neurometabolism in MS, e.g., [25–29]. In one study, Ge and colleagues [30] demonstrated decreases in brain-wide resting CMRO$_2$ for MS patients relative to HCs. Some neuroimaging studies have shown that neurometabolic alterations were related to white matter macrostructural (i.e., lesions, e.g., [30]) or microstructural damage in MS, e.g., [27,28]. For example, magnetic resonance spectroscopy in centrum semiovale white matter has shown that N-acetylaspertate (NAA) and NAA: creatine ratios were strongly related to diffusion-weighted indices of white matter structural integrity in MS patients [27].

It is intuitive that MS patients would show differences in in vivo neurometabolism when considering that postmortem analyses have revealed extensive alterations to the mitochondria in lesioned and non-lesioned MS neural tissue [31–33], see [34–36]. For instance, Singhal and colleagues [33] found decreases in postmortem NAA, a partial marker of neuronal respiratory capacity, and decreases in electron transport subunit proteins across lesioned and non-lesioned MS grey matter, relative to matched control participants' grey matter. Taken together, the results of postmortem and in vivo neuroimaging studies demonstrate that neurometabolic alterations are generally featured in MS.

Evaluating veCMRO$_2$ should also be particularly relevant as a diagnostic marker of MS because MS is marked by alterations to the neural substrate of the visual system, see [37–40] see also [5]. The use of advanced imaging techniques such as high-resolution structural brain imaging, optical coherence tomography (OCT), functional magnetic resonance imaging (fMRI), and diffusion tensor imaging (DTI) has revealed that visual system alterations exist even in MS patients without visual disturbances or a history of optic neuritis (a clinical syndrome closely linked to MS and marked by visual impairment and visual pathway insult). Indeed, there are MS-related structural alterations to both early (e.g., retinae) and later (e.g., optic radiations) portions of the afferent visual pathway, and alterations to visuocortical activity in patients without a history of optic neuritis see [39]. For instance, Alshowaier and colleagues [41] used electroencephalogram recordings to show that MS patients without a history of optic neuritis demonstrated delayed inion channel, multifocal visual-evoked electrical potentials relative to age- and sex-matched HCs. Previous work in our laboratory has also revealed alterations to visual cortex BOLD signal during visual stimulation in MS patients with normal or corrected-to-normal vision compared to matched HCs [42], see also [43]. Together, structural and functional imaging results suggest that changes to the visual system are a robust marker of MS pathology.

MS is associated with changes to neurometabolism and alterations to the neural substrate of the visual system. Thus, visual-evoked oxygen metabolism signals in visual cortex (i.e., $veCMRO_2$) should be a diagnostically relevant marker of MS. We assessed the extent to which $veCMRO_2$ signals could be used to discriminate between MS patients and HCs. The classification accuracy of $veCMRO_2$ was examined in the context of other variables commonly assayed in MS, including measures of neurological insult (e.g., gross lesion volume, parenchymal atrophy), neuropsychological change (e.g., Brief Repeatable Battery of Neuropsychological Tests [44]), and self-report symptom measures (e.g., subjective fatigue). We tested the extent to which $veCMRO_2$, and these other measures, could classify MS status using both within-sample and out-of-sample observations.

2. Materials and Methods

2.1. Participants

Participants between the ages of 18 and 65 were recruited for this study. Participants were required to be free of MR-contraindicators, concurrent substance abuse, have normal or corrected-to-normal vision, and speak fluent English. Because study procedures included a gas-inhalation challenge (see Section 2.4), participant selection was limited to non-smokers. Participants did not have histories of respiratory or pulmonary problems, cerebral vascular issues, or cardiac problems. Participants were required to have a score greater than 21 on the telephone interview for cognitive status [45]. Thirty-one participants in total met the inclusion criteria.

Twelve MS patients meeting the above criteria were recruited from the Clinical Center for Multiple Sclerosis at the University of Texas Southwestern Medical Center. Eleven patients had a diagnosis of relapsing-remitting MS and one patient had a diagnosis of secondary-progressive MS. Patients were required to be at least 1 month past their most recent exacerbation and their last corticosteroid treatment. Patients were recruited who did not report a history of optic neuritis. Patients without a history of optic neuritis were specifically selected so as to limit additional variability from attributed to severe, anterior visual pathway damage/dysfunction (e.g., such as that resulting from conduction block) and potential visual impairment. All MS patients' vision was normal or corrected-to-normal. Two patients withdrew or declined to undergo the gas challenge (total $n = 10$).

Nineteen HC participants were recruited from the Dallas-Fort Worth Metroplex via email, posted flyers, and word-of-mouth. These participants were evaluated for the general inclusion/exclusion criteria described above. Three HCs did not undergo the scanning protocol because of exclusions discovered after study enrollment (e.g., concussion history revealed after pre-screening, incidental MR finding). Two HCs withdrew or declined to undergo the gas challenge. During imaging processing (see Section 2.5), one HC's functional images failed to appropriately register to their anatomical image after multiple attempts, so this person was excluded. Thirteen HCs ($n = 13$) remained for subsequent analyses. These participants were closely age- and sex-matched to the MS patients (see Table 1).

Table 1. Group Characteristics.

	MS	HC	*p*
Age	50.10 (3.35)	50.77 (3.35)	0.885 [a]
MFIS	39.10 (7.62)	20.54 (4.57)	0.046 [a]
Sex (% female)	90.00%	84.62%	0.704 [b]
TICS Score	27.00 (0.82)	28.08 (1.43)	0.520 [a]
Age of MS Onset	38.67 (2.42)	-	-
Disease Duration	118.80 (19.32)	-	-
Last Flare-up	28.60 (11.32)	-	-
Neurological Disability Score	15.70 (3.71)	-	-
Disease Modifying Therapies			
Dalfampridine	50%	-	-
Dimethyl fumarate	10%	-	-
Fingolimod	20%	-	-
Glatiramer acetate	10%	-	-

Mean (SEM). Age in years. MFIS = modified fatigue impact score total. Sex in percent female. TICS score = telephone interview for cognitive status score. Age of MS onset in years. Disease duration and last flare-up in months. Neurological disability score measured by self-report [46]. Disease modifying therapies represent percent of participants reporting use of therapy. [a] *p*-value based upon independent samples *t*-test. [b] *p*-value based upon Pearson χ^2.

2.2. Study Procedures

Study procedures were approved by the University of Texas Southwestern Medical Center Institutional Review Board. Recruitment numbers were approximated based upon previous research showing sufficient power to demonstrate group changes in calibrated fMRI (cfMRI) contrasts with similar sample sizes [22,23]. Participants meeting inclusion criteria were asked to refrain from caffeine use at least two hours before their scheduled appointment time, e.g., [47]. They were also asked not to consume alcohol on the same calendar day before their scheduled appointment. Participants gave written informed consent before undergoing procedures and were compensated for their time. Participants underwent functional and structural neuroimaging on a Philips 3-Tesla magnet (Philips Medical Systems, Best, The Netherlands) with an 8-channel SENSE radiofrequency head coil. Foam padding was placed around the head to minimize motion during MRI scan acquisition. Participants completed standard neuropsychological tests (e.g., Brief Repeatable Battery of Neuropsychological tests [44]) and self-report measures regarding their general health and symptomology (i.e., SF-36 [48], Modified Fatigue Impact Scale (MFIS, [49]); see Table 2 for a complete list of model variables).

Table 2. Predictor Variables.

Predictor (Units if Available)	Predictor Category	What Predictor Measures
Normalized Grey Matter Volume (mm^3)	MR Image	Total grey matter volume normalized to skull
Normalized White Matter Volume (mm^3)	MR Image	Total white matter volume normalized to skull
Normalized Whole Brain Volume (mm^3)	MR Image	Total brain volume normalized to skull
Skeleton AD (mm^2/s)	MR Image	Diffusion along primary diffusion axis
Skeleton FA (proportion)	MR Image	Proportion of anisotropic diffusion
Skeleton MD (mm^2/s)	MR Image	Average Diffusion in primary diffusion axes
Skeleton RD (mm^2/s)	MR Image	Diffusion orthogonal to primary diffusion axis
T_2-FLAIR Lesion Burden-absolute lesion volume (mm^3)	MR Image	Total volume of lesioned brain tissue
T_2-FLAIR Lesion Burden-relative lesion volume (%)	MR Image	Total lesioned brain tissue relative to total white matter volume
T_2-FLAIR spatially distinct lesion count	MR Image	Total number of spatially distinct lesions
veBOLD (% signal change)	MR Image	Visual cortex BOLD response to visual stimulation task
veCBF (% signal change)	MR Image	Visual cortex CBF response to visual stimulation task
veCMRO$_2$ (% signal change)	MR Image	Visual cortex CMRO$_2$ response to visual stimulation task
ve$n$ (proportion)	MR Image	Visual cortex neural-vascular coupling
10/36 Delayed Recall (total correct after 15 min)	Neuropsych	Visuospatial memory/learning and delayed recall
10/36 Immediate Recall (total correct)	Neuropsych	Visuospatial memory/learning
25 Foot Walk (s)	Neuropsych	Walking ability and gait speed
9-Hole Peg Test-Dominant Hand (s)	Neuropsych	Finger and hand dexterity
9-Hole Peg Test-Non-dominant Hand (s)	Neuropsych	Finger and hand dexterity
Box Completion (items completed)	Neuropsych	Motor control
Controlled Oral Word Association Test (total correct)	Neuropsych	Verbal association fluency
Number Comparison (items completed)	Neuropsych	Processing speed

Table 2. *Cont.*

Predictor (Units if Available)	Predictor Category	What Predictor Measures
Paced Auditory Serial Addition Test 2 (% correct)	Neuropsych	Processing speed and selective/sustained attention
Paced Auditory Serial Addition Test 3 (% correct)	Neuropsych	Processing speed and selective/sustained attention
Selective Reminding Task Delayed (items recalled)	Neuropsych	Verbal learning and memory
Selective Reminding Task Long-term Storage (items recalled)	Neuropsych	Verbal learning and long-term memory
Symbol-digit Modalities Test (items completed)	Neuropsych	Sustained attention and concentration
Trail Making Task Form A (s)	Neuropsych	Visual search, attention, mental flexibility, and motor function
Trail Making Task Form B (s)	Neuropsych	Visual search, attention, mental flexibility, and motor function
Trail Making Task Form B-A (s)	Neuropsych	Visual search, attention, mental flexibility, and motor function
WAIS-III Digit Span Backward (items completed)	Neuropsych	Short-term, working memory
WAIS-III Digit Span Forward (items completed)	Neuropsych	Short-term, working memory
WAIS-III Digit Span Total (items completed)	Neuropsych	Short-term, working memory
WAIS-III Digit symbol coding (items completed)	Neuropsych	Performance subtest of WAIS
Modified Fatigue Impact Score	Symptoms	Fatigue symptomology
SF-36 Bodily Pain Scale	Symptoms	General measure of bodily pain
SF-36 Emotion	Symptoms	Role limitations due to emotional problems
SF-36 General Health Scale	Symptoms	General measure of health wellbeing
SF-36 Mental Health Scale	Symptoms	General measure of mental health
SF-36 Physical Functioning Scale	Symptoms	General measure of physical functioning
SF-36 Role Physical Function Scale	Symptoms	Role limitations due to physical problems
SF-36 Social Functioning Scale	Symptoms	General measure of social functioning
SF-36 Vitality Scale	Symptoms	General measure of energy/fatigue

FLAIR = Fluid-attenuated inversion recovery. WAIS = Wechsler adult intelligent scale. SF-36 = Short-form health survey. MR Image = magnetic resonance image; Neuropsych = neuropsychological test; Symptoms = self-report general health and symptom measures. Explanations of neuropsychological tests and symptom measures taken from [44,48,50,51].

2.3. cfMRI Parameters and Theory

Dual-echo pseudocontinuous arterial spin labeling (pCASL) and BOLD images (together referred to as dual-echo images) were acquired using an interleaved echo scanning protocol see [7,52]. Together, the perfusion (Echo 1) and BOLD-weighted (Echo 2) images along with biophysical modeling procedures allowed for estimation of CMRO$_2$ and a neural-vascular coupling coefficient (n, see [8]) associated with steady-state, neural stimulation [5,7]. One task run of dual-echo imaging data and one gas-challenge run of dual-echo imaging data were collected using the following parameters: Echo 1: labeling duration 1650 ms, labeling flip angle 18°, labeling gap = 63.5 mm, 3.44 × 3.44 × 5 mm voxel, repetition time (TR) = 4000 ms, echo time (TE) = 14 ms, 1525 ms post-label delay, 0 mm slice gap. Echo 2: 90° flip angle, 3.44 × 3.44 × 5 mm voxel, TR = 4000 ms, TE = 40 ms, 0 mm slice gap. Total scan time for the visual stimulation task = 600 s (72 dual-echo dynamics). Total scan time for the gas challenge = 624 s (75 dual-echo dynamics).

Estimations of CMRO2 and n were based upon the Davis model of BOLD signal change [6,7]:

$$\frac{\Delta S}{S_0} = M\left(\left(1 - \frac{\Delta CBF}{CBF_0}\right)^{\alpha-\beta}\left(\frac{\Delta CMRO_2}{CMRO_{2|0}}\right)^{\beta}\right) \tag{1}$$

where $\Delta x/x_0$ denotes a change from baseline, α is an empirically derived constant linking cerebral blood flow and cerebral blood volume, and β is an empirically derived constant related to vascular exchange and susceptibility of deoxyhemoglobin at specific field strengths (e.g., [53–55]). We assumed $\alpha = 0.38$ [56] and $\beta = 1.3$ [52]; these values were chosen because they have been shown to be sensitive to group differences in neurophysiology [22,23]. Also, these values have previously demonstrated group-equivalence in the estimation of M, e.g., [22,23]. M is a subject-specific scaling factor dependent upon the washout resting deoxyhemoglobin see [8]. M was estimated in each participant, using the gas challenge detailed below.

The measurement of BOLD, CBF, and M allows for the estimation of CMRO$_2$. Here, ΔCMRO$_2$ reflects the visual task-related change in neurometabolism of oxygen from resting baseline:

$$\frac{\Delta CMRO_2}{CMRO_{2|0}} = \left(1 - \frac{\frac{\Delta BOLD}{BOLD_0}}{M}\right)^{1/\beta}\left(\frac{\Delta CBF}{CBF_0}\right)^{1-\alpha/\beta} \tag{2}$$

where $\Delta x / x_0$ reflects percent change of signal during task compared to resting baseline. With the estimation of $\Delta CMRO_2$, n, may also be estimated:

$$n = \frac{\dfrac{\Delta CBF}{CBF_0}}{\dfrac{\Delta CMRO_2}{CMRO_{2|0}}} \qquad (3)$$

thus, n reflects per unit output of ΔCBF per unit input of $\Delta CMRO_2$ see [8].

2.4. cfMRI Task and Gas Challenge

Participants completed a visual stimulation task during dual-echo task imaging. This task was chosen for two reasons. First, differences in the functional response to visual stimulation have been observed in MS visual cortex see [42,57]. Second, because this task required minimal effort, group differences in performance were not expected to be a factor.

Participants were trained on the task before entering the MR environment. During the task, participants focused on a fixation cross at the center of their visual field. Participants were required to respond via bilateral, thumb-button press when a change in the luminance of the fixation cross occurred. This task was used in order to control the center of the participants' visual field [22,23,58]. Change in luminance was jittered and occurred every 2, 3, 4, or 6 s. Visual stimulation occurred in a block format. There were 6 visual stimulation task blocks consisting of 60 s of continual annulus flickering in the participants' near-foveal visual field. Annuli alternated at orthogonal orientations (0 to 90°) to avoid neural adaptation [58]. Alterations occurred at a constant frequency of 8 Hz because both electrochemical neural activity and BOLD signal have been shown to peak at this frequency, potentially yielding the greatest signal-to-noise estimates, e.g., [59,60]. Rest blocks were jittered at 32, 34, 36, 38, and 40 s intervals (see Figure 1).

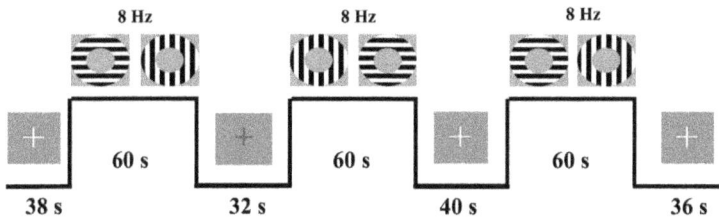

Figure 1. Example of three-trial visual stimulation task. Participants viewed a fixation cross at the center of the screen. This cross changed color at jittered intervals throughout task. Rest periods were also jittered. Continuous stimulation blocks lasted 60 s with 0° to 90° flickering annuli (at 8 Hz). *Note:* fixation cross was presented during task and rest periods however it cannot be seen in the task example periods here.

Participants also completed a gas-challenge in order to estimate *M*. Participants breathed 4 min of room air (~0.03% CO_2: 21% O_2: 78% N_2) and 6 min of an iso-oxic, CO_2 solution (5% CO_2: 21% O_2: 74% N_2) during dual-echo imaging. Each participant was fitted with a two-way, non-rebreathing valve/mouthpiece and a nose clip. Baseline end-tidal CO_2 (EtCO$_2$), O_2 saturation, breath rate, and heart rate measures were collected. After the 4 min of room air breathing, a valve was opened to release the CO_2 solution from a Douglas airbag which then flowed into the participants' breathing apparatus [22,23]. The CO_2 inhalation lasted 6 min.

Hypercapnic challenge, via the inhaled 5% CO_2 solution, increases global CBF, but probably has no or a minimal depressant effect on oxygen metabolism, e.g., [61–63]. Hypercapnia acts to wash out local baseline concentrations of deoxyhemoglobin, yielding a local maximum estimate of resting BOLD signal. Potential changes to oxygen metabolism due hypercapnic challenge have not been shown to appreciably alter the estimation of M as relationships between hypercapnia-derived M and M derived from non-hypercapnic techniques show high correspondence [64].

2.5. cfMRI Processing

Task and gas-challenge Echo 1 and Echo 2 data were processed in analysis of functional neuroimages (AFNI [65]) and the Functional MRI of the Brain Software Library (FSL [66]). Data were transformed into cardinal planes. Anomalous data points in each voxel time series were then attenuated using an interpolation method based upon the average signal. Data were volume registered to correct for motion to the fourth functional volume of each dataset's (task or gas challenge) Echo 2 sequence using a heptic polynomial interpolation method. CBF was estimated from Echo 1 images using the surround subtraction method [67]. Dual-echo BOLD data were also interpolated by pairwise averaging of temporally adjacent images.

For the visual stimulation task, Echo 2 data were linearly registered (12 degrees-of-freedom) to each participant's anatomical data using AFNI's *align_epi_anay.py* program. The transformation matrix from this registration was then applied to Echo 1 data, placing these two datasets in the same space. For gas-challenge data, a binary mask was created for functional voxels in Echo 2 to aid in co-registration. This mask was then registered to the respective participant's anatomical space using the *align_epi_anay.py* program. Gas-challenge Echo 2 and Echo 1 data were also aligned to the mask which was registered in native anatomical space. After alignment, Echoes 1 and 2 data from both the visual task and gas challenge were visually inspected for registration errors. One HC participant failed to register correctly after multiple attempts and was discarded from further analyses. Echoes 1 and 2 data from the visual task and gas challenge were then spatially smoothed using a Gaussian kernel (FWHM = 8 mm) and high-pass filtered (0.0039 Hz).

Preprocessed data from Echoes 1 and 2 in the visual stimulation task were analyzed via generalized linear modeling of task versus rest periods using a boxcar reference function. This modeling quantified task-related CBF and BOLD changes from baseline. BOLD and CBF beta-values were scaled to each voxel's resting baseline signal and were multiplied by 100, yielding percent signal change estimates from baseline (ΔBOLD and ΔCBF). Data were averaged from a visual (functional) region of interest (ROI) comprised of overlapping ΔBOLD and ΔCBF suprathreshold signals within occipital lobe (see Structural and Functional ROI; [22,23]). ΔBOLD, ΔCBF, ΔCMRO$_2$, and n results extracted from the functional region of interest were taken as the visual-evoked signals (i.e., veBOLD, veCBF, veCMRO$_2$, and ven).

For the gas challenge, resting baseline BOLD and CBF signals during room air breathing were averaged for each voxel time-series (BOLD$_0$ and CBF$_0$). The first two minutes of hypercapnia BOLD and CBF time-series were discarded to allow participants' blood flow to stabilize on the CO_2 solution, e.g., [22,23]. The last four minutes of hypercapnia BOLD and CBF time-series were averaged to yield BOLD$_{hc}$ and CBF$_{hc}$ respectively. Average values were extracted from a functional region of interest (see Structural and Functional ROI) using overlapping BOLD$_{hc}$ and CBF$_{hc}$ suprathreshold signals within occipital lobe, and were used to calculate M, using the following equation:

$$M = \frac{\dfrac{BOLD_{hc} - BOLD_0}{BOLD_0}}{\left(1 - \left(1 + \dfrac{CBF_{hc} - CBF_0}{CBF_0}\right)^{\alpha - \beta}\right)} \tag{4}$$

where $(x_{hc} - x_0)/x_0$ reflects percent change in signal from normocapnic to hypercapnic states, normalized by the signals during normocapnia and multiplied by 100. Once M was estimated, ΔCMRO$_2$ and n were also estimated (see Equations (2) and (3); see Figure 2) within a functional region of interest (see Structural and Functional ROI).

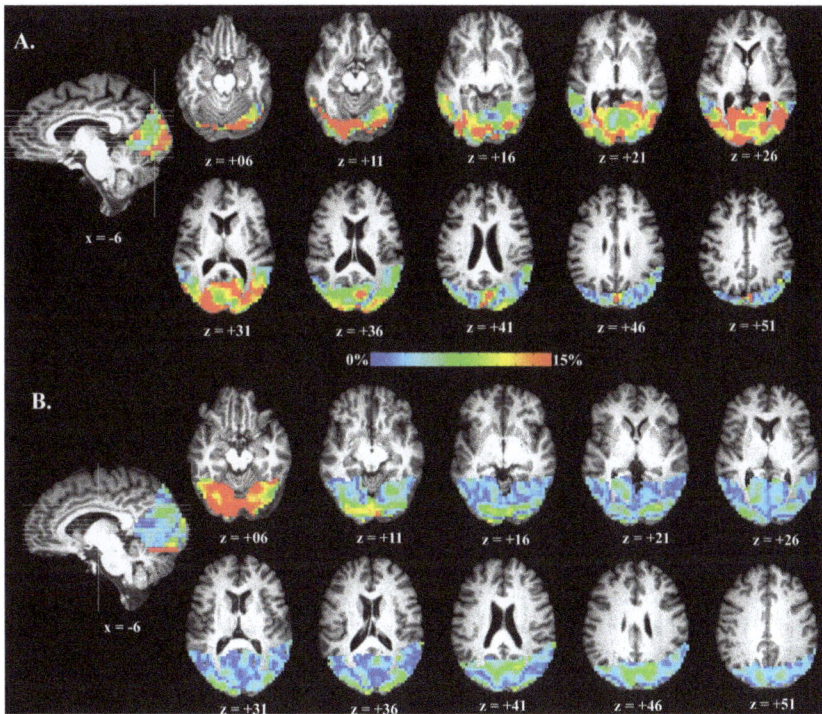

Figure 2. Examples of oxygen metabolism changes ($\Delta CMRO_2$) in occipital lobe. (**A**) HC $\Delta CMRO_2$; (**B**) MS patient $\Delta CMRO_2$. x = right-left, z = superior-inferior.

2.6. Structural and Functional ROIs

First, the magnetization-prepared rapid acquisition gradient-echo (MPRAGE) data were processed to create a native-space, occipital ROI. The skull was removed using an automated command, separating parenchyma and cerebral spinal fluid from the skull. An intensity based automated segmentation algorithm was used to delineate primarily white matter, grey matter, and cerebral spinal fluid voxels yielding a partial volume estimate of each tissue type, for each voxel. A grey matter mask was then created, retaining voxels with only a greater than or equal to grey matter partial volume estimate of 80%. A structural ROI of occipital lobe was manually delineated on each participant's MPRAGE image. These were drawn in native space because native space analyses tend to allow for more sensitive patient-control contrasts [68]. The structural ROI was drawn using gyral and sulcal landmarks and encompassed most of occipital cortex including calcarine sulcus, cuneus, and occipital portions of lingual gyrus. Several anatomical landmarks were used in the demarcation of this ROI (parieto-occipital sulcus, occipital pole, pre-occipital notch). Within the anatomically defined occipital lobe, only voxels with partial volume estimates of grey matter ($\geq 80\%$) were retained. These final masks were down-sampled to the functional voxel size.

A visual task functional ROI was created within the structural ROI described above to estimate veBOLD, veCBF, veCMRO$_2$, and ve$n$ (see Figure 3). This procedure eschewed noise from inactive voxels, e.g., [68]. Voxels comprising each participant's functional ROI were the overlapping top 5% of BOLD and top 5% of CBF t-values obtained from the generalized model, within the structural ROI. This ensured that average veBOLD and veCBF estimates were being derived from the same, task-responsive voxels and that veCMRO$_2$ and ve$n$ were derived in voxels with both CBF and BOLD task-related increases (see Figure 3).

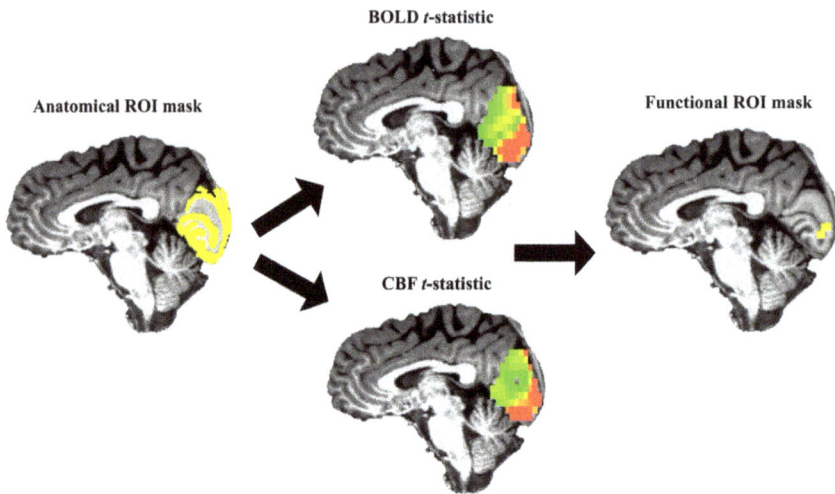

Figure 3. Graphical overview of masking procedure. For each participant, their top 5%, overlapping BOLD and CBF *t*-statistics (middle) within the anatomical ROI (left, yellow) were used to create the functional ROI mask (right, yellow). Functional measures (veBOLD, veCBF, veCMRO$_2$, and ve$n$) were extracted from each participant's functional ROI mask.

veCMRO$_2$ was calculated voxel-wise within the functional ROI using ΔBOLD, ΔCBF, M (which was extracted from functional ROI described below). ve$n$ was then calculated similarly. The final product of these analyses was average positive veBOLD, veCBF, and veCMRO$_2$, and ve$n$ extracted from the functional ROI (see Figure 3).

Because the gas challenge data differed in occipital coverage compared to the visual task data, M was estimated ex situ. To create a functional ROI for the gas challenge, ΔBOLD$_{hc}$/BOLD$_0$ and ΔCBF$_{hc}$/CBF$_0$ maps were thresholded and extracted from the structural ROI detailed above. The criteria for retention of a voxel within these maps required that the voxel was within the top 15% (top 20% for one participant) of ΔBOLD$_{hc}$/BOLD$_0$ and ΔCBF$_{hc}$/CBF$_0$ voxels in the structural ROI, and that these ΔBOLD$_{hc}$/BOLD$_0$ and ΔCBF$_{hc}$/CBF$_0$ voxels overlapped. This procedure ensured complementary maximum ΔBOLD$_{hc}$/BOLD$_0$ and ΔCBF$_{hc}$/CBF$_0$ signals in the retained voxels. Average ΔBOLD$_{hc}$/BOLD$_0$ and ΔCBF$_{hc}$/CBF$_0$ signals were extracted from this ROI and M was calculated (see Equation (4)).

2.7. Structural Images

One T$_1$-weighted MPRAGE image was acquired for each participant: 160 slices, TE = 3.7 ms, repetition time TR = 8.1 ms, sagittal slice orientation, $1 \times 1 \times 1$ mm^3 voxel, 12° flip angle. SIENAX [15,69] was used to obtain measures of grey matter, white matter, and total brain volume normalized by participant's head size. This technique uses partial volume estimation to calculate volume of differing tissue types (see Figure 4B,C). Further, this technique takes into account lesioned tissue, as demarcated by lesion masks (see below), in order to avoid misclassification of this tissue. The final products of these analyses were scaled estimates of each participant's grey matter, white matter, and total brain volume (mm^3).

A T$_2$ fluid attenuated inversion recovery (FLAIR) scan was also acquired for each participant: 33 slices, TE = 125 ms, TR = 11,000 ms, no slice gap, transverse slice orientation, $0.45 \times 0.45 \times 5.00$ mm^3 voxel, 120° refocusing angle. FLAIR images were used to estimate the extent of gross lesion burden for each participant. Hyperintense voxels were demarcated using in-house MATLAB code based upon slice-wise, signal intensity (i.e., voxels that were $\geq$1.25 SD over the slice mean intensity).

Next, lesions were manually delineated from the hyperintense tissue by two trained researchers (L.H., S.F.). Manual delineation ruled out false positives in lesion classification due to fat signals, motion, ventricular edge effects, skull, or signal inhomogeneites [70]. Lesion burden was estimated by extracting the number of voxels that were demarcated by the automated and manual procedures. Inter-rater agreement of lesion burden was calculated using a Dice ratio (κ) of the lesion burden estimates made by the two researchers on a sample of several subjects [71]. After the researchers were trained on lesion classification, inter-rater agreement was found to be high, $\kappa = 0.89$; where $\kappa > 0.70$ is generally thought to reflect excellent inter-rater agreement [72]. Lesion burden was quantified using absolute (total mm^3 of lesioned tissue; see Figure 4E) and relative scales (percent of total mm^3 of lesioned tissue scaled by uncorrected white matter volume in mm^3). Spatially distinct lesion count was also obtained by counting the number of non-touching lesions for each subject (see Figure 4F), e.g., [73]. A lesion was required to have at least 3 mm^3 volume in order to be added to the total lesion count. Thus, the final products of these analyses were absolute lesion volume, relative lesion volume, and spatially distinct lesion count.

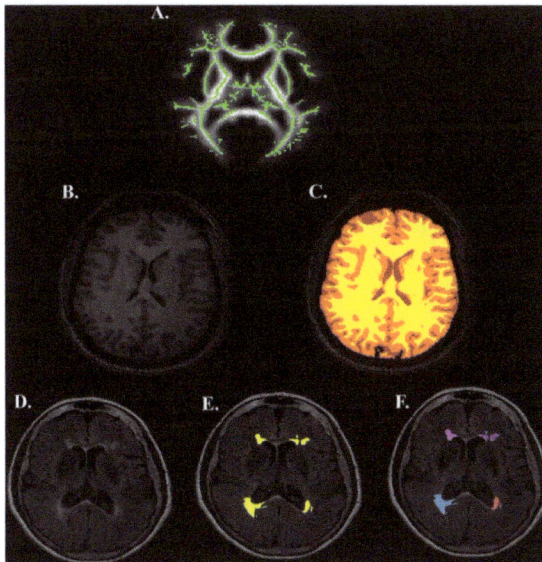

Figure 4. Diffusion and Structural Image Processing Examples. (**A**) Diffusion tensor imaging white matter skeleton. (**B**) T_1 image. (**C**) T_1 image segmented into white matter (yellow), grey matter (orange), and cerebral spinal fluid (red) using SIENAX. (**D**) T_2-FLAIR image. (**E**) Lesions demarcated (yellow) on T_2-FLAIR image used for calculating lesion burden. (**F**) Spatially distinct lesions demarcated on T_2-FLAIR image.

2.8. Diffusion Images

DTI images were acquired using a single-shot, echo-planar imaging sequence with a Sensitivity Encoding parallel imaging scheme (reduction factor = 2.3), 112 × 112 matrix, field of view = 224 × 224 mm^2 (nominal resolution of 2 mm), 65 slices (0 mm gap), slice thickness = 2 mm, TR = 7.78 s, TE = 97 ms. The diffusion weighting was encoded along 30 independent orientations [74] and the b value was 1000 s/mm^2. Imaging time was 5 min and 15 s. Two HCs did not undergo DTI (*n*HC = 11).

Automatic Image Registration [75] was performed on raw diffusion-weighted images to correct distortion caused by eddy currents. Six elements of the 3 × 3 diffusion tensor were determined by multivariate least-squares fitting. The tensor was diagonalized to obtain three eigenvalues (λ_{1-3})

and eigenvectors (v_{1-3}). Standard tensor fitting was conducted with DTIStudio [76] to generate the most common DTI-derived diffusion characteristics, fractional anisotropy (FA), axial diffusivity (AD), mean diffusivity (MD), and radial diffusivity (RD).

DTI measurements were obtained at the skeletons of the white matter using FSL [77] to alleviate partial volume effects with tract-based spatial statistics (see Figure 4F–H) [77]. Participant FA maps were registered nonlinearly to the EVE single-subject FA template [78–80] for better alignment with a digital white matter atlas (JHU ICBM-DTI-81) [81]. Registered FA maps of all subjects were averaged to generate a mean FA map, from which an FA skeleton mask was created. Skeletonized FA images of all subjects were obtained by projecting the registered FA images onto the mean FA skeleton mask. Skeletonized AD, MD, and RD metrics were obtained by applying the same registration, projection, and skeletonization procedures. We extracted skeleton-wide averages of each DTI metric (i.e., AD, FA, MD, RD), wherein an average of each metric is calculated across all voxels within the white matter skeleton (see Figure 4A).

2.9. Statistical Analyses

All analyses were performed on distributions free of outliers ($\geq \pm 2$ SD from group mean for simple group comparisons, $\geq \pm 3$ MAD from group median for classification modeling see [82]). Binary logistic regression was used for classifying MS status. A description of model variables can be found in Table 2. The accuracies of these models were computed as the proportion of correct classification outcomes over all outcomes. Accuracy was chosen as the metric of interest because it combines sensitivity and specificity in binary classification analysis by taking into account both true positives and true negatives relative to all outcomes. We used resampling-based hypothesis testing to examine both within-sample and out-of-sample classification of patient status see [83]. Because we used relatively conservative analytic techniques, inherently reducing the likelihood of Type I error and increasing the generalizability of our results, the criterion for a rejection of the null hypothesis was not corrected for multiple comparisons and all models were evaluated at the field-standard $\alpha = 0.05$. We also denote which hypothesis tests survived Benjamini-Hotchberg correction (Table 4; Figure 7).

Within-sample classification analyses obtained bias-corrected and accelerated (BCa) bootstrapped-resampled (B = 10,000) 95% confidence intervals of the accuracy of binary logistic regression models. The BCa procedure was used because it is robust to both skewness and sampling bias in the bootstrap distribution [84]. To avoid unstable classification, we stratified all resamples to match the original sample's constitution of patients and controls, 56.5% and 43.5%, respectively. If the BCa-derived 95% confidence interval did not contain a value at or below 0.50 (binary chance), this would demonstrate the measure's accuracy was significantly greater than chance to classify MS patients and HCs.

Out-of-sample classification analyses used a leave-one-out cross-validation approach [85]. This technique used training and sample iterations to test the ability of the model derived from the training set to predict an observation in the test (out-of-sample) set, thus, circumventing problems of sample bias, model over fitting, and lending a true predictive element to these analyses. Briefly, the leave-one-out cross validation (LOOCV) approach fitted N models, where N was proportional to our sample size. Each model was trained on N-1 samples and then the accuracy of the training model was assessed on the left-out sample. The N accuracies were then averaged to attain a representative and generalizable measure of the average out-of-sample classification accuracy. Permutation based p-values (5000 permutations) were computed to assess the significance of the LOOCV-derived accuracy statistics. The test permuted patient status labels and recomputed the accuracy of the model at each iteration, thus building the null distribution. The p-values were calculated from the percentage of the accuracy estimates of the permuted samples that were better than actual LOOCV-derived accuracy statistic of each model. This procedure was slightly modified according to Ojala and Garriga [86].

3. Results

3.1. Visual Task Performance

MS patients (92.75 ± 1.11%) did not significantly differ from HCs (94.86 ± 0.44%) on accuracy on the visual stimulation secondary task, $t(10.54) = -1.76$, $p = 0.108$. Patients (492.06 ms ± 31.15) also did not significantly differ from HCs (487.19 ms ± 24.10) on their average correct response time to press the button on the secondary task, $t(16.22) = 0.12$, $p = 0.903$.

3.2. Group Physiology, Cerebrovascular Response to Gas Challenge, and M

MS and HCs did not significantly differ in breath rate, end-tidal CO_2, heart rate, or O_2 saturation at baseline or during CO_2 solution breathing (all ps > 0.05; see Table 3). We tested whether MS patients differed in their CBF response to the CO_2 solution (($CBF_{hc} - CBF_0$)/CBF_0) and M in their respective gas challenge ROIs within occipital lobe see [87]. MS patients did not significantly differ in CBF response to the CO_2 solution (167.48 ± 19.8%) compared to HCs (146.90 ± 14.64%), $t(15.70) = 0.83$, $p = 0.417$. MS patients (3.88 ± 0.48%) did not significantly differ in M compared to HCs (5.11 ± 0.39%), $t(18.90) = -1.98$, $p = 0.062$.

Table 3. Sample Physiological Data.

	MS	HC	p
Baseline			
Breath Rate	11.20 (1.00)	10.25 (0.79)	0.747 [a]
EtCO$_2$	42.70 (1.81)	39.23 (0.74)	0.101 [b]
Heart Rate	66.90 (2.38)	72.08 (3.18)	0.207 [b]
SpO$_2$	98.10% (0.35%)	97.85% (0.32%)	0.596 [b]
5% CO$_2$			
Breath Rate	13.35 (1.28)	15.42 (1.07)	0.236 [c]
EtCO$_2$	48.95 (1.45)	49.06 (0.64)	0.950 [c]
Heart Rate	69.67 (2.38)	75.04 (2.60)	0.147 [d]
SpO$_2$	97.58% (0.39%)	98.20% (0.20%)	0.139 [d]

Mean (SEM). Breath Rate in breaths per minute. EtCO$_2$ = end-tidal CO_2 in mmHg. Heart Rate in beats per minute. SpO$_2$ = peripheral oxygen saturation in percent hemoglobin saturation. *p*-values were based on independent samples. [a] 22 degrees-of-freedom; [b] 21 degrees-of-freedom; [c] 16 degrees-of-freedom; [d] 17 degrees-of-freedom.

3.3. Group Comparisons on Visual Task cfMRI Measures

MS patients (1.12 ± 0.77%) did not significantly differ from HCs (1.18 ± 0.66%) on veBOLD response to visual stimulation, $t(19.18) = -0.60$, $p = 0.555$. MS patients (4.08 ± 0.35) did not show significant changes in ven compared to HCs (4.23 ± 0.23), $t(16.16) = -0.35$, $p = 0.731$. MS patients (48.06 ± 12.58%) had significant decreases in veCBF compared to HCs (92.68 ± 17.29%), $t(19.76) = -2.09$, $p = 0.050$. MS patients (9.59 ± 0.90%) also showed significant decreases in veCMRO$_2$ compared to HCs (17.85 ± 1.97%), $t(16.45) = -3.81$, $p = 0.002$ (see Figure 5).

3.4. Within-Sample Classification Analyses

Measures are ranked on original accuracy and presented in Table 4. Accuracy and smoothed density distributions for the significant and bottom 5 measures can be found in Figure 6.

3.5. Out-of-Sample Classification Analyses

Predictors presented in Figure 7 are ranked on LOOCV-derived accuracy.

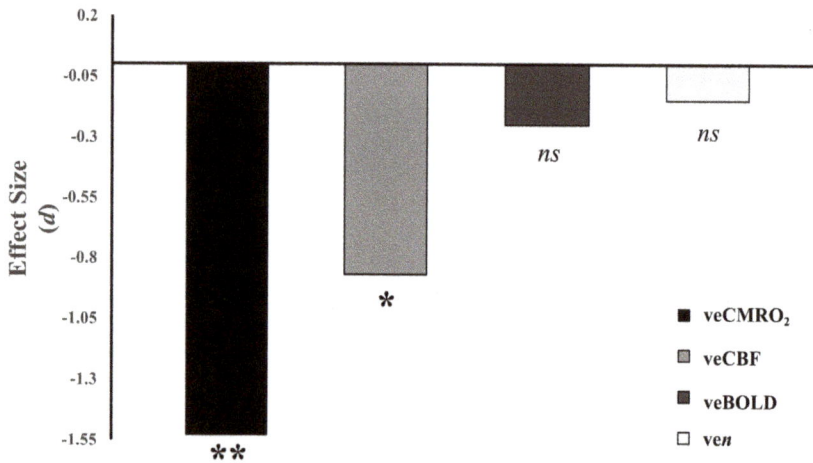

Figure 5. Effect sizes of group contrasts on calibrated functional magnetic imaging measures. Effect sizes reflect Cohen's *d*. *ns* = non-significant effect, $p > 0.05$; * $p < 0.05$; ** $p < 0.01$.

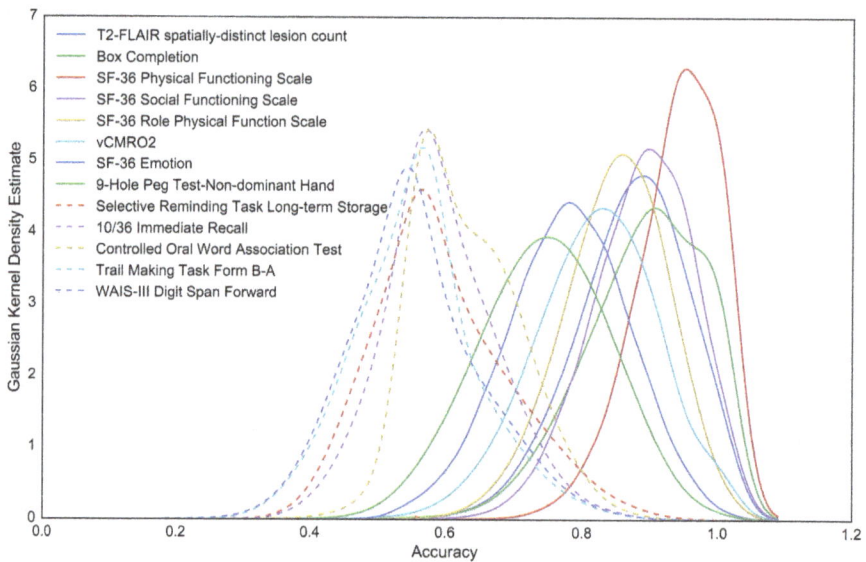

Figure 6. Smoothed density estimates of BCa-bootstap distributions. Distributions of significant (solid lines) and bottom 5 (dashed lines) within-sample predictors of MS status are illustrated. *Note*: because of smoothing, tails of distributions may exceed 1.

Table 4. Accuracy and 95% Confidence Limits of Within-Sample Classification Analyses.

Predictor	Predictor Accuracy	95% LCL	95% UCL	Significant
SF-36 Physical Functioning Scale	0.94	0.65	1.00	Yes †
SF-36 Social Functioning Scale	0.89	0.61	0.94	Yes †
T_2-FLAIR spatially distinct lesion count	0.86	0.57	0.95	Yes †
Box Completion	0.86	0.52	0.95	Yes †
SF-36 Role Physical Function Scale	0.85	0.60	0.95	Yes †
veCMRO$_2$	0.82	0.55	0.91	Yes ‡
Normalized Grey Matter Volume	0.81	0.43	0.95	No ‡
T_2-FLAIR Lesion Burden-absolute lesion volume	0.80	0.50	0.90	No ‡
T_2-FLAIR Lesion Burden-relative lesion volume	0.80	0.50	0.90	No ‡
SF-36 Emotion	0.78	0.56	0.89	Yes
9-Hole Peg Test-Non-dominant Hand	0.77	0.55	0.91	Yes ‡
SF-36 General Health Scale	0.77	0.50	0.86	No ‡
veCBF	0.75	0.45	0.85	No ‡
Normalized Whole Brain Volume	0.73	0.45	0.86	No
9-Hole Peg Test-Dominant Hand	0.73	0.50	0.82	No
SF-36 Bodily Pain Scale	0.73	0.45	0.86	No
Skeleton AD	0.71	0.43	0.81	No
Skeleton MD	0.71	0.48	0.86	No
Paced Auditory Serial Addition Test 2 s	0.71	0.48	0.86	No
Modified Fatigue Impact Score Total	0.70	0.43	0.78	No ‡
Normalized White Matter Volume	0.68	0.45	0.82	No
Paced Auditory Serial Addition Test 3 s	0.68	0.45	0.82	No
Skeleton RD	0.67	0.48	0.76	No
Trail Making Task Form A	0.65	0.43	0.78	No
SF-36 Vitality Scale	0.65	0.43	0.74	No
25 Foot Walk	0.64	0.50	0.77	No
WAIS-III Digit Span Backward	0.64	0.41	0.77	No
WAIS-III Digit Span Total	0.64	0.41	0.82	No
10/36 Delayed Recall	0.63	0.42	0.74	No
Trail Making Task Form B	0.62	0.33	0.76	No
SF-36 Mental Health Scale	0.62	0.38	0.62	No
veBOLD	0.61	0.48	0.78	No
Selective Reminding Task Delayed	0.60	0.35	0.60	No
Symbol-digit Modalities Test	0.60	0.30	0.70	No
Number Comparison	0.59	0.36	0.68	No
WAIS-III Digit symbol coding	0.58	0.37	0.58	No
Skeleton FA	0.57	0.37	0.67	No
ve*n*	0.57	0.39	0.52	No
Selective Reminding Task Long-term Storage	0.57	0.35	0.70	No
Controlled Oral Word Association Test	0.57	0.35	0.57	No
10/36 Immediate Recall	0.52	0.30	0.57	No
WAIS-III Digit Span Forward	0.50	0.27	0.55	No
Trail Making Task Form B-A	0.48	0.29	0.52	No

LCL = lower confidence limit. UCL = upper confidence limit. Confidence limits based upon 10,000 iteration BCa-corrected bootstrapping procedure. Yes = 95% confidence interval (CI) does not contain 0.50; No = 95% CI contains 0.50. Note: that the original parameter estimates do not necessarily need to lie within the 95% CI of the BCa-corrected, empirically derived distributions. † permutation *p*-value significant using Benjamini-Hotchberg correction ($p < 0.05$). ‡ permutation *p*-value marginally significant using Benjamini-Hotchberg correction ($p < 0.10$).

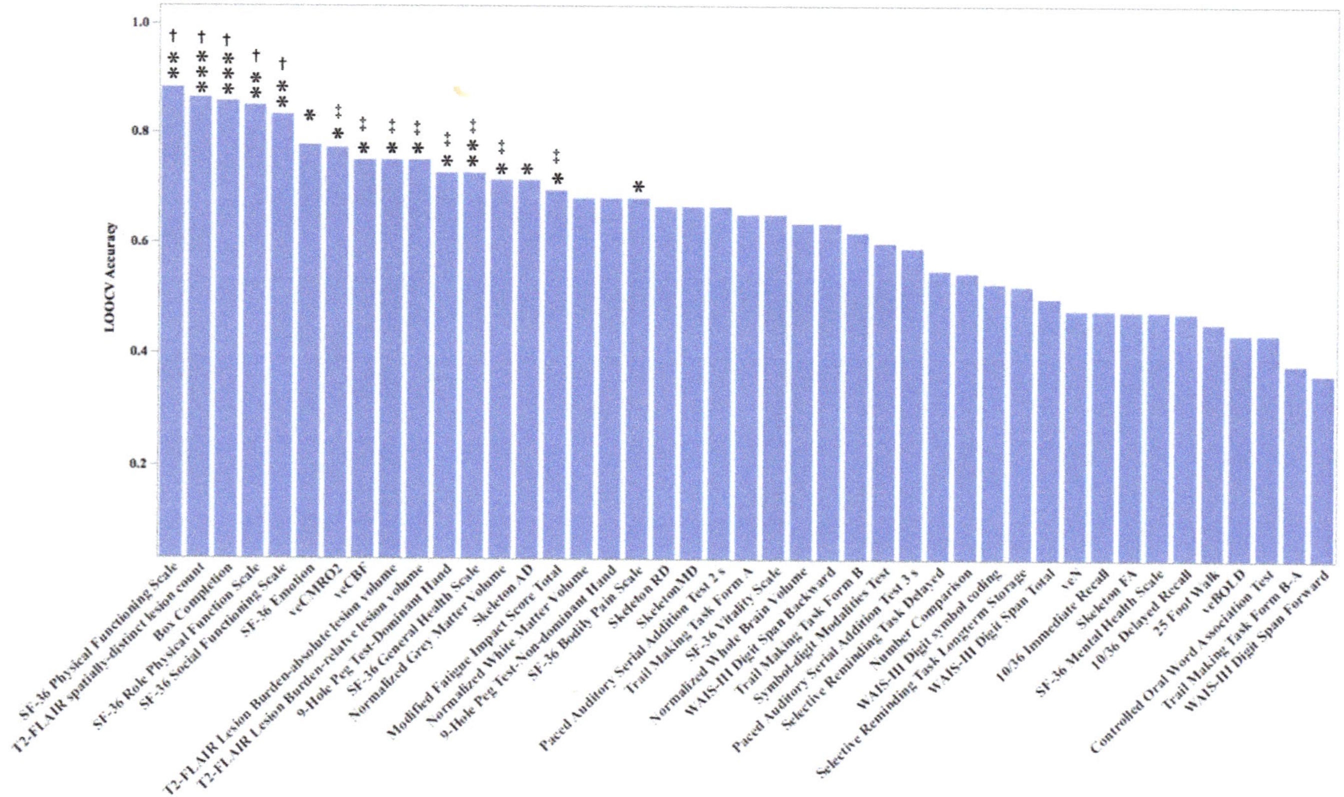

Figure 7. Leave-one-out cross-validation (LOOCV) out-of-sample classification accuracy of each model. * $p < 0.05$; ** = $p < 0.01$; *** = $p < 0.001$. † p-value also significant using Benjamini-Hotchberg correction ($p < 0.05$). ‡ p-value marginally significant using Benjamini-Hotchberg correction ($p < 0.10$).

4. Discussion

In the present study, we used a neuroimaging approach novel to MS research (cfMRI) to assess the accuracy of veCMRO$_2$ in classifying MS patients and closely age- and sex-matched HC participants. MS patients showed similar responses to HCs in veBOLD and ve$n$, however showed decreased veCBF and a pronounced decrease in veCMRO$_2$ relative to HCs. Groups were similar on visual task performance and on physiological measures pertaining to the CO$_2$ challenge, indicating that potential MS-related changes in physiological response to carbon dioxide, e.g., [87] or visual attention were not likely contributors to group CMRO$_2$ differences. Within-sample classification analyses demonstrated that veCMRO$_2$ was significant and one of the top measures to accurately classify MS status, discriminating between MS patients and HCs with exceptional accuracy (82%). Results also showed that within-sample classification accuracy by veCMRO$_2$ was comparable to neuroimaging measures often used to gauge MS pathology, such as T$_2$-FLAIR lesion burden (80% accuracy) and T$_1$ grey matter volume (81% accuracy). veCMRO$_2$ was also significantly accurate in MS classification using out-of-sample observations (77% accuracy). The use of such out-of-sample modeling afforded a predictive element to this study and demonstrated that veCMRO$_2$ can accurately classify new observations of MS and HC participants, offering support for its potential diagnostic utility.

One question that arises from these results is whether veCMRO$_2$ can add predictive value over other advanced imaging techniques not studied here. For instance, measurements of multifocal visual-evoked potentials have been of great interest to the MS research community. This technique, which uses visual stimulation and electroencephalogram signals in occipital channels proximal to the inion has been demonstrated to (1) more sensitively and specifically detect visual abnormalities in MS eyes relative to other visual-system measurements [88], (2) predict conversion to an MS diagnosis in persons with optic neuritis [89], and (3) relate to the extent of MS-related damage to visual white matter tracts [41]. Not surprisingly, this technique can also accurately discriminate between MS patients and HCs, e.g., [90]. For example, one study showed that measurements gathered from multifocal visual-evoked potentials were on average 74.76% accurate (range: 62.7%–96.1%) in classifying within-sample observations of MS patients without optic neuritis and HCs ([90], average calculated from Figures 5 and 6, pp. 910–911). We can compare these figures with the within-sample accuracy of veCMRO$_2$ observed here (82%). This suggests that veCMRO$_2$ accuracy is in about the same range as multifocal evoked potentials. However, it performs appreciably better than the average multifocal evoked potential measure. Future research directly comparing veCMRO$_2$ to electroencephalogram and other measures is necessary to more faithfully adjudicate claims about the relative performance of this technique.

A second avenue for future research could involve examining whether the integration of evoked CMRO$_2$ from other neural systems could maximize MS classification accuracy. Here, we showed significant decreases in MS patients' veCMRO$_2$ relative to HCs. This variable was also largely accurate in the prediction of MS status. We looked at veCMRO$_2$ specifically because of robust alterations to the visual system in MS see [37–40]. However, because (1) mitochondrial alterations are found in multiple forms of neural tissue in MS [31,33] and (2) global brain decreases in oxygen metabolism have been found in MS patients relative to HCs [30], it is likely that evoked CMRO$_2$ is affected in other neural systems as well. Our work and others' have shown altered patterns of brain activity in MS patients in motor, e.g., [42,91,92] and association cortices [43,93–95], see [96]. It is possible that the addition of measures of evoked CMRO$_2$ in these areas could lend improvements in the accuracy of MS classification. One advantage of the cfMRI approach over other advanced imaging approaches in MS, like OCT or visual-evoked potentials, is that this technique can specifically and simultaneously assay multiple neural systems. Work underway in our laboratories is examining the extent to which evoked motor and executive system CMRO$_2$ differs between MS patients and age- and sex-matched healthy HCs, and whether these changes, along with veCMRO$_2$, can help build optimal neurodiagnostic models of MS.

The utility of imaging biomarkers in MS is not limited to assisting in diagnosis see [97]. For instance, OCT measures have been shown to be effective in predicting brain atrophy and visual acuity loss in MS see [38]. The retinal nerve fiber thickness and macular volume measures from OCT might also be useful in differentiating different subtypes of MS [98]. Other imaging-based measures, such as T_2-lesion burden, have shown prognostic ability by prediction of future MS disability, e.g., [99], see also [100–102]. One potential avenue for future research is to evaluate the use of oxygen metabolism signals in MS prognosis. For example, Ge and colleagues' [27] research showed that lower resting brain-wide levels of oxygen metabolism were associated with both increased neurological disability and increased lesion burden in MS patients. Although these findings were cross-sectional, they suggested that oxygen metabolism could be a marker of the trajectory of disease course. To wit, future longitudinal work should examine whether measures of oxygen metabolism in early MS can predict future disease progression cf. [89]. veCMRO$_2$ or resting oxygen metabolic markers could also be evaluated for their abilities to predict the transition from risk states (such as clinically or radiologically isolated syndrome) to clinically definite MS see [100,102,103].

A recent wave of findings related to metabolic dysfunction in MS has led to metabolic hypotheses to explain the pathophysiology of MS see [34–36]. For instance, Paling and colleagues furthered an energy failure hypothesis of the pathophysiology of MS [35,104]. These authors postulated a link between white matter damage and energy demand in MS, wherein this damage causes neuroenergetic demand to exceed the supply of metabolic substrate. This hypothesis is largely consistent with the findings of the present study, wherein the observed relative decrease in veCBF (the supply of oxygen and glucose) in MS might have limited the neurometabolic response (veCMRO$_2$) relative to HCs. Further, issues of oxygen extraction due to mitochondrial damage/dysfunction could have also contributed to the relative decrease in veCMRO$_2$ for MS patients relative to HCs see [34–36].

Imaging techniques here and elsewhere have produced convincing biomarkers of MS see [38,97,100]. However, MS is a complex, multifaceted disease. Thus, it is not surprising that our results revealed a diverse array of measures that were accurate in classifying MS patients and HCs. The goal of this work was to examine the ability of a new marker (veCMRO$_2$) to accurately classify MS. However, a truly prodigious advance in MS diagnostics will likely evolve from models that combine many relevant factors. It is possible that a "gold-standard" model of MS diagnostics would contain information about evoked CMRO$_2$, along with other information like lesion count, self-reported symptomology, neuropsychological performance, and potentially other strong associates of MS not examined here (e.g., low-contrast letter acuity performance see [105], oligoclonal band status [106], retinal nerve fiber layer thickness see [38]). For instance, research from the Alzheimer's Disease Neuroimaging Initiative showed that a complement of multimodal neuroimaging, cerebrospinal fluid proteins, along with standard clinical evaluations allow for optimal prediction of conversion from mild cognitive impairment to Alzheimer's disease [107]; see also [108] for application in psychiatry.

5. Conclusions

This study was the first to apply cfMRI in an MS sample. Presently, the intricacies of cfMRI acquisition and post-acquisition processing probably hinder it from having an immediate impact upon routine diagnosis or tracking of MS. However, acquisition continues to be optimized and research is showing promise toward eliminating the gas-challenge component of this method, see [8], which should increase the ease of cfMRI administration and the diversity of patients in which it can be applied. With contemporary research highlighting the importance of neurometabolism in the pathophysiology of MS and continued optimization of this technique, cfMRI shows promise as a translational diagnostic/prognostic tool for MS.

Our findings demonstrated that veCMRO$_2$ was accurate in classifying both within- and out-of-sample observations of MS patients and HCs. Out-of-sample analyses suggested that predictive models using veCMRO$_2$ could be useful in MS diagnostics and potentially new cases of MS. Although out-of-sample analyses provide confidence in the generalizability of our findings, larger, independent samples are desirable to confirm the robustness of these effects. However, the present findings represent an encouraging first step in realizing the diagnostic relevance of veCMRO$_2$ in MS.

Acknowledgments: This work was supported by grants from the National Multiple Sclerosis Society (to DTO and BR; number RG-1507-04951; and to BR RG-1510-06687) and from the National Institutes of Health (to BR and HL; number 5RO1AG047972-02). The authors wish to thank Hannah Grotzinger and Judith Gallagher for their contributions to manuscript preparation.

Author Contributions: N.A.H. contributed to study design, data analysis, and wrote the manuscript. Y.S.A. and C.C. contributed to statistical analysis. M.P.T., L.H. and B.P.T. contributed to study design, data collection and analyses, and manuscript writing. M.O. and H.H., contributed to diffusion tensor imaging analyses. S.F. contributed to data collection and data processing. J.H., Jr., D.T.O. and B.R. contributed to study design, conceptualization, and manuscript writing.

Conflicts of Interest: N.A.H., Y.S.A., C.C., M.O., M.P.T., L.H., S.F., B.P.T., J.H., Jr., H.H. and B.R. declare no perceived conflicts of interests. D.T.O. received lecture fees from Acorda, Genzyme, and TEVA Neuroscience, consulting and advisory board fees from EMD Serono, Genentech, Genzyme, Novartis and TEVA Neuroscience, and research support from Biogen not related to this study.

References

1. Polman, C.H.; Reingold, S.C.; Banwell, B.; Clanet, M.; Cohen, J.A.; Filippi, M.; Fujihara, K.; Havrdova, E.; Hutchison, M.; Kappos, L.; et al. Diagnostic Criteria for Multiple Sclerosis: 2010 Revisions to the McDonald Criteria. *Ann. Neurol.* **2011**, *69*, 292–302. [CrossRef] [PubMed]
2. George, I.C.; Sati, P.; Absinta, M.; Cortese, I.C.M.; Sweeney, E.M.; Shea, C.D.; Reich, D.S. Clinical 3-Tesla FLAIR* MRI Improves Diagnostic Accuracy in Multiple Sclerosis. *Mult. Scler.* **2016**, *22*, 1578–1586. [CrossRef] [PubMed]
3. Wattjes, M.P.; Barkhof, F. High Field MRI in the Diagnosis of Multiple Sclerosis: High Field-High Yield? *Neuroradiology* **2009**, *51*, 279–292. [CrossRef] [PubMed]
4. Metcalf, M.; Xu, D.; Okuda, D.T.; Carvajal, L.; Srinivasan, R.; Kelley, D.A.C.; Mukherjee, P.; Nelson, S.J.; Vigneron, D.B.; Pelletier, D. High-Resolution Phased-Array MRI of the Human Brain and 7 Tesla: Initial Experience in Multiple Sclerosis Patients. *J. Neuroimaging* **2010**, *20*, 141–147. [CrossRef] [PubMed]
5. Oberwahrenbrok, T.; Ringelstein, M.; Jentschke, S.; Deuschle, K.; Klumbies, K.; Bellmann-Strobl, J.; Harmel, J.; Ruprecht, K.; Schippling, S.; Hartung, H.P.; et al. Retinal Ganglion Cell and Inner Plexiform Layer Thinning in Clinically Isolated Syndrome. *Mult. Scler.* **2013**, *19*, 1887–1895. [CrossRef] [PubMed]
6. Davis, T.L.; Kwong, K.K.; Weiskoff, R.M.; Rosen, B.R. Calibrated Functional MRI: Mapping the Dynamics of Oxidative Metabolism. *Proc. Natl. Acad. Sci. USA* **1998**, *95*, 1834–1839. [CrossRef] [PubMed]
7. Hoge, S.A.; Atkinson, J.; Gill, B.; Crelier, G.R.; Marrett, S.; Pike, G.B. Linear Coupling Between Cerebral Blood Flow and Oxygen Consumption in Activated Human Cortex. *Proc. Natl. Acad. Sci. USA* **1999**, *96*, 9403–9408. [CrossRef] [PubMed]
8. Hoge, R.D. Calibrated fMRI. *NeuroImage* **2012**, *62*, 930–937. [CrossRef] [PubMed]
9. Herman, P.; Sanganahalli, B.G.; Blumenfeld, H.; Hyder, F. Cerebral Oxygen Demand for Short-Lived and Steady-State Events. *J. Neurochem.* **2009**, *109*, 73–79. [CrossRef] [PubMed]
10. Herman, P.; Sanganahalli, B.G.; Blumenfeld, H.; Rothman, D.L.; Hyder, F. Quantitative Basis for Neuroimaging of Cortical Laminae with Calibrated Functional MRI. *Proc. Natl. Acad. Sci. USA* **2013**, *110*, 15115–15120. [CrossRef] [PubMed]
11. Hyder, F.; Kida, I.; Behar, K.L.; Kennan, R.P.; Maciejewski, P.K.; Rothman, D.L. Quantitative Functional Imaging of the Brain: Towards Mapping Neuronal Activity by BOLD fMRI. *NMR Biomed.* **2001**, *14*, 413–431. [CrossRef] [PubMed]
12. Hyder, F.; Rothman, D.L.; Shulman, R.G. Total Neuroenergetics Support Localized Brain Activity: Implications for the Interpretation of fMRI. *Proc. Natl. Acad. Sci. USA* **2002**, *99*, 10771–10776. [CrossRef] [PubMed]
13. Hyder, F. Neuroimaging with Calibrated FMRI. *Stroke* **2004**, *35*, 2635–2641. [CrossRef] [PubMed]
14. Lin, A.-L.; Fox, P.T.; Hardies, J.; Duong, T.Q.; Gao, J.H. Nonlinear Coupling Between Cerebral Blood Flow, Oxygen Consumption, and ATP Production in Human Visual Cortex. *Proc. Natl. Acad. Sci. USA* **2010**, *107*, 8446–8451. [CrossRef] [PubMed]

15. Smith, A.J.; Blumenfeld, H.; Behar, K.J.; Rothman, D.L.; Shulman, R.G.; Hyder, F. Cerebral Energetics and Spiking Frequency: The Neurophysiological Basis of fMRI. *Proc. Natl. Acad. Sci. USA* **2002**, *99*, 10765–10770. [CrossRef] [PubMed]

16. He, B.J.; Snyder, A.Z.; Zempel, J.M.; Smyth, M.D.; Raichle, M.E. Electrophysiological Correlates of the Brain's Intrinsic Large-Scale Functional Architecture. *Proc. Natl. Acad. Sci. USA* **2008**, *105*, 16039–16044. [CrossRef] [PubMed]

17. Leopold, D.A.; Maier, A. Ongoing Physiological Processes in the Cerebral Cortex. *NeuroImage* **2012**, *62*, 2190–2200. [CrossRef] [PubMed]

18. Logothetis, N.K.; Pauls, J.; Augath, M.; Trinath, T.; Oeltermann, A. Neurophysiological Investigation of the Basis of the fMRI Signal. *Nature* **2001**, *412*, 150–157. [CrossRef] [PubMed]

19. Lu, H.; Zuo, Y.; Gu, H.; Waltz, J.A.; Zhan, W.; Scholl, C.A.; Rea, W.; Yang, W.; Stein, E.A. Synchronized Delta Oscillations Correlate with the Resting-State Functional MRI Signal. *Proc. Natl. Acad. Sci. USA* **2007**, *104*, 18265–18269. [CrossRef] [PubMed]

20. Zhu, Z.; Johnson, N.F.; Kim, C.; Gold, B.T. Reduced Frontal Cortex Efficiency is Associated with Lower White Matter Integrity in Aging. *Cereb. Cortex* **2015**, *25*, 138–146. [CrossRef] [PubMed]

21. Mark, C.I.; Mazerolle, E.L.; Chen, J.J. Metabolic and Vascular Origins of the BOLD Effect: Implications for Imaging Pathology and Resting-State Brain Function. *J. Magn. Reson. Imaging* **2015**, *42*, 231–246. [CrossRef] [PubMed]

22. Hutchison, J.L.; Lu, H.; Rypma, B. Neural Mechanisms of Age-Related Slowing: The $\Delta CBF/\Delta CMRO_2$ Ratio Mediates Age-Differences in BOLD Signal and Human Performance. *Cereb. Cortex* **2013**, *23*, 2337–2346. [CrossRef] [PubMed]

23. Hutchison, J.L.; Shokri-Kojori, E.; Lu, H.; Rypma, B. A BOLD Perspective on Age-Related Neurometabolic-Flow Coupling and Neural Efficiency Changes in Human Visual Cortex. *Front. Psychol.* **2013**, *4*, 1–12. [CrossRef] [PubMed]

24. Iannetti, G.D.; Wise, R.G. BOLD Functional MRI in Disease and Pharmacological Studies: Room for Improvement? *Magn. Reson. Imaging* **2007**, *25*, 978–988. [CrossRef] [PubMed]

25. Cader, S.; Johansen-Berg, H.; Wylezinska, M.; Palace, J.; Behrens, T.E.; Smith, S.; Matthews, P.M. Discordant White Matter N-acetylasparate and Diffusion MRI Measure Suggest that Chronic Metabolic Dysfunction Contributes to Axonal Pathology in Multiple Sclerosis. *NeuroImage* **2007**, *36*, 19–27. [CrossRef] [PubMed]

26. Pfueller, C.F.; Brandt, A.U.; Schubert, F.; Bock, M.; Walaszek, B.; Waiszies, H.; Schwenteck, T.; Dörr, J.; Bellmann-Strobl, J.; Mohr, C.; et al. Metabolic Changes in the Visual Cortex are Linked to Retinal Nerve Fiber Layer Thinning in Multiple Sclerosis. *PLoS ONE* **2011**, *6*, e18019. [CrossRef] [PubMed]

27. Hannoun, S.; Bagory, M.; Durand-Dubief, F.; Ibarrola, D.; Comte, J.C.; Confavreux, C.; Cotton, F.; Sappey-Marinier, D. Correlation of diffusion and Metabolic Alterations in Different Clinical Forms of Multiple Sclerosis. *PLoS ONE* **2012**, *7*, e32525. [CrossRef]

28. Sijens, P.E.; Irwan, R.; Potze, J.H.; Mostert, J.P.; De Keyser, J.; Ouderk, M. Analysis of the Human Brain in Primary Progressive Multiple Sclerosis with Mapping of the Spatial Distributions Using 1H MR Spectroscopy and Diffusion Tensor Imaging. *Eur. Radiol.* **2005**, *15*, 1686–1693. [CrossRef] [PubMed]

29. Sun, X.; Tanaka, M.; Kondo, S. Clinical Significance of Reduced Cerebral Metabolism in Multiple Sclerosis: A Combined PET and MRI Study. *Ann. Nucl. Med.* **1998**, *12*, 89–94. [CrossRef] [PubMed]

30. Ge, Y.; Zhang, Z.; Lu, H.; Tang, L.; Jaggi, H.; Herbert, J.; Babb, J.S.; Rusinek, H.; Grossman, R.I. Characterizing Brain Oxygen Metabolism in Patients with Multiple Sclerosis with T2-Relaxation-Under-Spin-Tagging MRI. *J. Cereb. Blood Flow Metab.* **2012**, *32*, 403–412. [CrossRef] [PubMed]

31. Dutta, R.; McDonough, J.; Yin, X.; Peterson, J.; Chang, A.; Torres, T.; Gudz, T.; Macklin, W.B.; Lewis, D.A.; Fox, R.J.; et al. Mitochondrial Dysfunction as a Cause of Axonal Degeneration in Multiple Sclerosis Patients. *Ann. Neurol.* **2006**, *59*, 478–489. [CrossRef] [PubMed]

32. Mahad, D.J.; Ziabreva, I.; Campbell, G.; Lax, N.; White, K.; Hanson, P.S.; Lassmann, H.; Turnbull, D.M. Mitochondrial Changes Within Axons in Multiple Sclerosis. *Brain* **2009**, *132*, 1161–1174. [CrossRef] [PubMed]

33. Singhal, N.K.; Li, S.; Arning, E.; Alkhayer, K.; Clements, R.; Sarcyk, Z.; Dassanayake, R.S.; Brasch, N.E.; Freeman, E.J.; Bottiglieri, T.; et al. Changes in Methionine Metabolism and Histone H3 Trimethylation are Linked to Mitochondrial Defects in Multiple Sclerosis. *J. Neurosci.* **2015**, *35*, 15170–15186. [CrossRef] [PubMed]

34. Cambron, M.; D'haeseleer, M.; Laureys, G.; Clinckers, R.; Debruyne, J.; De Keyser, J. White-Matter Astrocytes, Axonal Energy Metabolism, and Axonal Degeneration in Multiple Sclerosis. *J. Cereb. Blood Flow Metab.* **2012**, *32*, 413–424. [CrossRef] [PubMed]

35. Paling, D.; Golay, X.; Wheeler-Kingshott, C.; Kapoor, R.; Miller, D. Energy Failure in Multiple Sclerosis and its Investigation Using MR Techniques. *J. Neurol.* **2011**, *258*, 2113–2127. [CrossRef] [PubMed]

36. Su, K.; Bourdette, D.; Forte, M. Mitochondrial Dysfunction and Neurodegeneration in Multiple Sclerosis. *Front. Physiol.* **2013**, *4*, 1–10. [CrossRef] [PubMed]

37. Frohman, E.M.; Frohman, T.C.; Zee, D.S.; McColl, R.; Galetta, S. The Neuro-Ophthalmology of Multiple Sclerosis. *Lancet Neurol.* **2005**, *4*, 111–121. [CrossRef]

38. Frohman, E.M.; Fujimoto, J.G.; Frohman, T.C.; Calabresi, P.A.; Cutter, G.; Balcer, L.J. Optical Coherence Tomography: A Window Into the Mechanisms of Multiple Sclerosis. *Nat. Clin. Pract. Neurol.* **2008**, *4*, 664–675. [CrossRef] [PubMed]

39. Graham, S.L.; Klistorner, A. Afferent Visual Pathways in Multiple Sclerosis: A Review. *Clin. Exp. Ophthalmol.* **2017**, *45*, 62–72. [CrossRef] [PubMed]

40. Kolappan, M.; Henderson, A.P.D.; Jenkins, T.M.; Wheeler-Kingshott, C.A.; Plant, G.T.; Miller, D.H. Assessing Structure and Function of the Afferent Visual Pathway in Multiple Sclerosis and Associated Optic Neuritis. *J. Neurol.* **2009**, *256*, 305–319. [CrossRef] [PubMed]

41. Alshowaeir, D.; Yiannikas, C.; Garrick, R.; Paratt, J.; Barnett, M.H.; Graham, S.L.; Klistorner, A. Latency of Multifocal Visual Evoked Potentials in Nonoptic Neuritis Eyes of Multiple Sclerosis Patients Associated with Optic Radiation Lesions. *Investig. Ophthalmol. Vis. Sci.* **2014**, *55*, 3758–3764. [CrossRef] [PubMed]

42. Hubbard, N.A.; Turner, M.; Hutchison, J.L.; Ouyang, A.; Strain, J.; Oasay, L.; Sundaram, S.; Davis, S.; Remington, G.; Brigante, R.; et al. Multiple Sclerosis-Related White Matter Microstructural Change Alters the BOLD Hemodynamic Response. *J. Cereb. Blood Flow Metab.* **2016**, *36*, 1872–1884. [CrossRef] [PubMed]

43. Hubbard, N.A.; Hutchison, J.L.; Turner, M.P.; Sundaram, S.; Oasay, L.; Robinson, D.; Strain, J.; Weaver, T.; Davis, S.L.; Remington, G.M.; et al. Asynchrony in Executive Networks Predicts Cognitive Slowing in Multiple Sclerosis. *Neuropsychology* **2016**, *30*, 75. [CrossRef] [PubMed]

44. Rao, S.M. Cognitive Function Study Group of the National Multiple Sclerosis Society. In *A Manual for the Brief Repeatable Battery of Neuropsychological Tests in Multiple Sclerosis*; Medical College of Wisconsin: Milwaukee, WI, USA, 1990.

45. Brandt, J.; Spencer, M.; Folstein, M. The Telephone Interview for Cognitive Status. *Neuropsychiatry Neuropsychol. Behav. Neurol.* **1988**, *1*, 111–117.

46. Verdier-Taillefer, M.H.; Roullet, E.; Cesaro, P.; Alpérovitch, A. Validation of Self-Reported Neurological Disability in Multiple Sclerosis. *Int. J. Epidemiol.* **1994**, *23*, 148–154. [CrossRef] [PubMed]

47. Perthen, J.E.; Lansing, A.E.; Liau, J.; Liu, T.T.; Buxton, R.B. Caffeine-Induced Uncoupling of Cerebral Blood Flow and Oxygen Metabolism: A Calibrated BOLD fMRI Study. *NeuroImage* **2008**, *40*, 237–247. [CrossRef] [PubMed]

48. Ware, J.E.; Kosinski, M.; Keller, S.D. *SF-36 Physical and Mental Health Summary Scales: A Users' Manual*; The Health Institute: Scarborough, ON, Canada; New England Medical Center: Boston, MA, USA, 1994.

49. Fisk, J.D.; Pontefract, A.; Ritvo, P.G.; Archibald, C.J.; Muarray, T.J. The impact of fatigue on patients with multiple sclerosis. *Can. J. Neurol. Sci.* **1994**, *21*, 9–14. [CrossRef] [PubMed]

50. Boringa, J.B.; Lazeron, R.H.C.; Reuling, I.E.W.; Adèr, H.J.; Pfennings, L.E.M.A.; Lindeboom, J.; de Sonneville, L.M.J.; Kalkers, N.F.; Polman, C.H. The Brief Repeatable Battery of Neuropsychological Tests: Normative Values Allow Application in Multiple Sclerosis Clinical Practice. *Mult. Scler.* **2001**, *7*, 263–267. [CrossRef] [PubMed]

51. Strauss, E.; Sherman, E.M.S.; Spreen, O. *A Compendium of Neuropsychological Tests: Administration, Norms, and Commentary*; American Chemical Society: Washington, DC, USA, 2006.

52. Lu, H.; van Zijl, P. Experimental Measurement of Extravascular Parenchymal BOLD Effects and Tissue Oxygen Extraction Fractions Using Multi-Echo VASO fMRI at 1.5 and 3.0 T. *Magn. Reson. Med.* **2005**, *53*, 808–816. [CrossRef] [PubMed]

53. Ances, B.M.; Liang, C.L.; Leontiev, O.; Perthen, J.E.; Fleisher, A.S.; Lansing, A.E.; Buxton, R.B. Effects of Aging on Cerebral Blood Flow, Oxygen Metabolism, and Blood Oxygen Level Dependent Responses to Visual Stimulation. *Hum. Brain Mapp.* **2009**, *30*, 1120–1132. [CrossRef] [PubMed]

54. Buxton, R.B. Interpreting Oxygenation-Based Neuroimaging Signals: The Importance and the Challenge of Understanding Brain Oxygen Metabolism. *Front. Neuroenerg.* **2010**. [CrossRef] [PubMed]

55. Leontiev, O.; Buxton, R.B. Reproducibility of BOLD, Perfusion, and CMRO2 Measurements with Calibrated-BOLD fMRI. *NeuroImage* **2007**, *35*, 175–184. [CrossRef] [PubMed]

56. Grubb, R.L.; Raichle, M.E.; Eichling, J.O.; Ter-Pogossian, M.M. The Effects of Changes in PaCO2 Cerebral Blood Volume, Blood Flow, and Vascular Mean Transit Time. *Stroke* **1974**, *5*, 630–639. [CrossRef] [PubMed]

57. Hubbard, N.A.; Turner, M.P.; Robinson, D.M.; Sundaram, S.; Oasay, L.; Hutchison, J.L.; Ouyang, A.; Huang, H.; Rypma, B. Attenuated BOLD Hemodynamic Response Predicted by Degree of White Matter Insult, Slows Cognition in Multiple Sclerosis. *Mult. Scler. J.* **2014**, *20*, 267.

58. Pasley, B.N.; Inglis, B.A.; Freeman, R.D. Analysis of Oxygen Metabolism Implies a Neural Origin for the Negative BOLD Response in Human Visual Cortex. *NeuroImage* **2007**, *36*, 269–276. [CrossRef] [PubMed]

59. Lin, A.; Fox, P.T.; Yang, Y.; Lu, J.; Tan, L.H.; Gao, J.H. Evaluation of MRI Models in the Measurement of CMRO2 and Its Relationship with CBF. *Magn. Reson. Med.* **2008**, *60*, 380–389. [CrossRef] [PubMed]

60. Singh, M.; Kim, S.; Kim, T. Correlation Between BOLD-fMRI and EEG Signal Changes in Response to Visual Stimulus Frequency in Humans. *Magn. Reson. Med.* **2003**, *49*, 108–114. [CrossRef] [PubMed]

61. Peng, S.L.; Ravi, H.; Sheng, M.; Thomas, B.P.; Lu, H. Searching for a Truly "Iso-Metabolic" Gas Challenge in Physiological MRI. *J. Cereb. Blood Flow Metab.* **2017**, *37*, 715–725. [CrossRef] [PubMed]

62. Xu, F.; Uh, J.; Brier, M.R.; Hart, J., Jr.; Yezhuvath, U.S.; Gu, H.; Yang, Y.; Lu, H. The Influence of Carbon Dioxide on Brain Activity and Metabolism in Conscious Humans. *J. Cereb. Blood Flow Metab.* **2011**, *31*, 58–67. [CrossRef] [PubMed]

63. Zappe, A.C.; Uludağ, K.; Oeltermann, A.; Uğurbil, K.; Logothetis, N.L. The Influence of Moderate Hypercapnia on Neural Activity in the Anesthetized Nonhuman Primate. *Cereb. Cortex* **2008**, *18*, 2666–2673. [CrossRef] [PubMed]

64. Yucel, M.A.; Evans, K.C.; Selb, J.; Huppert, T.J.; Boas, D.A.; Gagnon, L. Validation of the Hypercapnic Calibrated fMRI Method Using DOT-fMRI Fusion Imaging. *NeuroImage* **2014**, *102*, 729–735. [CrossRef] [PubMed]

65. Cox, R.W. AFNI: Software for Analysis and Visualization of Functional Magnetic Resonance Neuroimages. *Comput. Biomed. Res.* **1996**, *29*, 162–173. [CrossRef] [PubMed]

66. FMRIB Analysis Group. FMRIB Software Library v5.0. Available online: https://fsl.fmrib.ox.ac.uk/fsl/fslwiki (accessed on 3 June 2017).

67. Liu, T.T.; Wong, E.C. A Signal Processing Model for Arterial Spin Labeling Functional MRI. *NeuroImage* **2005**, *24*, 207–215. [CrossRef] [PubMed]

68. Hutchison, J.L.; Hubbard, N.A.; Brigante, R.M.; Turner, M.; Sandoval, T.I.; Hillis, G.A.; Weaver, T.; Rypma, B. The Efficiency of fMRI Region of Interest Analysis Methods for Detecting Group Differences. *J. Neurosci. Methods* **2014**, *226*, 57–65. [CrossRef] [PubMed]

69. Smith, S.M.; De Stefano, N.; Jenkinson, M.; Matthews, P.M. Normalised Accurate Measurement of Longitudinal Brain Change. *J. Comput. Assist. Tomogr.* **2001**, *25*, 466–475. [CrossRef] [PubMed]

70. Hart, J., Jr.; Kraut, M.A.; Womack, K.B.; Strain, J.; Didehbani, N.; Bartz, E.; Conover, H.; Mansinghani, S.; Lu, H.; Cullum, C.M. Neuroimaging of Cognitive Dysfunction and Depression in Aging Retired National Football League Players. *JAMA Neurol.* **2013**, *70*, 326–335. [CrossRef] [PubMed]

71. Dice, L.R. Measures of the Amount of Ecologic Association between Species. *Ecology* **1945**, *26*, 297–302. [CrossRef]

72. Zhang, L.; Dean, D.; Liu, J.Z.; Sahgal, V.; Wang, X.; Yue, G.H. Quantifying Degeneration of White Matter in Normal Aging Using Fractal Dimension. *Neurobiol. Aging* **2007**, *28*, 1543–1555. [CrossRef] [PubMed]

73. Ghassemi, R.; Narayana, S.; Banwell, B.; Sled, J.G.; Shroff, M.; Arnold, D.L. Quantitative Determination of Regional Lesion Volume and distribution in Children and Adults with Relapsing-Remitting Multiple Sclerosis. *PLoS ONE* **2014**, *9*, e85741. [CrossRef] [PubMed]

74. Jones, D.K.; Simmons, A.; Williams, S.C.; Horsfield, M.A. Non-invasive Assessment of Axonal Fiber Connectivity in the Human Brain via Diffusion Tensor MRI. *Magn. Reson. Med.* **1999**, *42*, 37–41. [CrossRef]

75. Woods, R.P.; Grafton, S.T.; Holmes, C.J.; Cherry, S.R.; Mazziotta, J.C. Automated Image Registration: I. General Methods and Intrasubject, Intramodality Validation. *J. Comput. Assist. Tomogr.* **1998**, *22*, 139–152. [CrossRef] [PubMed]

76. Jiang, H.; van Zijl, P.C.J.K.; Pearlson, G.D.; Mori, S. DtiStudio: Resource Program for Diffusion Tensor Computation and Fiber Bundle Tracking. *Comput. Methods Programs Biomed.* **2006**, *81*, 106–116. [CrossRef] [PubMed]

77. Smith, S.M.; Jenkinson, M.; Johansen-Berg, H.; Rueckert, D.; Nichols, T.E.; Mackay, C.E.; Watkins, K.E.; Ciccarelli, O.; Cader, M.Z.; Matthews, P.M.; et al. Tract-Based Spatial Statistics: Voxelwise Analysis of Multi-Subject Diffusion Data. *NeuroImage* **2006**, *3*, 1487–1505. [CrossRef] [PubMed]

78. Huang, H.; Gundapuneedi, T.; Rao, U. White Matter Disruptions in Adolescents Exposed to Childhood Maltreatment and Vulnerability to Psychopathology. *Neuropsychopharmacology* **2012**, *37*, 2693–2701. [CrossRef] [PubMed]

79. Huang, H.; Fan, X.; Weiner, M.; Martin-Cook, K.; Xiao, G.; Davis, J.; Devous, M.; Rosenberg, R. Distinctive Disruption Patterns of White Matter Tracts in Alzheimer's Disease with Full Diffusion Tensor Characterization. *Neurobiol. Aging* **2012**, *33*, 2029–2045. [CrossRef] [PubMed]

80. Ouyang, M.; Cheng, H.; Mishra, V.; Gong, G.; Mosconi, M.; Sweeney, J.; Peng, Y.; Huang, H. Atypical age-dependent effects of autism on white matter microstructure in children of 2–7 years. *Hum. Brain Mapp.* **2016**, *37*, 819–832. [CrossRef] [PubMed]

81. Mori, S.; Oishi, K.; Jiang, H.; Jiang, L.; Li, X.; Akhter, K.; Hua, K.; Faria, A.V.; Mahmood, A.; Woods, R.; et al. Stereotaxic White Matter Atlas Based on Diffusion Tensor Imaging in an ICBM Template. *NeuroImage* **2008**, *40*, 570–582. [CrossRef] [PubMed]

82. Iglewicz, B.; Hoaglin, D. Volume 16: How to Detect and Handle Outliers. In *The ASQC Basic References in Quality Control: Statistical Techniques*; Mykytka, E.F., Ed.; American Society for Quality Control, Statistics Division: Milwaukee, WI, USA, 1993.

83. Gabrieli, J.D.E.; Ghosh, S.S.; Whitfield-Gabrieli, S. Prediction as a Humanitarian and Pragmatic Contribution from Human Cognitive Neuroscience. *Neuron* **2015**, *85*, 11–26. [CrossRef] [PubMed]

84. Efron, B. Better Bootstrap Confidence Intervals. *J. Am. Stat. Assoc.* **1987**, *82*, 171–185. [CrossRef]

85. Kohavi, R. A Study of Cross-Validation and Bootstrap for Accuracy Estimation and Model Selection. In Proceedings of the Fourteenth International Joint Conference on Artificial Intelligence, Montreal, QC, Canada, 20–25 August 1995; Morgan Kaufmann Publishers: San Mateo, CA, USA, 1995; Volume 2, pp. 1137–1143.

86. Ojala, M.; Garriga, G.C. Permutation Tests for Studying Classifier Performance. *J. Mach. Learn. Res.* **2010**, *11*, 1833–1863.

87. Marshall, O.; Lu, H.; Brisset, J.C.; Xu, F.; Liu, P.; Herbert, J.; Grossman, R.I.; Ge, Y. Impaired Cerebrovascular Reactivity in Multiple Sclerosis. *JAMA Neurol.* **2014**, *71*, 1275–1281. [CrossRef] [PubMed]

88. Laron, M.; Cheng, H.; Zhang, B.; Schiffman, J.S.; Tang, R.A.; Frishman, L.J. Comparison of Multifocal Visual Evoked Potential, Standard Automated Perimetry and Optical Coherence Tomography in Assessing Visual Pathways in Multiple Sclerosis Patients. *Mult. Scler.* **2010**, *16*, 412–426. [CrossRef] [PubMed]

89. Fraser, C.; Klistorner, A.; Graham, S.; Garrick, R.; Billson, F.; Grigg, J. Multifocal Visual Evoked Potential Latency Analysis: Predicting Progression to Multiple Sclerosis. *Arch. Neurol.* **2006**, *63*, 847–850. [CrossRef] [PubMed]

90. Ruseckaite, R.; Maddess, T.; Danta, G.; Lueck, C.J.; James, A.C. Sparse Multifocal Stimuli for the Detection of Multiple Sclerosis. *Ann. Neurol.* **2005**, *57*, 904–913. [CrossRef] [PubMed]

91. Pantano, P.; Mainero, C.; Caramia, F. Functional Brain Reorganization in Multiple Sclerosis: Evidence from fMRI Studies. *J. Neuroimaging* **2006**, *16*, 104–114. [CrossRef] [PubMed]

92. White, A.T.; Lee, J.N.; Light, A.R.; Light, K.C. Brain Activation in Multiple Sclerosis: A BOLD fMRI Study of the Effects of Fatiguing Hand Exercise. *Mult. Scler.* **2009**, *15*, 580–586. [CrossRef] [PubMed]

93. Chiaravalloti, N.D.; Hillary, F.G.; Ricker, J.H.; Christodoulou, C.; Kalnin, A.J.; Liu, W.C.; Steffener, J.; DeLuca, J. Cerebral Activation Patterns During Working Memory Performance in Multiple Sclerosis using fMRI. *J. Clin. Exp. Neuropsychol.* **2005**, *27*, 33–54. [CrossRef] [PubMed]

94. Genova, H.M.; Sumowski, J.F.; Chiaravalloti, N.; Voelbel, G.T.; DeLuca, J. Cognition in Multiple Sclerosis: A Review of Neuropsychological and fMRI Research. *Front. Biosci.* **2009**, *14*, 1730–1744. [CrossRef]

95. Sweet, L.H.; Rao, S.M.; Primeau, M.; Durgerian, S.; Cohen, R.A. Functional Magnetic Resonance Imaging Response to Increased Verbal Working Memory Demands Among Patients with Multiple Sclerosis. *Hum. Brain. Mapp.* **2006**, *27*, 28–36. [CrossRef] [PubMed]

96. Genova, H.M.; Hillary, F.G.; Wylie, G.; Rypma, B.; DeLuca, J. Examination of Processing Speed Deficits in Multiple Sclerosis Using Functional Magnetic Resonance Imaging. *J. Int. Neuropsychol. Soc.* **2009**, *15*, 383–393. [CrossRef] [PubMed]

97. Comabella, M.; Sastre-Garriga, J.; Montalban, X. Precision Medicine in Multiple Sclerosis: Biomarkers for Diagnosis, Prognosis, and Treatment Response. *Curr. Opin. Neurol.* **2016**, *29*, 254–262. [CrossRef] [PubMed]

98. Pulicken, M.; Gordon-Lipkin, E.; Balcer, L.J.; Frohman, E.; Cutter, G.; Calabresi, P.A. Optical Coherence Tomography and Disease Subtype in Multiple Sclerosis. *Neurology* **2007**, *69*, 2085–2092. [CrossRef] [PubMed]

99. Fisniku, L.K.; Brex, P.A.; Altmann, D.R.; Miszkiel, K.A.; Benton, C.E.; Lanyon, R.; Thompson, A.J.; Miller, D.H. Disability and T2 MRI Lesions: A 20-Year Follow-Up of Patients with Relapse Onset of Multiple Sclerosis. *Brain* **2008**, *131*, 808–817. [CrossRef] [PubMed]

100. Lebrun, C.; Bensa, C.; Debouverie, M.; Wiertlevski, S.; Brassat, D.; de Seze, J.; Rumbach, L.; Pelletier, J.; Labauge, P.; Brochet, B.; et al. Association Between Clinical Conversion to Multiple Sclerosis in Radiologically Isolated Syndrome and Magnetic Resonance Imaging, Cerebrospinal Fluid, and Visual Evoked Potential. *Arch. Neurol.* **2009**, *66*, 841–846. [CrossRef] [PubMed]

101. Leocanti, L.; Rocca, M.A.; Comi, G. MRI and Neurophysiological Measures to Predict Course, Disability and Treatment Response in Multiple Sclerosis. *Curr. Opin. Neurol.* **2016**, *29*, 243–253. [CrossRef] [PubMed]

102. Okuda, D.T.; Mowry, E.M.; Cree, B.A.C.; Crabtree, E.C.; Goodin, D.S.; Waubant, E.; Pelletier, D. Asymptomatic Spinal Cord Lesions Predict Disease Progression in Radiologically Isolated Syndrome. *Neurology* **2011**, *76*, 686–692. [CrossRef] [PubMed]

103. Stromillo, M.L.; Giorgio, A.; Rossi, F.; Battaglini, M.; Hakiki, B.; Malentacchi, G.; Santangelo, M.; Gasperini, C.; Bartolozzi, M.L.; Portaccio, E.; et al. Brain metabolic changes suggestive of axonal damage in radiologically isolated syndrome. *Neurology* **2013**, *80*, 2090–2094. [CrossRef] [PubMed]

104. Campbell, G.R.; Worrall, J.T.; Mahad, D.J. The Central Role of Mitochondrial in Axonal Degeneration in Multiple Sclerosis. *Mult. Scler.* **2014**, *20*, 1806–1813. [CrossRef] [PubMed]

105. Balcer, L.J.; Raynowska, J.; Nolan, R.; Galetta, S.L.; Kapoor, R.; Benedict, R.; Phillips, G.; LaRocca, N.; Hudson, L.; Rudick, R.; et al. Validity of Low-Contrast Letter Acuity as a Visual Performance Outcome Measure for Multiple Sclerosis. *Mult. Scler.* **2017**, *23*, 734–747. [CrossRef] [PubMed]

106. Link, H.; Huang, Y.-M. Oligoclonal Bands in Multiple Sclerosis Cerebrospinal Fluid: An Update on Methodology and Clinical Usefulness. *J. Immunol.* **2006**, *180*, 17–28. [CrossRef] [PubMed]

107. Shaffer, J.L.; Petrella, J.R.; Sheldon, F.C.; Choudhury, K.R.; Calhoun, V.D.; Coleman, R.E.; Doraiswamy, P.M. Predicting Cognitive Decline in Subjects at Risk for Alzheimer Disease by Using Combined Cerebrospinal Fluid, MR Imaging, and PET Biomarkers. *Radiology* **2013**, *266*, 583–591. [CrossRef] [PubMed]

108. Whitfield-Gabrieli, S.; Ghosh, S.S.; Nieto-Castanon, A.; Saygin, Z.; Doehrmann, O.; Chai, X.J.; Reynolds, G.O.; Hofmann, S.G.; Pollack, M.H.; Gabrieli, J.D.E. Brain Connectomics Predict Response to Treatment in Social Anxiety Disorder. *Mol. Psychiatry* **2015**, *21*, 680–685. [CrossRef] [PubMed]

brain
sciences

MDPI

Review

Contribution of the Degeneration of the Neuro-Axonal Unit to the Pathogenesis of Multiple Sclerosis

Hannah E. Salapa [1], Sangmin Lee [2,3], Yoojin Shin [2,3] and Michael C. Levin [1,2,3,4,*]

[1]   Department of Anatomy and Cell Biology, CMSNRC (Cameco MS Neuroscience Research Center), University of Saskatchewan, Saskatoon, SK S7N0Z1, Canada; hes763@mail.usask.ca
[2]   Veterans Administration Medical Center, Memphis, TN 38104, USA; salee@uthsc.edu (S.L.); yshin2@uthsc.edu (Y.S.)
[3]   Department of Neurology, University of Tennessee Health Science Center, Memphis, TN 38104, USA
[4]   Department of Neurology, University of Saskatchewan, Saskatoon, SK S7N0Z1, Canada
*   Correspondence: michael.levin@usask.ca

Academic Editor: Evanthia Bernitsas
Received: 5 May 2017; Accepted: 14 June 2017; Published: 18 June 2017

Abstract: Multiple sclerosis (MS) is a demyelinating, autoimmune disease of the central nervous system. In recent years, it has become more evident that neurodegeneration, including neuronal damage and axonal injury, underlies permanent disability in MS. This manuscript reviews some of the mechanisms that could be responsible for neurodegeneration and axonal damage in MS and highlights the potential role that dysfunctional heterogeneous nuclear ribonucleoprotein A1 (hnRNP A1) and antibodies to hnRNP A1 may play in MS pathogenesis.

Keywords: multiple sclerosis; RNA binding protein; neurodegeneration; axonal damage; hnRNP A1

1. Multiple Sclerosis (MS)

Multiple sclerosis (MS) is a demyelinating, autoimmune disease of the central nervous system (CNS). Over two million affected individuals world-wide, typically diagnosed in young adulthood, makes MS the most debilitating neurological disease in this population. Symptoms associated with MS, such as fatigue, impaired coordination, and spasticity, limit the ability of people to function properly, which endows a financial burden on both the patient and their family. The majority of patients are initially diagnosed with relapsing remitting MS (RRMS) where symptoms develop and are followed by a period of recovery or remission with no symptoms [1–3]. A total of 50% of RRMS patients gradually advance to having secondary progressive MS (SPMS) where symptoms steadily increase and worsen with rare periods of recovery [1–4]. Approximately 5% of people develop primary progressive MS (PPMS) where patients experience gradual disease worsening with no periods of recovery [1,3]. Additional clinical subtypes include radiologically isolated syndrome (RIS) where lesions are discovered on MRI without the presentation of any symptoms and clinically isolated syndrome (CIS) where a patient experiences an initial clinical episode consistent with MS concurrent with an MRI suggestive of MS [4]. Despite the differences in onset, relapse rate, and initial subtype diagnosis, many patients with MS will progress to a stage of irreversible disability [5].

One of the most recognizable pathological features of MS is the plaque or lesion that is evident in vivo on MRI brain scans. These plaques, depending on their stage, can contain activated lymphocytes, microglia, and myelin debris from macrophage degradation [6]. Oligodendrocytes often attempt to remyelinate damaged axons, which leads to shadow plaques containing partially remyelinated axons [1]. Inflammation, plaques, and disease progression vary between individuals and may not necessarily correlate with disease severity. For example, white matter atrophy rates have been shown to be similar across each subtype of disease; however, there are vast differences in grey matter atrophy across disease subtypes [7]. Significantly more grey matter atrophy is observed in SPMS patients and those with higher expanded disability status scale (EDSS) scores as opposed to RRMS patients or those with lower EDSS scores [8]. These findings and other incongruences between disease state and pathology suggest that inflammation and neurodegeneration, including axonal damage, likely transpire simultaneously but independently during disease [9,10]. Axonal and neuronal injury exists in the absence of demyelination, with no correlation between plaque location and axonal loss in spinal cord long tracts (e.g. corticospinal tract and posterior columns) [11]. Additionally, focal axonal degeneration starts with focal swellings, which are observed in myelinated axons [12]. These differences suggest that neurodegeneration is a better correlate of disease severity than demyelination. Clearly, neurodegeneration in MS takes place throughout disease [13–15] and although neurodegeneration is known to cause permanent disability in MS, research regarding the underlying mechanisms is in its infancy.

The degeneration of axons is generally classified as either Wallerian degeneration or "dying back" degeneration. Wallerian degeneration occurs when the axon located distally to a site of injury begins to degenerate in an anterograde fashion. These injuries result in a disconnect between the neuronal cell body and distal portion of the axon. Dying back, on the other hand, is axonal degeneration from the distal end in a retrograde manner with degeneration of the axon happening before cell body loss. Both Wallerian and "dying back" degeneration are characterized by axonal dystrophy, which is visualized as axonal spheroids and varicosities, which have been found in MS lesions [16]. Axonal degeneration observed in MS has been most commonly categorized as Wallerian degeneration [17]. "Dying back" axonopathy has been observed in hereditary spastic paraplegia (HSP), a disease clinically similar to MS [18–20].

A number of mechanisms underlying neurodegeneration in neurologic diseases have been proposed. For instance, in amyotrophic lateral sclerosis (ALS), axonal transport deficits and mitochondrial dysfunction have been observed in ALS-mutant mice [21,22]. Additional evidence suggests that the interplay between mitochondria and impaired axonal transport leads to degeneration in Alzheimer's disease [23]. Dysfunctional mitochondria have also been shown to play a role in the pathogenesis of Parkinson's Disease [24]. More recent findings suggest a role for RNA binding proteins in neurologic disease [25–29]. RNA binding proteins are responsible for regulating RNA homeostasis, also known as "ribostasis". The mislocalization of these proteins from nucleus to cytoplasm and the formation of stress granules are key pathological features of RNA binding proteins in ALS, frontotemporal dementia (FTD), and spinal muscular atrophy (SMA) [25–28,30,31]. In MS, disrupted sodium and calcium ion channel dynamics [32,33], axonal transport deficiencies [9,34–36], mitochondrial dysfunction [37–40], and oxidative stress [40,41] contribute to neuronal and axonal damage. As in other diseases, these mechanisms may work in tandem to contribute to neurodegeneration and axonal damage as opposed to independently (Figure 1).

Axonal transport is an essential function for maintenance of neuronal health, and has long been implicated in neurodegenerative conditions. The neuronal cytoskeleton is composed of microtubules, neurofilaments, and actin filaments. Microtubules provide a track system for the movement of cargo along the axons by the motor proteins kinesin and dynein. Kinesins mediate anterograde transport, moving organelles or vesicles from the soma to the synapse or membrane while dynein molecules are involved in retrograde axonal transport to move cargo toward the cell body (Figure 1). Evidence of disrupted axonal transport and axonal damage is observed in MS post-mortem tissue through staining with amyloid precursor protein (APP), a protein involved in "fast" anterograde transport due to its ability to mediate interactions between cargo and kinesin proteins [42]. APP staining can be seen in acute MS brain lesions [34], axonal swellings in demyelinated plaques [35], and in normal appearing white matter of acute MS cases [9]. Furthermore, impaired transport of organelles is a prominent early feature in inflammatory MS-like lesions [16]. There is further support for this concept in experimental autoimmune encephalomyelitis (EAE) mice where compromised mitochondrial transport is an early event in EAE lesions [35]. Disruption of axonal transport, both retrograde and anterograde, has also been observed prior to demyelination in EAE mice [36]. Researchers used genetically engineered mice expressing fluorescent markers in mitochondria and peroxisomes under the *Thy1* promoter. After inducing EAE, researchers employed two-photon imaging to measure retrograde and anterograde transport of both organelles along axons and found decreased transport in swollen axons as well as in normal appearing axons around lesion areas [36]. These findings in the EAE model as well as those with APP staining in the MS cortex are evidence that loss of axonal transport could be an early first step towards axonal degeneration prior to demyelination.

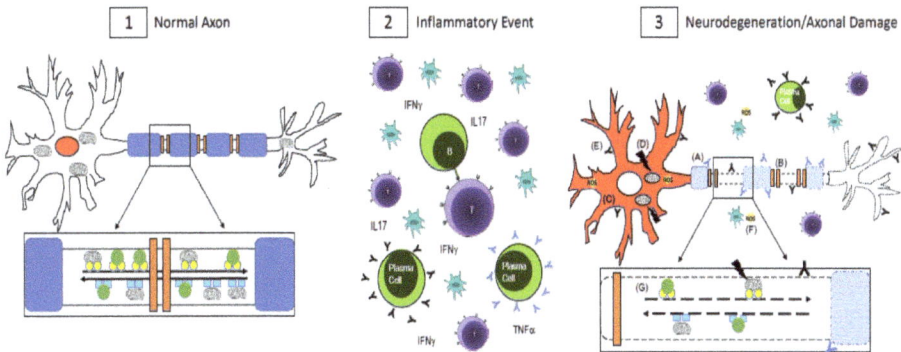

Figure 1. Axonal Damage in multiple sclerosis (MS). (**1**) In a normal, healthy axon, myelin (blue) wraps around the axon and ion channels (orange) are clustered in the unmyelinated nodes of Ranvier. This enables saltatory conduction for fast signal transmission down the axon. RNA binding proteins, which maintain RNA homeostasis, are localized to the nucleus (red). Mitochondria and other cargo (green circles) are transported retrogradely along the axon by dynein (turquoise squares, inset) while anterograde transport is done by kinesin motor proteins (yellow circles, inset). Transport is fast and uninterrupted because the axon is undamaged, has no energy shortage, and the motor proteins are intact; (**2**) In MS, there is central nervous system infiltration of T-cells, B-cells, plasma cells, and macrophages, which lead to a cascade of events including the release of pro-inflammatory cytokines and antibodies, which are thought to be harmful to both myelin and neurons and axons. [43–49]; (**3**) Axonal damage and neurodegeneration occur simultaneously with inflammation. There is ongoing demyelination due to antibodies against myelin antigens (blue Y, A) [43–45]. Demyelination leads to the redistribution of ion channels (B), which impairs conduction along the axon [32,33]. The redistribution of RNA binding proteins from their normal nuclear location (panel 1, red) to the cytoplasm (panel 3, C, red) is a pathological feature of neuronal degeneration in neurological diseases [25–28,30,31]. Mutations in mitochondrial DNA (D) can impair the cell's ability to generate enough ATP while antibodies to non-myelin antigens (black Y, E) damage axons [46–49]. Reactive oxygen species (yellow, F) could be released from activated microglia or as a result of dysfunctional mitochondria [37–41]. Impaired fast axonal transport (G) is also evident in MS [9,34–36]. A combination of these events, as opposed to one in particular, contributes to neurodegeneration and axonal damage in MS.

An essential component for axonal transport is energy, in the form of ATP, which is produced by mitochondria. There is an increasing body of work supporting a role for mitochondrial dysfunction in neurodegenerative diseases [50]. Mitochondria are responsible for maintaining a cell's energy production in the form of ATP generated by the respiratory chain complex. In addition to producing ATP, mitochondria also function to produce amino acids, maintain calcium homeostasis, and the modulation of reactive oxygen species (ROS). It is therefore understandable that perturbation of mitochondrial processes could result in neuronal dysfunction, decreased viability, and even apoptosis leading to neuronal loss and degeneration. Loss of neurons in the MS patient cortex is commonly observed [51,52]. Incidentally, a decrease in mitochondrial electron transport gene expression is also observed in MS brain tissues, suggesting that mitochondrial dysfunction could be contributing neuronal loss in patients [38]. Furthermore, mitochondrial DNA (mtDNA) accumulates deletions in the grey matter of SPMS patients irrespective of lesions [12,39]. These damaged mitochondria could influence anterograde transport leading to axonal transport deficits [37] and compromised mitochondrial transport is an early change in inflammatory EAE lesions [35]. In addition to compromising axonal transport, dysmorphic swollen mitochondria lead to increased ROS and reactive nitrogen species (RNS) concentrations and therefore, a release of proapoptotic mediators [12].

In addition to these distinctive hypotheses regarding neurodegeneration, there is also evidence for the contribution of antibodies to neurodegeneration in MS. The inflammatory environment in MS primarily consists of an initial T cell infiltrate along with activated macrophages and microglia with another T- and B-cell infiltrate after myelin has been broken apart [13]. The invading T-cells consist of both CD4+ and CD8+ T-cells, however, data suggests that CD8+ may have a more profound effect especially during the later phases of disease [53,54]. Natural killer (NK) cells may also be present in the inflammatory milieu and have been shown to play both protective and deleterious roles in MS [55]. Lymphocytes are present during active demyelination, however, antibody producing plasma cells are more evident in SPMS and PPMS patients, suggesting a role for autoantibodies in disease progression [14]. Autoantibodies to myelin antigens such as myelin oligodendrocyte glycoprotein (MOG), myelin basic protein (MBP), and myelin proteolipid protein (PLP) have been identified. MOG antibodies have been shown to play a primary role in the demyelination of axons as opposed to degeneration [43] through complement cascade activation [44,45] as two myelin proteins bind C1 directly to activate the complement cascade [45]. For example, C3d immunoreactivity is seen in areas of partly demyelinated axons as well as in active lesions [56]. On the other hand, antibodies to non-myelin antigens such as neurofascin, neurofilament, and KIR 4.1 (a glial potassium channel) have been shown to contribute to axonal and neuronal injury [46–49]. Furthermore, the injection of neurofascin antibodies into EAE mice also leads to axonal injury [48]. Although antibodies to myelin and non-myelin antigens seem to have different effects on myelination and axonal damage, respectively, they may have a common mechanism of action through activation of the complement system [46–48] and could both be responsible for the pathology observed in MS tissue. Furthermore, progressive patients have IgG-containing plasma cells in their meninges and throughout the brain, which remain even after T- and B-cell levels decrease [14,57]. Clearly, the humoral response is an important contributing factor to axonal damage but may be particularly influential in patients with progressive disease.

2. hnRNP A1 and RNA Metabolism

Support for a role of antibodies to non-myelin, neuronal antigens in MS pathogenesis, specifically neurodegeneration, is strong [14,46,48,49,57,58]. Data from our lab also provides compelling evidence to strengthen this hypothesis. Initial experiments from our lab showed that IgG from human T-lymphotrophic virus type 1 associated myelopathy/tropical spastic paraparesis (HAM/TSP) patients, a disease clinically similar to progressive MS, immunoreacted with a 33 kDa protein from isolated human brain neurons on a Western blot [59]. This protein was identified as heterogeneous nuclear ribonucleoprotein A1 (hnRNP A1) [59]. hnRNP A1 is an RNA binding protein that performs a multitude of functions related to ribostasis, including mRNA transport, pre-mRNA processing, and translation [60]. IgG from HAM/TSP patients was shown to preferentially react with areas commonly damaged in HAM/TSP, such as neurons and axons throughout the corticospinal system [61]. The immunodominant epitope of hnRNP A1 recognized by HAM/TSP IgG was identified as an amino acid sequence (AA 293-GQYFAKPRNQGG-304) within the M9 nuclear localization sequence [62]. The "M9" area is required for the nucleocytoplasmic transport of hnRNP A1 [63]. HAM/TSP and progressive forms of MS show similarities and as such, it was hypothesized that MS patient IgG would also react with hnRNP A1, indicating the development of antibodies against this RNA binding protein

(RBP). MS patients were found to make antibodies to hnRNP A1, specifically to the same M9 epitope as HAM/TSP patients [64,65]. Healthy controls and patients with Alzheimer's disease were examined as controls and were found to show no immunoreactivity to hnRNP A1 [64].

Because antibodies to other non-myelin antigens, such as neurofascin, have been shown to worsen EAE and lead to axonal damage, we hypothesized that anti-hnRNP A1-M9 antibodies, which recognize the same immunodominant epitope as MS patient IgG, might show similar effects. Neurons were exposed to control antibodies as well as hnRNP A1-M9 antibodies. Anti-hnRNP A1-M9 exposure led to neurodegeneration and neuronal death [64]. Microarray analyses comparing anti-hnRNP A1-M9 antibodies to both control IgG and untouched neuronal cells revealed altered RNA expression in the anti-hnRNP A1-M9 antibody condition [64]. Interestingly, some of the genes affected by the anti-hnRNP A1-M9 antibodies included the spinal paraplegia genes (SPGs) implicated in the pathogenesis of hereditary spastic paraplegia (HSP), which clinically mimics HAM/TSP and progressive MS. Specifically, it identified spastin (SPG4), paraplegin (SPG7), and spartin (SPG20) [64]. Furthermore, anti-hnRNP A1-M9 antibodies also altered expression of axonal transport RNAs, including kinesin family member 5 (KIF5C) as well as a number of genes related to hnRNP A1's function in "ribostasis" [64].

We sought to determine whether anti-hnRNP A1-M9 antibodies had an effect on hnRNP A1's ability to bind its target RNA (which is bound via the RNA binding domains). Using the RNA Binding Protein Data Base (RBPDB.com), we determined that spastin (SPG4) contains a 100% binding sequence (NM_014946, b.3282-3288) match with the hnRNP A1 binding sequence while SPG7 and SPG20 showed lesser degrees of RNA sequence alignment. By using RNA immunoprecipitation, we found that SP4 and SPG7 bound hnRNP A1 while SPG20 did not [3,66]. Furthermore, SPG4 and SPG7 levels were remarkably decreased in neuronal cells that had been exposed to anti-hnRNP A1-M9 antibodies but not in cells exposed to control antibodies [66]. Because anti-hnRNP A1-M9 antibodies had led to neuronal death, loss of neuronal processes, apoptosis, and mislocazation of hnRNP A1 from the nucleus to the cytoplasm [64,66,67], we hypothesized that the immunodominant anti-hnRNP A1-M9 antibodies might impact the EAE disease course in mice.

To test this hypothesis, we induced EAE in mice and upon the first sign of disease, we injected anti-hnRNP A1-M9 antibodies, control antibodies, and phosphate buffered saline (PBS) three times for a total of 300 micrograms of antibody. We clinically scored animals and approximately 11 days following the first injection, animals injected with anti-hnRNP A1-M9 antibodies showed significantly higher clinical scores, indicating worsened disease [68]. Subsequent staining with Fluoro Jade C showed preferential neurodegeneration in the anti-hnRNP A1-M9 animals in the deep white matter of the cerebellum and the distal ventral spinocerebellar tract (VSCT) as it enters the cerebellum [68]. The cell bodies of the VSCT, an afferent pathway, lie in laminae VII, VIII, and IX of the lumbosacral spinal cord. Recent studies show entry of T-cells happens in this region early in EAE [69,70]. This pattern of neurodegeneration, in which axonal injury follows a distal to proximal pattern, suggests a "dying back" axonopathy, which is commonly observed in HSP [18,19]. Furthermore, these animals developed spasticity (a major clinical feature in MS patients) in their hind limbs whereas those injected with control antibodies or PBS did not.

The development of spasticity, the interaction between hnRNP A1 and SPG4 and SPG7, and the "dying back" axonopathy suggest a similar or shared mechanism of pathology between HSP and progressive MS. In HSP, mutations within SPG4 account for the majority of cases. Spastin is a member of the AAA protein family with multiple isoforms (M1, M87) that has microtubule severing functions. The severing of microtubules by spastin is crucial for efficient microtubule transport and mutations lead to loss of microtubule-severing activity and distal axonal end degeneration [18]. The M1 spastin isoform (68 kDa) is only detectable in the adult mouse spinal cord whereas the M87 (60 kDa) isoform is more widely distributed and more abundant [18]. The presence of the M1 isoform strongly correlates with axonal degeneration in HSP, suggesting a gain of function mechanism due to perturbed alternative splicing mechanisms [71]. Mutations in either isoform alter microtubule dynamics and lead to the formation of toxic aggregates [18]. Mutations within SPG7 also account for a smaller portion of HSP cases. SPG7 plays a role in the inner mitochondrial membrane and cultured myoblasts from patients with SPG7 mutations show defects in respiratory chain function [72].

Because mutations in RNA binding proteins have been shown to lead to other neurological diseases [26,29], we wanted to determine whether MS patients had mutations in hnRNP A1. DNA was isolated from peripheral blood monocytes (PMBC) from each subtype of MS patient and PCR was performed to isolate human hnRNP A1 genomic DNA containing exons 8 and 9. Following amplification, genomic DNA was cloned into vectors and sequenced. PPMS patients had a greater number of novel somatic nucleotide variants, which when translated into protein resulted in more amino acid substitutions than RRMS, SPMS, or healthy controls [73]. These mutations were cloned into expression vectors and transfected into SKNSH cells and stained for hnRNP A1 as well as stress granules. Mutant forms of hnRNP A1, as opposed to wildtype, showed mislocalization from the nucleus to the cytoplasm as well as the formation of stress granules [73], which is similar to pathogenic features observed in other neurologic diseases involving RNA binding proteins [24,27]. Furthermore, exposing SKNSH cells to anti-hnRNP A1-M9 antibodies also leads to mislocalization and the formation of stress granules [66,67]. This suggests that hnRNP A1 dysfunction, either related to an autoimmune or genetic mechanism, may contribute to MS pathogenesis in a manner similar to other diseases.

Taken together, these studies suggest that hnRNP A1, an RNA binding protein, may show pathogenic features, such as mislocalization and stress granule formation, in MS that are similar to other neurologic diseases. After exposure to anti-hnRNP A1 antibodies, hnRNP A1 mislocalizes to the cytoplasm of cells. This mislocalization is also observed in [74]. The effect of anti-hnRNP A1-M9 antibodies on neuronal cell lines and RNA targets in vitro suggests that antibodies may be altering endogenous hnRNP A1 functions by interrupting normal ribostasis such as mRNA binding. If hnRNP A1 does not properly bind RNA targets, such as SPG4 or KIF5C in vivo, this could disrupt normal functioning of these proteins resulting in impaired axonal transport or the development of spasticity. Additionally, anti-hnRNP A1-M9 antibodies, in the setting of a pro-inflammatory environment (EAE), lead to the development of hind limb spasticity and neurodegeneration, including axonal dying back, phenotypes both observed in HSP and MS [20,75,76]. Further understanding the mechanisms and consequences of dysfunctional hnRNP A1 and anti-hnRNP A1 antibodies could lead to the development of better therapies that alleviate symptoms, such as spasticity.

3. Conclusions

Several factors contribute to axonal damage and neurodegeneration in MS. Impaired fast axonal transport, the release of reactive oxygen species, mitochondria dysfunction, the redistribution of RNA binding proteins from their normal nuclear location, and antibodies to non-myelin antigens along with pro-inflammatory events occur during MS. A combination of these events, as opposed to one in particular, may be responsible for neuronal and axonal damage observed in disease.

Acknowledgments: This work is based upon work supported by the Office of Research and Development, Medical Research Service, Department of Veterans Affairs (MCL grant #I01 BX001996) and the Multiple Sclerosis Research Fund and Neuroscience Institute at the University of Tennessee Health Science Center.

Author Contributions: H.E.S. contributed experimental data and was the primary writer of the manuscript. S.L. and Y.S. contributed experimental data and expertise to the manuscript. M.C.L. wrote, reviewed, and edited the manuscript and approved the final version including figures.

Conflicts of Interest: The authors declare no conflicts of interest.

References

1. Lassmann, H.; Brück, W.; Lucchinetti, C.F. The immunopathology of multiple sclerosis: An overview. *Brain Pathol.* **2007**, *17*, 210–218. [CrossRef] [PubMed]
2. Noseworthy, J.H.; Lucchinetti, C.; Rodriguez, M.; Weinshenker, B.G. Multiple sclerosis. *N. Engl. J. Med.* **2000**, *343*, 938–952. [CrossRef] [PubMed]

3.   Levin, M.C.; Lee, S.; Gardner, L.A.; Shin, Y.; Douglas, J.N.; Groover, C.J. Pathogenic mechanisms of neurodegeneration based on the phenotypic expression of progressive forms of immune-mediated neurologic disease. *Degener. Neurol. Neuromuscul. Dis.* **2012**, *2*, 175–187. [CrossRef]

4.   Milo, R.; Miller, A. Revised diagnostic criteria of multiple sclerosis. *Autoimmun. Rev.* **2014**, *13*, 518–524. [CrossRef] [PubMed]

5.   Confavreux, C.; Vukusic, S.; Moreau, T.; Adeleine, P. Relapses and progression of disability in multiple sclerosis. *N. Engl. J. Med.* **2000**, *343*, 1430–1438. [CrossRef] [PubMed]

6.   Love, S. Demyelinating diseases. *J. Clin. Pathol.* **2006**, *59*, 1151–1159. [CrossRef] [PubMed]

7.   Fisher, E.; Lee, J.C.; Nakamura, K.; Rudick, R.A. Gray matter atrophy in multiple sclerosis: A longitudinal study. *Ann. Neurol.* **2008**, *64*, 255–265. [CrossRef] [PubMed]

8.   Fisniku, L.K.; Chard, D.T.; Jackson, J.S.; Anderson, V.M.; Altmann, D.R.; Miszkiel, K.A.; Thompson, A.J.; Miller, D.H. Gray matter atrophy is related to long-term disability in multiple sclerosis. *Ann. Neurol.* **2008**, *64*, 247–254. [CrossRef] [PubMed]

9.   Kornek, B.; Storch, M.K.; Weissert, R.; Wallstroem, E.; Stefferl, A.; Olsson, T.; Linington, C.; Schmidbauer, M.; Lassmann, H. Multiple sclerosis and chronic autoimmune encephalomyelitis: A comparative quantitative study of axonal injury in active, inactive, and remyelinated lesions. *Am. J. Pathol.* **2000**, *157*, 267–276. [CrossRef]

10. Kutzelnigg, A.; Lucchinetti, C.F.; Stadelmann, C.; Brück, W.; Rauschka, H.; Bergmann, M.; Schmidbauer, M.; Parisi, J.E.; Lassmann, H. Cortical demyelination and diffuse white matter injury in multiple sclerosis. *Brain* **2005**, *128*, 2705–2712. [CrossRef] [PubMed]

11. DeLuca, G.C.; Williams, K.; Evangelou, N.; Ebers, G.C.; Esiri, M.M. The contribution of demyelination to axonal loss in multiple sclerosis. *Brain* **2006**, *129*, 1507–1516. [CrossRef] [PubMed]

12. Nikić, I.; Merkler, D.; Sorbara, C.; Brinkoetter, M.; Kreutzfeldt, M.; Bareyre, F.M.; Brück, W.; Bishop, D.; Misgeld, T.; Kerschensteiner, M. A reversible form of axon damage in experimental autoimmune encephalomyelitis and multiple sclerosis. *Nat. Med.* **2011**, *17*, 495–499. [CrossRef] [PubMed]

13. Lassmann, H.; van Horssen, J. The molecular basis of neurodegeneration in multiple sclerosis. *FEBS Lett.* **2011**, *585*, 3715–3723. [CrossRef] [PubMed]

14. Frischer, J.M.; Bramow, S.; Dal-Bianco, A.; Lucchinetti, C.F.; Rauschka, H.; Schmidbauer, M.; Laursen, H.; Sorensen, P.S.; Lassmann, H. The relation between inflammation and neurodegeneration in multiple sclerosis brains. *Brain* **2009**, *132*, 1175–1189. [CrossRef] [PubMed]

15. Trapp, B.D.; Nave, K.A. Multiple sclerosis: An immune or neurodegenerative disorder? *Annu. Rev. Neurosci.* **2008**, *31*, 247–269. [CrossRef] [PubMed]

16. Benarroch, E.E. Acquired axonal degeneration and regeneration. *Neurology* **2015**, *84*, 2076–2085. [CrossRef] [PubMed]

17. Dziedzic, T.; Metz, I.; Dallenga, T.; König, F.B.; Müller, S.; Stadelmann, C.; Brück, W. Wallerian degeneration: A major component of early axonal pathology in multiple sclerosis. *Brain Pathol.* **2010**, *20*, 976–985. [CrossRef] [PubMed]

18. Solowska, J.M.; Baas, P.W. Hereditary spastic paraplegia SPG4: What is known and not known about the disease. *Brain* **2015**, *138*, 2471–2484. [CrossRef] [PubMed]

19. DeLuca, G.C.; Ebers, G.C.; Esiri, M.M. The extent of axonal loss in the long tracts in hereditary spastic paraplegia. *Neuropathol. Appl. Neurobiol.* **2004**, *30*, 576–584. [CrossRef] [PubMed]

20. Denton, K.R.; Lei, L.; Grenier, J.; Rodionov, V.; Blackstone, C.; Li, X. Loss of spastin function results in disease-specific axonal defects in human pluripotent stem cell-based models of hereditary spastic paraplegia. *Stem Cells* **2014**, *32*, 414–423. [CrossRef] [PubMed]

21. Bilsland, L.G.; Sahai, E.; Kelly, G.; Golding, M.; Greensmith, L.; Schiavo, G. Deficits in axonal transport precede ALS symptoms in vivo. *Proc. Natl. Acad. Sci. USA* **2010**, *107*, 20523–20528. [CrossRef] [PubMed]

22. Shi, P.; Gal, J.; Kwinter, D.M.; Liu, X.; Zhu, H. Mitochondrial dysfunction in amyotrophic lateral sclerosis. *Biochim. Biophys. Acta* **2010**, *1802*, 45–51. [CrossRef] [PubMed]

23. Calkins, M.J.; Reddy, P.H. Amyloid beta impairs mitochondrial anterograde transport and degenerates synapses in Alzheimer's disease neurons. *Biochim. Biophys. Acta* **2011**, *1812*, 507–513. [CrossRef] [PubMed]

24. Schapira, A.H.V. Mitochondria in the aetiology and pathogenesis in Parkinson's disease. *Lancet Neurol.* **2008**, *7*, 97–109. [CrossRef]

25. Alami, N.H.; Smith, R.B.; Carrasco, M.A.; Williams, L.A.; Winborn, C.S.; Han, S.S.; Kiskinis, E.; Winborn, B.; Freibaum, B.D.; Kanagaraj, A.; et al. Axonal transport of TDP-43 mRNA granules is impaired by ALS-causing mutations. *Neuron* **2014**, *81*, 536–543. [CrossRef] [PubMed]

26. Kim, H.J.; Kim, N.C.; Wang, Y.D.; Scarboough, E.A.; Moore, J.; Diaz, Z.; MacLea, K.S.; Freibaum, B.; Li, S.; Molliex, A.; et al. Mutations in prion-like domains in hnRNP A2B1 and hnRNPA1 cause multisystem proteinopathy and ALS. *Nature* **2013**, *495*, 467–473. [CrossRef] [PubMed]

27. Ling, S.C.; Polymenidou, M.; Cleveland, D.W. Converging mechanisms in ALS and FTD: Disrupted RNA and protein homeostasis. *Neuron* **2013**, *79*, 416–438. [CrossRef] [PubMed]

28. Ranaswami, M.; Taylor, J.P.; Parker, R. Altered "ribostasis": RNA-protein granule formation or persistence in the development of degenerative disorders. *Cell* **2013**, *154*, 727–735. [CrossRef] [PubMed]

29. Ugras, S.E.; Shorter, J. RNA-binding proteins in amyotrophic lateral sclerosis and neurodegeneration. *Neurol. Res. Int.* **2012**, *2012*, 432780. [CrossRef] [PubMed]

30. Lukong, K.E.; Chang, K.W.; Khandjian, E.W.; Richard, S. RNA-binding proteins in human genetic disease. *Trends Genet.* **2008**, *24*, 416–425. [CrossRef] [PubMed]

31. Orr, H.T. FTD and ALS: Genetic ties that bind. *Neuron* **2011**, *72*, 189–190. [CrossRef] [PubMed]

32. Black, J.A.; Newcombe, J.; Trapp, B.D.; Waxman, B.D. Sodium channel expression within chronic multiple sclerosis plaques. *J. Neuropathol. Exp. Neurol.* **2007**, *66*, 828–837. [CrossRef] [PubMed]

33. Craner, M.J.; Newcombe, J.; Black, J.A.; Hartle, C.; Cuzner, M.L.; Waxman, S.G. Molecular changes in neurons in multiple sclerosis: Altered axonal expression of Nav1.2 and Nav1.6 sodium channels and Na$^+$/Ca2$^+$ exchanger. *Proc. Natl. Acad. Sci. USA* **2004**, *101*, 8168–8173. [CrossRef] [PubMed]

34. Ferguson, B.; Matyszak, M.K.; Esiri, M.M.; Perry, V.H. Axonal damage in acute multiple sclerosis lesions. *Brain* **1997**, *120*, 393–399. [CrossRef] [PubMed]

35. Friese, M.A.; Schattling, B.; Fugger, L. Mechanisms of neurodegeneration and axonal dysfuction in multiple sclerosis. *Nat. Rev. Neurol.* **2014**, *10*, 225–238. [CrossRef] [PubMed]

36. Sorbara, C.D.; Wagner, N.E.; Ladwig, A.; Nikić, I.; Merkler, D.; Kleele, T.; Marinković, P.; Naumann, R.; Godinho, L.; Bareyre, F.M.; et al. Pervasive axonal transport deficits in multiple sclerosis models. *Neuron* **2014**, *84*, 1183–1190. [CrossRef] [PubMed]

37. De Vos, K.J.; Grierson, A.J.; Ackerley, S.; Miller, C.C.J. Role of axonal transport in neurodegenerative diseases. *Annu. Rev. Neurosci.* **2008**, *31*, 151–173. [CrossRef] [PubMed]

38. Dutta, R.; McDonough, J.; Yin, X.; Peterson, J.; Chang, A.; Torres, T.; Gudz, T.; Macklin, W.B.; Lewis, D.A.; Fox, R.J.; et al. Mitochondrial dysfunction as a cause of axonal degeneration in multiple sclerosis patients. *Ann. Neurol.* **2006**, *59*, 478–489. [CrossRef] [PubMed]

39. Campbell, G.R.; Ziabreva, I.; Reeve, A.K.; Krishnan, K.J.; Reynolds, R.; Howell, O.; Lassmann, H.; Turnbull, D.M.; Mahad, D.J. Mitochondrial DNA deletions and neurodegeneration in multiple sclerosis. *Ann. Neurol.* **2011**, *69*, 481–492. [CrossRef] [PubMed]

40. Fischer, M.T.; Sharma, R.; Lim, J.L.; Haider, L.; Frischer, J.M.; Drexhage, J.; Mahad, D.; Bradl, M.; van Horssen, J.; Lassmann, H. NADPH oxidase expression in active multiple sclerosis lesions in relation to oxidative tissue damage and mitochondrial injury. *Brain* **2012**, *135*, 886–899. [CrossRef] [PubMed]

41. Haider, L.; Fischer, M.T.; Frischer, J.M.; Bauer, J.; Höftberger, R.; Botond, G.; Esterbauer, H.; Binder, C.J.; Witztum, J.L.; Lassmann, H. Oxidative damage in multiple sclerosis lesions. *Brain* **2011**, *134*, 1914–1924. [CrossRef] [PubMed]

42. Satpute-Krishnan, P.; DeGiorgis, J.A.; Conley, M.P.; Jang, M.; Bearer, E.L. A peptide zipcode sufficient for anterograde transport within amyloid precursor protein. *Proc. Natl. Acad. Sci. USA* **2006**, *103*, 16532–16537. [CrossRef] [PubMed]

43. Schluesener, H.J.; Sobel, R.A.; Linington, C.; Weiner, H.L. A monoclonal antibody against a myelin oligodendrocyte glycoprotein induces relapses and demyelination in central nervous system autoimmune disease. *J. Immunol.* **1987**, *139*, 4016–4021. [PubMed]

44. Vanguri, P.; Koski, C.L.; Silverman, B.; Shin, M.L. Complement activation by isolated myelin: Activation of the classical pathway in the absence of myelin-specific antibodies. *Proc. Natl. Acad. Sci. USA* **1982**, *79*, 3290–3294. [CrossRef] [PubMed]

45. Vanguri, P.; Shin, M.L. Activation of complement by myelin: Identification of C1-binding proteins of human myelin from central nervous tissue. *J. Neurochem.* **1986**, *46*, 1535–1541. [CrossRef] [PubMed]

46. Huizinga, R.; Heijmans, N.; Schubert, P.; Gschmeissner, S.; 't Hart, B.A.; Herrmann, H.; Amor, S. Immunization with neurofilament light protein induces spastic paresis and axonal degeneration in Biozzi ABH mice. *J. Neuropathol. Exp. Neurol.* **2007**, *66*, 295–304. [CrossRef] [PubMed]

47. Huizinga, R.; Gerritsen, W.; Heijmans, N.; Amor, S. Axonal loss and gray matter pathology as a direct result of autoimmunity to neurofilaments. *Neurobiol. Dis.* **2008**, *32*, 461–470. [CrossRef] [PubMed]

48. Mathey, E.K.; Derfuss, T.; Storch, M.K.; Williams, K.R.; Hales, K.; Woolley, D.R.; Al-Hayani, A.; Davies, S.N.; Rasband, M.N.; Olsson, T.; et al. Neurofascin as a novel target for autoantibody-mediated axonal injury. *J. Exp. Med.* **2007**, *204*, 2363–2372. [CrossRef] [PubMed]

49. Levin, M.C.; Lee, S.; Gardner, L.A.; Shin, Y.; Douglas, J.N.; Cooper, C. Autoantibodies to non-myelin antigens as contributors to the pathogenesis of multiple sclerosis. *J. Clin. Cell Immunol.* **2013**, *30*, 4. [CrossRef] [PubMed]

50. Johri, A.; Beal, M.F. Mitochondrial dysfunction in neurodegenerative diseases. *J. Pharmacol. Exp. Ther.* **2012**, *342*, 619–630. [CrossRef] [PubMed]

51. Peterson, J.W.; Bö, L.; Mörk, S.; Chang, A.; Trapp, B.D. Transected neurites, apoptotic neurons, and reduced inflammation in cortical multiple sclerosis lesions. *Ann. Neurol.* **2001**, *50*, 389–400. [CrossRef] [PubMed]

52. Vercellino, M.; Plano, F.; Votta, B.; Mutani, R.; Giordana, M.T.; Cavalla, P. Grey matter pathology in multiple sclerosis. *J. Neuropathol. Exp. Neurol.* **2005**, *64*, 1101–1107. [CrossRef] [PubMed]

53. Booss, J.; Esiri, M.M.; Tourtellotte, W.W.; Mason, D.Y. Immunohistological analaysis of T lymphocyte subsets in the central nervous system in chronic progressive multiple sclerosis. *J. Neurol. Sci.* **1983**, *62*, 219–232. [CrossRef]

54. Lassmann, H.; Bradl, M. Multiple sclerosis: Experimental models and reality. *Acta Neuropathol.* **2017**, *133*, 223–244. [CrossRef] [PubMed]

55. Morandi, B.; Bramanti, P.; Bonaccorsi, I.; Montalto, E.; Oliveri, D.; Pezzino, G.; Navarra, M.; Ferlazzo, G. Role of natural killer cells in the pathogenesis and progression of multiple sclerosis. *Pharmacol. Res.* **2008**, *57*, 1–5. [CrossRef] [PubMed]

56. Barnett, M.H.; Parratt, J.D.; Cho, E.S.; Prineas, J.W. Immunoglobulins and complement in postmortem multiple sclerosis tissue. *Ann. Neurol.* **2009**, *65*, 32–46. [CrossRef] [PubMed]

57. Meinl, E.; Derfuss, T.; Krumbholz, M.; Pröbstel, A.K.; Hohlfeld, R. Humoral autoimmunity in multiple sclerosis. *J. Neurol. Sci.* **2010**, *306*, 180–182. [CrossRef] [PubMed]

58. Mead, R.J.; Neal, J.W.; Griffiths, M.R.; Linington, C.; Botto, M.; Lassmann, H.; Morgan, B.M. Deficiency of the complement regulator CD59a enhances disease severity, demyelination and axonal injury in murine acute experimental allergic encephalomyelitis. *Lab. Investig.* **2004**, *84*, 21–28. [CrossRef] [PubMed]

59. Levin, M.C.; Lee, S.; Kalume, F.; Morcos, Y.; Dohan, F.C.; Hasty, K.A.; Callaway, J.C.; Zunt, J.; Desiderio, D.M.; Stuart, J.M. Autoimmunity due to molecular mimicry as a cause of neurological disease. *Nat. Med.* **2002**, *8*, 509–513. [CrossRef] [PubMed]

60. Jean-Philippe, J.; Paz, S.; Caputi, M. hnRNP A1: The Swiss army knife of gene expression. *Int. J. Mol. Sci.* **2013**, *14*, 18999–19024. [CrossRef] [PubMed]

61. Jernigan, M.; Morcos, Y.; Lee, S.M.; Dohan, F.C., Jr.; Raine, C.; Levin, M.C. IgG in brain correlates with clinicopathological damage in HTLV-1 associated neurologic disease. *Neurology* **2003**, *60*, 1320–1327. [CrossRef] [PubMed]

62. Lee, S.M.; Dunnavant, F.D.; Jang, H.; Zunt, J.; Levin, M.C. Autoantibodies that recognize functional domains of hnRNPA1 implicate molecular mimicry in the pathogenesis of neurologic disease. *Neurosci. Lett.* **2006**, *401*, 188–193. [CrossRef] [PubMed]

63. Jean-Philippe, J.; Paz, S.; Lu, M.L.; Caputi, M. A truncated hnRNP A1 isoform, lacking the RGG-box RNA binding domain, can efficiently regulate HIV-1 splicing and replication. *Biochim. Biophys. Acta* **2014**, *1839*, 251–258. [CrossRef] [PubMed]

64. Lee, S.; Xu, L.; Shin, Y.; Gardner, L.; Hartzes, A.; Dohan, F.C.; Raine, C.; Homayouni, R.; Levin, M.C. A potential link between autoimmunity and neurodegeneration in immune-mediated neurological disease. *J. Neuroimmunol.* **2011**, *235*, 56–69. [CrossRef] [PubMed]

65. Yukitake, M.; Sueoka, E.; Sueoka-Aragane, N.; Sato, A.; Ohashi, H.; Yakushiji, Y.; Saito, M.; Osame, M.; Izumo, S.; Kuroda, Y. Significantly increased antibody response to heterogeneous nuclear ribonucleoproteins in cerebrospinal fluid of multiple sclerosis patients not in patients with human T-lymphotropic virus I-associated myelopathy/tropical spastic paraparesis. *J. Neurovirol.* **2008**, *14*, 130–135. [CrossRef] [PubMed]

66. Douglas, J.N.; Gardner, L.A.; Salapa, H.E.; Levin, M.C. Antibodies to the RNA binding protein heterogeneous nuclear ribonucleoprotein A1 colocalize to stress granules resulting in altered RNA and protein levels in a model of neurodegeneration in multiple sclerosis. *J. Clin. Cell. Immunol.* **2016**, *7*, 402. [PubMed]

67. Douglas, J.N.; Gardner, L.A.; Levin, M.C. Antibodies to an intracellular antigen penetrate neuronal cells and cause deleterious effects. *J. Clin. Cell. Immunol.* **2013**, *4*, 134. [CrossRef]

68. Douglas, J.N.; Gardner, L.A.; Salapa, H.E.; Lalor, S.J.; Lee, S.; Segal, B.M.; Sawchenko, P.E.; Levin, M.C. Antibodies to the RNA-binding protein hnRNP A1 contribute to neurodegeneration in a model of central nervous system autoimmune inflammatory disease. *J. Neuroinflamm.* **2016**, *13*, 178. [CrossRef] [PubMed]

69. Arima, Y.; Harada, M.; Kamimura, D.; Park, J.H.; Kawano, F.; Yull, F.E.; Kawamoto, T.; Iwakura, Y.; Betz, U.A.; Márquez, G.; et al. Regional neural activation defines a gateway for autoreactive T cells to cross the blood-brain barrier. *Cell* **2012**, *148*, 447–457. [CrossRef] [PubMed]

70. Kamimura, D.; Yamada, M.; Harada, M.; Sabharwal, L.; Meng, J.; Bando, H.; Ogura, H.; Atsumi, T.; Arima, Y.; Murakami, M. The gateway theory: Bridging neural and immune interactions in the CNS. *Front. Neurosci.* **2013**, *29*, 204. [CrossRef] [PubMed]

71. Solowska, J.M.; Morfini, G.; Falnikar, A.; Himes, B.T.; Brady, S.T.; Huang, D.; Baas, P.W. Quantitative and functional analyses of spastin in the nervous system: Implications for hereditary spastic paraplegia. *J. Neurosci.* **2008**, *28*, 2147–2157. [CrossRef] [PubMed]

72. Wilkinson, P.A.; Crosby, A.H.; Turner, C.; Bradley, L.J.; Ginsberg, L.; Wood, N.W.; Schapira, A.H.; Warner, T.T. A clinical, genetic, and biochemical study of SPG7 mutations in hereditary spastic paraplegia. *Brain* **2004**, *127*, 973–980. [CrossRef] [PubMed]

73. Lee, S.; Levin, M.C. Novel somatic single nucleotide variants within the RNA binding protein hnRNP A1 in multiple sclerosis patient. *F1000Research* **2014**, *3*, 132. [CrossRef] [PubMed]

74. Lee, S.L.; Shin, Y.; Levin, M.C.; University of Tennessee Health Science Center, Memphis, TN. Unpublished work, 2017.

75. Kremenchutzky, M.; Rice, G.P.A.; Baskerville, J.; Wingerchuk, D.M.; Ebers, G.C. The natural history of multiple sclerosis: A geographically based study 9: Observations on the progressive phase of the disease. *Brain* **2006**, *129*, 584–594. [CrossRef] [PubMed]

76. Miller, D.H.; Leary, S.M. Primary-progressive multiple sclerosis. *Lancet Neurol.* **2007**, *6*, 903–912. [CrossRef]

brain sciences

MDPI

Review

The Role of Peripheral CNS-Directed Antibodies in Promoting Inflammatory CNS Demyelination

Silke Kinzel [1] and Martin S. Weber [1,2,*]

1   Institute of Neuropathology, University Medical Center, 37099 Göttingen, Germany;
    silke.kinzel@med.uni-goettingen.de
2   Institute of Neuropathology and Department of Neurology, University Medical Center,
    Georg August University, Robert-Koch-Str. 40, 37099 Göttingen, Germany
*   Correspondence: martin.weber@med.uni-goettingen.de; Tel.: +49-551-39-7706; Fax: +49-551-39-10800

Academic Editor: Evanthia Bernitsas
Received: 18 May 2017; Accepted: 17 June 2017; Published: 22 June 2017

Abstract: In central nervous system (CNS) demyelinating disorders, such as multiple sclerosis (MS), neuromyelitis optica (NMO) and related NMO-spectrum disorders (NMO-SD), a pathogenic role for antibodies is primarily projected into enhancing ongoing CNS inflammation by directly binding to target antigens within the CNS. This scenario is supported at least in part, by antibodies in conjunction with complement activation in the majority of MS lesions and by deposition of anti-aquaporin-4 (AQP-4) antibodies in areas of astrocyte loss in patients with classical NMO. A currently emerging subgroup of AQP-4 negative NMO-SD patients expresses antibodies against myelin oligodendrocyte glycoprotein (MOG), again suggestive of their direct binding to CNS myelin. However, both known entities of anti-CNS antibodies, anti-AQP-4- as well as anti-MOG antibodies, are predominantly found in the serum, which raises the questions why and how a humoral response against CNS antigens is raised in the periphery, and in a related manner, what pathogenic role these antibodies may exert outside the CNS. In this regard, recent experimental and clinical evidence suggests that peripheral CNS-specific antibodies may indirectly activate peripheral CNS-autoreactive T cells by opsonization of otherwise unrecognized traces of CNS antigen in peripheral compartments, presumably drained from the CNS by its newly recognized lymphatic system. In this review, we will summarize all currently available data on both possible roles of antibodies in CNS demyelinating disorders, first, directly enhancing damage within the CNS, and second, promoting a peripheral immune response against the CNS. By elaborating on the latter scenario, we will develop the hypothesis that peripheral CNS-recognizing antibodies may have a powerful role in initiating acute flares of CNS demyelinating disease and that these humoral responses may represent a therapeutic target in its own right.

Keywords: multiple sclerosis; neuromyelitis optica; aquaporin-4; myelin oligodendrocyte glycoprotein; opsonization; autoantibody; central nervous system; CNS-draining lymphatics

---

1. Introduction

Several recent investigations highlight that B cells and antibodies can be crucially involved in the pathogenesis of central nervous system (CNS) demyelinating disorders, such as multiple sclerosis (MS), neuromyelitis optica (NMO) and NMO-spectrum disorders (NMO-SD) [1,2]. In particular the empirical success of clinical trials testing B cell-depleting anti-CD20 antibodies as therapeutic approach in MS and NMO substantiate this notion [3–6]. In these conditions, B cells are assumed to equally contribute to the inflammatory process by providing pro-inflammatory cytokines [7] and by acting as professional antigen-presenting cells (APC) [8], leading to the activation and propagation of autoreactive T cells (Figure 1). In contrast to these cellular B cell functions, the pathomechanistic involvement of antibodies may substantially differ in MS, NMO and NMO-SD.

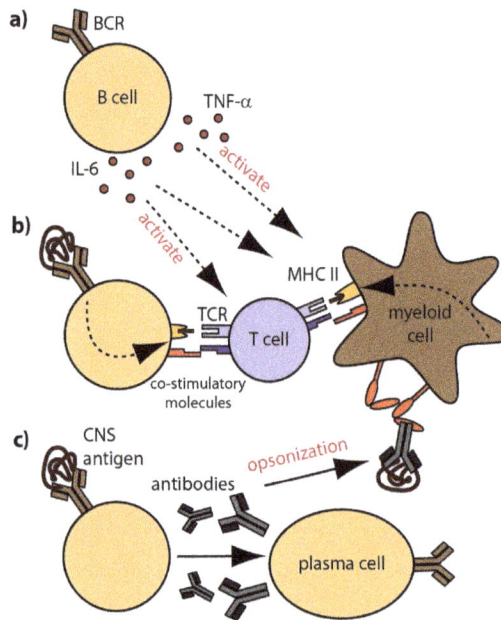

Figure 1. Cellular and molecular B cell properties in MS; (**a**) B cells modulate the activation and differentiation of immune cells by secretion of pro- and anti-inflammatory cytokines; (**b**) Antigen-specific B cells recognize CNS antigen via their BCR and internalize, process and present linearized antigens to responding T cells. Ligation of co-stimulatory molecules and secretion of pro-inflammatory cytokines foster the generation of effector T cells; (**c**) B cells differentiate into antibody-producing plasma cells. Secreted CNS-reactive antibodies that reach the CNS contribute to demyelination and inflammation by complement-mediated cytotoxicity. In the periphery, opsonization of rare CNS antigen by antibodies fosters the generation of auto-reactive T cells; Fc receptors on myeloid APC recognize antibody-antigen complexes and trigger internalization, processing and presentation of opsonized antigen to responding T cells. Definitions: APC = antigen-presenting cells; BCR = B cell receptor; CNS = central nervous system.

Due to some clinical, radiological and histopathological similarities, NMO was for decades considered to be a variant of MS. The discovery of antibodies against aquaporin-4 (AQP-4), a water channel expressed on astrocytes demonstrated in an impressive manner that it is a disease in its own right [9]. The presence of these autoantibodies in the serum of patients with CNS demyelination applies now as a unique feature separating NMO from MS [10]. Although initially introduced as a diagnostic marker, more recent investigations emphasize that anti-AQP-4 antibodies are critically involved in NMO pathogenesis [11,12]. In our current understanding, classical NMO is an autoimmune astrocytopathy, where AQP-4-directed antibodies directly destroy astrocytes and demyelination occurs only as a consequence of astrocyte loss [13]. It is important to note that in NMO patients, autoantibodies are mainly detectable in the serum, but not in the cerebrospinal fluid [14,15] suggesting that NMO is a peripheral humoral autoimmune disorder. In MS in contrast, no distinct humoral immune response could be identified so far unequivocally in the periphery, but most patients present oligoclonal immunoglobulins (Ig) termed oligoclonal bands (OCB) in the cerebrospinal fluid (CSF) [16], which were mostly absent in NMO patients [17]. These OCB originate from locally supported plasma cells [18,19]. Although it is still elusive whether intrathecal Ig are pathogenic or not, they are of important diagnostic value. In addition to OCB, in a subgroup of MS patients antibody depositions are found to co-localize with complement accumulation in areas of ongoing CNS demyelination [20,21], while astrocytes remain preserved. These findings indicate that in MS lesions, myelin and/or oligodendrocytes may be directly affected.

Based on the histopathology of MS and NMO, the role of CNS-reactive antibodies was primarily projected into enhancing ongoing CNS destruction during acute disease flares [22]; at that time, the blood-brain barrier is compromised due to immune cell infiltration, and peripheral antibodies have access to the CNS. However, novel findings suggest that CNS-directed autoantibodies may be of significance even before they enter the CNS. In rodents, antibodies directed against the myelin component, myelin oligodendrocyte glycoprotein (MOG), are capable of opsonizing antigen [23] in the periphery and thereby trigger its uptake, presentation and subsequent activation of encephalitogenic T cells, resulting in experimental autoimmune encephalomyelitis (EAE) [24], an animal model for CNS demyelinating disorders.

Deciphering the pathogenic function of such autoantibodies may be of particular interest and best to study in a recently emerging group of patients with CNS demyelinating disorder in which antibodies against MOG can be detected in the serum. MOG is an extracellular component of the myelin sheath, which is exclusively expressed in the CNS and assumed to be a prime candidate autoantigen in CNS demyelinating disorders. In this context, the question arises how a peripheral immune response can be raised against antigens that are exclusively present in the CNS. In the following sections, we will summarize the current knowledge about autoantibodies in CNS demyelination. Furthermore, we will discuss how a humoral immune response against CNS antigens may be raised in the periphery and by which mechanisms CNS-reactive antibodies potentially contribute to the development of CNS demyelinating disorders.

2. Towards a Mechanistic Understating on the Role of Autoantibodies in CNS-Demyelinating Disorders

B cells can drive inflammation on the one hand by the secretion of pro-inflammatory cytokines and by exerting antigen-presenting function on the other hand (Figure 1a,b). Especially due to the ability of B cells to bind conformational protein-antigen specifically via their B cell receptor, they are highly competent in recognizing small amounts of protein antigen [25]. Hence, B cells are very efficient APC when they share antigen recognition with T cells [26,27] and the mere co-existence of CNS-specific B and T cells in mice is sufficient to induce EAE [28–30]. Another approach to induce experimental demyelination in various species is immunization with CNS antigens. One well established autoantigen activating T and B cells is MOG [31]; in mice, active immunization with MOG-derived peptides leads to the development of encephalitogenic T cells without the involvement of B cells, while immunization with conformational MOG protein additionally activates B cells in a pathogenic manner [32,33]. As a consequence, B cells differentiate into plasma cells which, in an appropriate induction regime, secrete pathogenic antibodies. Those antibodies represent the soluble counterpart of the B cell receptor and may mediate similar properties in recognizing protein antigens with a low prevalence. In EAE, MOG-specific antibodies have been shown to exacerbate ongoing disease [34,35] by promoting CNS inflammation and demyelination [36,37]. It is assumed that encephalitogenic T cells compromise the integrity of the blood-brain barrier allowing peripheral autoantibodies to enter the CNS. There, CNS-reactive antibodies can mediate myelin destruction directly by fixation of the complement system [38] and/or by increasing the uptake and intracellular metabolism of myelin by macrophages [23,39]. Lyons, Ramsbottom and Cross [40], however, suggest a more fundamental role of myelin-specific antibodies for the initiation of EAE in mice than just enhancing demyelination. They demonstrate that adoptive transfer of anti-MOG antibodies restores the ability of B cell deficient mice to develop clinical and histological EAE upon active immunization, indicating that antibodies are required for T cell (re-)activation. In the same line, studies of our group demonstrated that anti- MOG antibodies opsonize traces of otherwise undetected soluble MOG in the periphery and thereby trigger and amplify a respective pathogenic T cell response (Figure 1c). Based on this mechanism, myelin-specific antibodies in combination with myelin-reactive T cells were sufficient to induce spontaneous EAE in mice. Importantly, prior to EAE, no antibody deposition was detected within the CNS, which indicates that peripherally applied MOG-specific antibodies do not directly bind to CNS-located myelin but trigger activation of T cells outside the CNS. In conclusion, these findings indicate that peripheral CNS-specific antibodies cannot only contribute to myelin destruction by direct binding to the CNS, but additionally, promote the development of encephalitogenic T cells in the periphery.

3. Stratifying Human Demyelinating Disorders by the Involvement of Distinct Autoantibodies

Neuroinflammation and subsequent demyelination can be caused by a variety of extrinsic factors, but also by the immune system itself attacking endogenous molecules as it is presumably the case in MS, NMO and NMO-SD. Although the targeted antigens and pathological mechanism may differ, these diseases result in the loss of CNS myelin, neuronal damage and axonal impairment as a consequence. For decades, T cells were considered to be the major effector cell type in autoimmune CNS demyelinating disorders. However, more recent observations highlight that B cells and B cell-derived products are equally important key players in their pathogenesis. In MS, the first indication for this assumption was the discovery of OCB in the CSF of 90% of MS patients [16], which originate from locally supported plasma cells [18,19]. The particular antigen(s) recognized by these autoantibodies remain elusive and concomitantly, it is still under debate whether OCB are of pathogenic relevance for MS. However, in a subgroup of MS patients with so-called type II lesions, antibody depositions are found to co-localize with complement accumulation in areas of ongoing CNS demyelination [20,21]. This observation together with the aforementioned findings in EAE, highlight a potential role of autoantibodies for the complement-mediated destruction of myelin in acute demyelinating MS lesions [41]. Within the intrathecal humoral immune response, many potential CNS targets have been suggested in recent years, including neuroglial and astrocytic antigens such as neurofascin and contactin-2 [42,43], and also myelin antigens such as myelin basic protein and MOG [44,45], but could not be confirmed. The fact that the expression pattern of OCB in MS patients have no apparent reflection in the blood suggests that at least some of the antibodies present in the CSF are produced within the CNS. However, this comparison does not formally exclude that individual antibody entities may originate from the periphery. In line with the assumption of intrathecal IgG production, ectopic B cell follicle-like structures in the meninges of secondary progressive MS patients suggest that B cell function gradually shift from the periphery into the inflamed CNS during disease progression [46]. Furthermore, patients with primary- or secondary-progressive MS only rarely show the formation of new inflammatory spots, but rather, a gradual expansion of consisting lesions pointing towards a CNS intrinsic pathogenic mechanism that is independent of the peripheral immune system.

The heterogeneous appearance of MS, and the fact that no common autoantigen could be identified so far, suggests that MS may consist of different disease entities. Based on the discovery of anti-AQP-4 antibodies, NMO was the first condition that has been separated from the "core disorder" MS [47]. AQP-4-directed antibodies are suggested to directly destroy astrocytes [10], while demyelination occurs only as a consequence of astrocyte loss in later stages of the disease [13]. Interestingly, antibody producing plasma cells are only infrequently found in the CSF of NMO patients [48], while AQP-4 positive plasmablasts are selectively increased in the blood and shown to peak at relapses [49]. Further, OCB are only present in 15–30% of NMO patients and disappear mostly during disease progression [50]. These findings together indicate that anti-AQP-4 antibodies are mainly generated in the periphery and suggest that NMO is a peripheral humoral autoimmune disorder. However, it is poorly understood how anti-AQP-4 Ab enter the CNS and induce lesion formation. Normally, in the absence of ongoing inflammation, the CNS is assumed to be an immune-privileged organ and antibodies should not be able to cross the intact blood–brain barrier. Several observations in NMO patients support such prerequisites. Signs of viral infections prior to NMO relapses were observed in 15–30% of NMO patients, in which the virally provoked immune response may trigger an autoimmune reaction [51,52]. Further, relatively high anti-AQP-4 antibody titers can be detected in many patients during remission phase or, in individual cases, even years before disease onset, indicating the necessity of inflammatory conditions accompanying the peripheral autoantibody response [53,54]. Therefore, it is assumed that anti-AQP-4 antibodies reach the CNS only after inflammation-induced opening of the blood–brain barrier. In support of this hypothesis, disease progression occurs in NMO patients only during acute disease flares, when peripheral immune cells penetrate the CNS [55]. Apart from that, recently published in vitro data point towards the possibility that anti-AQP-4 antibodies themselves may contribute to blood–brain barrier penetration by interaction with astrocytes, which in turn prompt endothelial cells to decrease barrier function [56].

It is important to note that not all patients with a NMO-suggestive condition show anti-AQP-4 antibodies; an observation that led to the introduction of the broader category NMO-SD [57]. Within this category, a small proportion of patients show antibodies against MOG in the serum [58]. MOG is a transmembrane protein expressed on the surface of oligodendrocytes [59] and the outermost lamella of the myelin sheath. Although its exact function remains unclear, it is assumed to mediate the adhesion of neighboring myelinated fibers acting as an adhesive glue between them [60]. Due to its extracellular localization and its lack of expression in the thymus, MOG represents a plausible target for autoimmune responses [61,62]. Interestingly, patients negative for anti-AQP-4, but positive for anti-MOG antibodies fulfil many of the clinical and radiological criteria for NMO, but their relapse biology and prognosis rather resembles MS [63]. Especially the fact that the respective CNS biopsy/autopsy material showed no astrocytopathy, but demyelination as primary destructive mechanism, differentiates these patients sharply from classical NMO [64–66]. Consequently, these findings initiated the debate whether anti-MOG antibody-positive encephalomyelitis should be continued to be considered as a part of NMO-SD or may be defined as separate disease entity [67].

Notwithstanding these differences, the most crucial similarity between anti-MOG positive NMO-SD and anti-AQP positive NMO is that autoantibodies are predominantly found in the blood, but not in the CSF. Thus, for both conditions the question arises why and how a humoral response against CNS antigens is raised in the periphery. Furthermore, it remains elusive what pathogenic role these antibodies may exert outside the CNS and whether opsonization of endogenous CNS antigens, as described above in mice, can also occur in this human condition as a prime disease-driving mechanism. Indeed, in vitro experiments demonstrated that anti-MOG antibodies isolated from the blood of NMO-SD patients were, in principle, capable of opsonizing human MOG protein, resulting in an increased uptake of antigen by macrophages [24]. However, it remains elusive, where antibodies confer antigen recognition to myeloid APC, given that MOG is solely expressed in the CNS. A recently recognized lymphatic system which drains the CNS may represent this so far missing anatomical link between the CNS and the peripheral immune system. These newly appreciated lymphatic vessels have been shown to drain molecules from the CSF into cervical lymph nodes [68,69]. Furthermore, traces of myelin have been found there in MS patients as well as healthy controls [70,71] (Figure 2). This implies that even under non-pathologic conditions, CNS antigens can (occasionally) be transported to cervical lymph nodes. A first indication for the clinical relevance of cervical lymph has been given by a chronic-relapsing EAE model where lymphadenectomized mice show a reduced relapse severity [72].

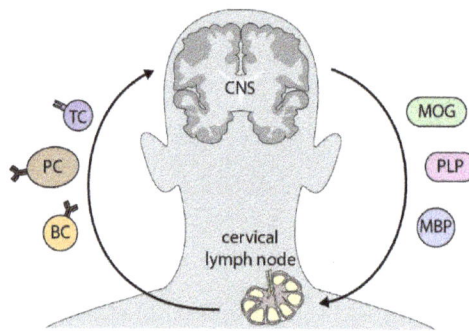

Figure 2. CNS antigens potentially activate peripheral immune cells in CNS-draining lymph nodes. Cerebrospinal fluid, which occasionally contains CNS antigen, such as myelin components, is drained by lymphatic vessels into deep cervical lymph nodes. There, antigen is encountered and processed by professional APC and subsequently presented to autoreactive antigen-specific T cells. By this interaction, immune cells are activated and in turn, migrate to the CNS, where they contribute to inflammation. Definitions: MOG = myelin oligodendrocyte glycoprotein; PLP = proteolipid-Protein 1; MBP = myelin basic protein; TC = T cell, PC = plasma cell; BC = B cell.

4. Conclusions

B cells and antibodies have been recognized to be important players in the pathogenesis of MS, NMO and NMO-SD. While it emerges that cellular properties of B cells such as cytokine secretion and antigen presentation can drive inflammation in all of these conditions, the exact role of antibodies is still under discussion. In NMO, anti-AQP-4 antibodies are assumed to bind directly to astrocytes and induce a complement-mediated astrocytopathy. Similarly, in a subgroup of MS patients, lesions are characterized by the co-localization of antibodies and complement in areas of ongoing demyelination pointing towards antibody-mediated degradation of myelin. These mechanisms likely require, though, that the blood-brain barrier is compromised during acute disease flares and peripheral antibodies have access to the CNS, or that pathogenic antibodies are produced locally within the CNS. Notwithstanding these observations, we highlight that antibodies harbor additional pathogenic properties that may contribute to the pathogenesis of demyelinating disorders even before autoantibodies reach the CNS. Studies in mice and first transitional experiments with Ig isolated from NMO-SD patients demonstrated that CNS-reactive antibodies are capable of opsonizing soluble antigen, resulting in an increased uptake by macrophages. Subsequently, T cells differentiated in an encephalitogenic manner and induced spontaneous EAE in mice. Based on these findings, it is plausible that antibodies can contribute to the generation of auto-reactive T cells by increasing the uptake of available antigen; a mechanism probably primarily important in NMO and NMO-SD patients, where CNS-directed autoantibodies accumulate in the serum. Thus, if it can be confirmed that CNS-draining lymphoid organs are the site where antibody-mediated opsonization triggers de novo recognition of CNS antigen, both the peripheral humoral response as well as CNS draining lymph nodes may be promising targets for future therapeutic interventions in CNS demyelinating disorders. Furthermore, understanding the relative importance of antibodies for the pathogenesis of each disease entity may offer the possibility to refine treatment options in the most suitable way.

Author Contributions: S.K. wrote the manuscript and drafted the figures. M.S.W. conceptualized and wrote the manuscript.

Conflicts of Interest: The authors declare no conflict of interest.

References

1. Kinzel, S.; Weber, M.S.B. Cell-Directed Therapeutics in Multiple Sclerosis: Rationale and Clinical Evidence. *CNS Drug* **2016**, *30*, 1137–1148. [CrossRef] [PubMed]
2. Bennett, J.L.; O'Connor, K.C.; Bar-Or, A.; Zamvil, S.S.; Hemmer, B.; Tedder, T.F.; von Büdingen, H.C.; Stuve, O.; Yeaman, M.R.; Smith, T.J.; et al. B lymphocytes in neuromyelitis optica. *Neurol. Neuroimmunol. Neuroinflam.* **2015**, *2*, e104. [CrossRef] [PubMed]
3. Hauser, S.L.; Bar-Or, A.; Comi, G.; Giovannoni, G.; Hartung, H.P.; Hemmer, B.; Lublin, F.; Montalban, X.; Rammohan, K.W.; Selmaj, K.; et al. Ocrelizumab versus Interferon Beta-1a in Relapsing Multiple Sclerosis. *N. Engl. J. Med.* **2017**, *376*, 221–234. [CrossRef] [PubMed]
4. Montalban, X.; Hauser, S.L.; Kappos, L.; Arnold, D.L.; Bar-Or, A.; Comi, G.; de Seze, J.; Giovannoni, G.; Hartung, H.-P.; Hemmer, B.; et al. Ocrelizumab versus Placebo in Primary Progressive Multiple Sclerosis. *N. Engl. J. Med.* **2017**, *376*, 209–220. [CrossRef] [PubMed]
5. Cree, B.A.; Lamb, S.; Morgan, K.; Chen, A.; Waubant, E.; Genain, C. An open label study of the effects of rituximab in neuromyelitis optica. *Neurology* **2005**, *64*, 1270–1272. [CrossRef] [PubMed]
6. Lehmann-Horn, K.; Kronsbein, H.C.; Weber, M.S. Targeting B cells in the treatment of multiple sclerosis: Recent advances and remaining challenges. *Ther. Adv. Neurol. Disord.* **2013**, *6*, 161–173. [CrossRef] [PubMed]
7. Duddy, M.; Niino, M.; Adatia, F.; Hebert, S.; Freedman, M.; Atkins, H.; Kim, H.J.; Bar-Or, A. Distinct effector cytokine profiles of memory and naive human B cell subsets and implication in multiple sclerosis. *J. Immunol.* **2007**, *178*, 6092–6099. [CrossRef] [PubMed]
8. Constant, S.; Schweitzer, N.; West, J.; Ranney, P.; Bottomly, K.B. lymphocytes can be competent antigen-presenting cells for priming CD4+ T cells to protein antigens in vivo. *J. Immunol.* **1995**, *155*, 3734–3741. [PubMed]
9. Lennon, V.A.; Wingerchuk, D.M.; Kryzer, T.J.; Pittock, S.J.; Lucchinetti, C.F.; Fujihara, K.; Nakashima, I.; Weinshenker, B.G. A serum autoantibody marker of neuromyelitis optica: Distinction from multiple sclerosis. *Lancet* **2004**, *364*, 2106–2112. [CrossRef]
10. Misu, T.; Fujihara, K.; Kakita, A.; Konno, H.; Nakamura, M.; Watanabe, S.; Takahashi, T.; Nakashima, I.; Takahashi, H.; Itoyama, Y. Loss of aquaporin 4 in lesions of neuromyelitis optica: Distinction from multiple sclerosis. *Brain* **2007**, *130*, 1224–1234. [CrossRef] [PubMed]
11. Lucchinetti, C.F.; Mandler, R.N.; McGavern, D.; Bruck, W.; Gleich, G.; Ransohoff, R.M.; Trebst, C.; Weinshenker, B.; Wingerchuk, D.; Parisi, J.E.; et al. A role for humoral mechanisms in the pathogenesis of Devic's neuromyelitis optica. *Brain* **2002**, *125*, 1450–1461. [CrossRef] [PubMed]
12. Hinson, S.R.; Pittock, S.J.; Lucchinetti, C.F.; Roemer, S.F.; Fryer, J.P.; Kryzer, T.J.; Lennon, V.A. Pathogenic potential of IgG binding to water channel extracellular domain in neuromyelitis optica. *Neurology* **2007**, *69*, 2221–2231. [CrossRef] [PubMed]
13. Levy, M.; Wildemann, B.; Jarius, S.; Roemer, S.F.; Fryer, J.P.; Kryzer, T.J.; Lennon, V.A. Immunopathogenesis of neuromyelitis optica. *Adv. Immunol.* **2014**, *121*, 213–242. [PubMed]
14. Waters, P.J.; McKeon, A.; Leite, M.I.; Rajasekharan, S.; Lennon, V.A.; Villalobos, A.; Palace, J.; Mandrekar, J.N.; Vincent, A.; Bar-Or, A.; et al. Serologic diagnosis of NMO: A multicenter comparison of aquaporin-4-IgG assays. *Neurology* **2012**, *78*, 665–671. [CrossRef] [PubMed]
15. Majed, M.; Fryer, J.P.; McKeon, A.; Lennon, V.A.; Pittock, S.J. Clinical utility of testing AQP4-IgG in CSF: Guidance for physicians. *Neurol. Neuroimmunol. Neuroinflam.* **2016**, *3*, e231. [CrossRef] [PubMed]
16. Kabat, E.A.; Freedman, D.A. A study of the crystalline albumin, gamma globulin and total protein in the cerebrospinal fluid of 100 cases of multiple sclerosis and in other diseases. *Am. J. Med. Sci.* **1950**, *I*, 55–64. [CrossRef]
17. Jarius, S.; Paul, F.; Franciotta, D.; Ruprecht, K.; Ringelstein, M.; Bergamaschi, R.; Rommer, P.; Kleiter, I.; Stich, O.; Reuss, R.; et al. Cerebrospinal fluid findings in aquaporin-4 antibody positive neuromyelitis optica: Results from 211 lumbar punctures. *J. Neurol. Sci.* **2011**, *306*, 82–90. [CrossRef] [PubMed]
18. Obermeier, B.; Mentele, R.; Malotka, J.; Kellermann, J.; Kümpfel, T.; Wekerle, H.; Lottspeich, F.; Hohlfeld, R.; Dornmair, K. Matching of oligoclonal immunoglobulin transcriptomes and proteomes of cerebrospinal fluid in multiple sclerosis. *Nat. Med.* **2008**, *14*, 688–693. [CrossRef] [PubMed]

19. Von Budingen, H.C.; Gulati, M.; Kuenzle, S.; Fischer, K.; Rupprecht, T.A.; Goebels, N. Clonally expanded plasma cells in the cerebrospinal fluid of patients with central nervous system autoimmune demyelination produce "oligoclonal bands". *J. Neuroimmunol.* **2010**, *218*, 134–139. [CrossRef] [PubMed]

20. Genain, C.P.; Cannella, B.; Hauser, S.L.; Raine, C.S. Identification of autoantibodies associated with myelin damage in multiple sclerosis. *Nat. Med.* **1999**, *5*, 170–175. [CrossRef] [PubMed]

21. Storch, M.K.; Piddlesden, S.; Haltia, M.; Iivanainen, M.; Morgan, P.; Lassmann, H. Multiple sclerosis: In situ evidence for antibody- and complement-mediated demyelination. *Ann. Neurol.* **1998**, *43*, 465–471.

22. Weber, M.S.; Hemmer, B.; Cepok, S. The role of antibodies in multiple sclerosis. *Biochim. Biophys. Acta* **2011**, *1812*, 239–245. [CrossRef] [PubMed]

23. Van der Goes, A.; Kortekaas, M.; Hoekstra, K.; Dijkstra, C.D.; Amor, S. The role of anti-myelin (auto)-antibodies in the phagocytosis of myelin by macrophages. *J. Neuroimmunol.* **1999**, *101*, 61–67. [CrossRef]

24. Kinzel, S.; Lehmann-Horn, K.; Torke, S.; Häusler, D.; Winkler, A.; Stadelmann, C.; Payne, N.; Feldmann, L.; Saiz, A.; Reindl, M.; et al. Myelin-reactive antibodies initiate T cell-mediated CNS autoimmune disease by opsonization of endogenous antigen. *Acta Neuropathol.* **2016**, *132*, 43–58. [CrossRef] [PubMed]

25. Constant, S.; Sant'Angelo, D.; Pasqualini, T.; Taylor, T.; Levin, D.; Flavell, R.; Bottomly, K. Peptide and protein antigens require distinct antigen-presenting cell subsets for the priming of CD4+ T cells. *J. Immunol.* **1995**, *154*, 4915–4923. [PubMed]

26. Lanzavecchia, A. Antigen-specific interaction between T and B cells. *Nature* **1985**, *314*, 537–539. [CrossRef] [PubMed]

27. Weber, M.S.; Hemmer, B. Cooperation of B cells and T cells in the pathogenesis of multiple sclerosis. *Results Probl. Cell Differ.* **2010**, *51*, 115–126. [PubMed]

28. Krishnamoorthy, G.; Lassmann, H.; Wekerle, H.; Holz, A. Spontaneous opticospinal encephalomyelitis in a double-transgenic mouse model of autoimmune T cell/B cell cooperation. *J. Clin. Investig.* **2006**, *116*, 2385–2392. [CrossRef] [PubMed]

29. Pollinger, B.; Krishnamoorthy, G.; Berer, K.; Lassmann, H.; Bösl, M.R.; Dunn, R.; Domingues, H.S.; Holz, A.; Kurschus, F.C.; Wekerle, H. Spontaneous relapsing-remitting EAE in the SJL/J mouse: MOG-reactive transgenic T cells recruit endogenous MOG-specific B cells. *J. Exp. Med.* **2009**, *206*, 1303–1316. [CrossRef] [PubMed]

30. Molnarfi, N.; Schulze-Topphoff, U.; Weber, M.S.; Patarroyo, J.C.; Prod'homme, T.; Varrin-Doyer, M.; Shetty, A.; Linington, C.; Slavin, A.J.; Hidalgo, J.; et al. MHC class II-dependent B cell APC function is required for induction of CNS autoimmunity independent of myelin-specific antibodies. *J. Exp. Med.* **2013**, *210*, 2921–2937. [CrossRef] [PubMed]

31. Constantinescu, C.S.; Farooqi, N.; O'Brien, K.; Gran, B. Experimental autoimmune encephalomyelitis (EAE) as a model for multiple sclerosis (MS). *Br. J. Pharmacol.* **2011**, *164*, 1079–1106. [CrossRef] [PubMed]

32. Weber, M.S.; Prod'homme, T.; Patarroyo, J.C.; Molnarfi, N.; Karnezis, T.; Lehmann-Horn, K.; Danilenko, D.M.; Eastham-Anderson, J.; Slavin, A.J.; Linington, C.; et al. B-cell activation influences T-cell polarization and outcome of anti-CD20 B-cell depletion in central nervous system autoimmunity. *Ann. Neurol.* **2010**, *68*, 369–383. [CrossRef] [PubMed]

33. Lalive, P.H.; Molnarfi, N.; Benkhoucha, M.; Weber, M.S.; Santiago-Raber, M.L. Antibody response in MOG35–55 induced EAE. *J. Neuroimmunol.* **2011**, *240–241*, 28–33. [CrossRef] [PubMed]

34. Schluesener, H.J.; Sobel, R.A.; Linington, C.; Weiner, H.L. A monoclonal antibody against a myelin oligodendrocyte glycoprotein induces relapses and demyelination in central nervous system autoimmune disease. *J. Immunol.* **1987**, *139*, 4016–4021. [PubMed]

35. Flach, A.C.; Litke, T.; Strauss, J.; Haberl, M.; Gómez, C.C.; Reindl, M.; Saiz, A.; Fehling, H.J.; Wienands, J.; Odoardi, F.; et al. Autoantibody-boosted T-cell reactivation in the target organ triggers manifestation of autoimmune CNS disease. *Proc. Natl. Acad. Sci. USA* **2016**, *113*, 3323–3328. [CrossRef] [PubMed]

36. Oliver, A.R.; Lyon, G.M.; Ruddle, N.H. Rat and human myelin oligodendrocyte glycoproteins induce experimental autoimmune encephalomyelitis by different mechanisms in C57BL/6 mice. *J. Immunol.* **2003**, *171*, 462–468. [CrossRef] [PubMed]

37. Storch, M.K.; Stefferl, A.; Brehm, U.; Weissert, R.; Wallström, E.; Kerschensteiner, M.; Olsson, T.; Linington, C.; Lassmann, H. Autoimmunity to myelin oligodendrocyte glycoprotein in rats mimics the spectrum of multiple sclerosis pathology. *Brain Pathol.* **1998**, *8*, 681–694. [CrossRef] [PubMed]

38. Urich, E.; Gutcher, I.; Prinz, M.; Becher, B. Autoantibody-mediated demyelination depends on complement activation but not activatory Fc-receptors. *Proc. Natl. Acad. Sci. USA* **2006**, *103*, 18697–18702. [CrossRef] [PubMed]

39. Trotter, J.; DeJong, L.J.; Smith, M.E. Opsonization with antimyelin antibody increases the uptake and intracellular metabolism of myelin in inflammatory macrophages. *J. Neurochem.* **1986**, *47*, 779–789. [CrossRef] [PubMed]

40. Lyons, J.A.; Ramsbottom, M.J.; Cross, A.H. Critical role of antigen-specific antibody in experimental autoimmune encephalomyelitis induced by recombinant myelin oligodendrocyte glycoprotein. *Eur. J. Immunol.* **2002**, *32*, 1905–1913. [CrossRef]

41. Seifert, C.L.; Wegner, C.; Sprenger, T.; Weber, M.S.; Brück, W.; Hemmer, B.; Sellner, J. Favourable response to plasma exchange in tumefactive CNS demyelination with delayed B-cell response. *Mult. Scler.* **2012**, *18*, 1045–1049. [CrossRef] [PubMed]

42. Mathey, E.K.; Derfuss, T.; Storch, M.K.; Williams, K.R.; Hales, K.; Woolley, D.R.; Al-Hayani, A.; Davies, S.N.; Rasband, M.N.; Olsson, T.; et al. Neurofascin as a novel target for autoantibody-mediated axonal injury. *J. Exp. Med.* **2007**, *204*, 2363–2372. [CrossRef] [PubMed]

43. Derfuss, T.; Parikh, K.; Velhin, S.; Braun, M.; Mathey, E.; Krumbholz, M.; Kümpfel, T.; Moldenhauer, A.; Rader, C.; Sonderegger, P.; et al. Contactin-2/TAG-1-directed autoimmunity is identified in multiple sclerosis patients and mediates gray matter pathology in animals. *Proc. Natl. Acad. Sci. USA* **2009**, *106*, 8302–8307. [CrossRef] [PubMed]

44. Warren, K.G.; Catz, I. Relative frequency of autoantibodies to myelin basic protein and proteolipid protein in optic neuritis and multiple sclerosis cerebrospinal fluid. *J. Neurol. Sci.* **1994**, *121*, 66–73. [CrossRef]

45. Lalive, P.H.; Menge, T.; Delarasse, C.; Williams, K.R.; Hales, K.; Woolley, D.R.; Al-Hayani, A.; Davies, S.N.; Rasband, M.N.; Olsson, T.; et al. Antibodies to native myelin oligodendrocyte glycoprotein are serologic markers of early inflammation in multiple sclerosis. *Proc. Natl Acad. Sci. USA* **2006**, *103*, 2280–2285. [CrossRef] [PubMed]

46. Serafini, B.; Rosicarelli, B.; Magliozzi, R.; Stigliano, E.; Aloisi, F. Detection of ectopic B-cell follicles with germinal centers in the meninges of patients with secondary progressive multiple sclerosis. *Brain Pathol.* **2004**, *14*, 164–174. [CrossRef] [PubMed]

47. Lennon, V.A.; Kryzer, T.J.; Pittock, S.J.; Verkman, A.S.; Hinson, S.R. IgG marker of optic-spinal multiple sclerosis binds to the aquaporin-4 water channel. *J. Exp. Med.* **2005**, *202*, 473–477. [CrossRef] [PubMed]

48. Bennett, J.L.; Lam, C.; Kalluri, S.R.; Saikali, P.; Bautista, K.; Dupree, C.; Glogowska, M.; Case, D.; Antel, J.P.; Owens, G.P.; et al. Intrathecal pathogenic anti–aquaporin-4 antibodies in early neuromyelitis optica. *Ann. Neurol.* **2009**, *66*, 617–629. [CrossRef] [PubMed]

49. Chihara, N.; Aranami, T.; Sato, W.; Miyazaki, Y.; Miyake, S.; Okamoto, T.; Ogawa, M.; Toda, T.; Yamamura, T. Interleukin 6 signaling promotes anti-aquaporin 4 autoantibody production from plasmablasts in neuromyelitis optica. *Proc. Natl. Acad. Sci. USA* **2011**, *108*, 3701–3706. [CrossRef] [PubMed]

50. Jarius, S.; Paul, F.; Franciotta, D.; Waters, P.; Zipp, F.; Hohlfeld, R.; Vincent, A.; Wildemann, B. Mechanisms of Disease, aquaporin-4 antibodies in neuromyelitis optica. *Nat. Clin. Pract. Neurol.* **2008**, *4*, 202–214. [CrossRef] [PubMed]

51. Ghezzi, A.; Bergamaschi, R.; Martinelli, V.; Trojano, M.; Tola, M.R.; Merelli, E.; Mancardi, L.; Gallo, P.; Filippi, M.; Zaffaroni, M.; et al. Clinical characteristics, course and prognosis of relapsing Devic's Neuromyelitis Optica. *J. Neurol.* **2004**, *251*, 47–52. [CrossRef] [PubMed]

52. Jarius, S.; Ruprecht, K.; Wildemann, B.; Kuempfel, T.; Ringelstein, M.; Geis, C.; Kleiter, I.; Kleinschnitz, C.; Berthele, A.; Brettschneider, J.; et al. Contrasting disease patterns in seropositive and seronegative neuromyelitis optica, A multicentre study of 175 patients. *J. Neuroinflam.* **2012**, *9*, 14. [CrossRef] [PubMed]

53. Jarius, S.; Franciotta, D.; Paul, F.; Bergamaschi, R.; Rommer, P.S.; Ruprecht, K.; Ringelstein, M.; Aktas, O.; Kristoferitsch, W.; Wildemann, B. Testing for antibodies to human aquaporin-4 by ELISA, Sensitivity, specificity, and direct comparison with immunohistochemistry. *J. Neurol. Sci.* **2012**, *320*, 32–37. [CrossRef] [PubMed]

54. Nishiyama, S.; Ito, T.; Misu, T.; Takahashi, T.; Kikuchi, A.; Suzuki, N.; Jin, K.; Aoki, M.; Fujihara, K.; Itoyama, Y. A case of NMO seropositive for aquaporin-4 antibody more than 10 years before onset. *Neurology* **2009**, *72*, 1960–1961. [CrossRef] [PubMed]

55. Kleiter, I.; Gahlen, A.; Borisow, N.; Fischer, K.; Wernecke, K.D.; Wegner, B.; Hellwig, K.; Pache, F.; Ruprecht, K.; Havla, J.; et al. Neuromyelitis optica, Evaluation of 871 attacks and 1,153 treatment courses. *Ann. Neurol.* **2016**, *79*, 206–216. [CrossRef] [PubMed]

56. Takeshita, Y.; Obermeier, B.; Cotleur, A.C.; Spampinato, S.F.; Shimizu, F.; Yamamoto, E.; Sano, Y.; Kryzer, T.J.; Lennon, V.A.; Kanda, T.; et al. Effects of neuromyelitis optica–IgG at the blood–brain barrier in vitro. *Neurol. Neuroimmunol. Neuroinflam.* **2017**, *4*, e311. [CrossRef] [PubMed]

57. Wingerchuk, D.M.; Banwell, B.; Bennett, J.L.; Cabre, P.; Carroll, W.; Chitnis, T.; de Seze, J.; Fujihara, K.; Greenberg, B.; Jacob, A.; et al. International consensus diagnostic criteria for neuromyelitis optica spectrum disorders. *Neurology* **2015**, *85*, 177–189. [CrossRef] [PubMed]

58. Probstel, A.K.; Rudolf, G.; Dornmair, K.; Collongues, N.; Chanson, J.-B.; Sanderson, N.S.R.; Lindberg, R.L.P.; Kappos, L.; de Seze, J.; Derfuss, T. Anti-MOG antibodies are present in a subgroup of patients with a neuromyelitis optica phenotype. *J. Neuroinflam.* **2015**, *12*, 46. [CrossRef] [PubMed]

59. Varrin-Doyer, M.; Shetty, A.; Spencer, C.M.; Schulze-Topphoff, U.; Weber, M.S.; Bernard, C.C.; Forsthuber, T.; Cree, B.A.; Slavin, A.J.; Zamvil, S.S. MOG transmembrane and cytoplasmic domains contain highly stimulatory T-cell epitopes in MS. *Neurol. Neuroimmunol. Neuroinflam.* **2014**, *1*, e20. [CrossRef] [PubMed]

60. Clements, C.S.; Reid, H.H.; Beddoe, T.; Tynan, F.E.; Perugini, M.A.; Johns, T.G.; Bernard, C.C.; Rossjohn, J. The crystal structure of myelin oligodendrocyte glycoprotein, a key autoantigen in multiple sclerosis. *Proc. Natl. Acad. Sci. USA* **2003**, *100*, 11059–11064. [CrossRef] [PubMed]

61. Bruno, R.; Sabater, L.; Sospedra, M.; Ferrer-Francesch, X.; Escudero, D.; Martínez-Cáceres, E.; Pujol-Borrell, R. Multiple sclerosis candidate autoantigens except myelin oligodendrocyte glycoprotein are transcribed in human thymus. *Eur. J. Immunol.* **2002**, *32*, 2737–2747. [CrossRef]

62. Shetty, A.; Gupta, S.G.; Varrin-Doyer, M.; Weber, M.S.; Prod'homme, T.; Molnarfi, N.; Ji, N.; Nelson, P.A.; Patarroyo, J.C.; Schulze-Topphoff, U.; et al. Immunodominant T-cell epitopes of MOG reside in its transmembrane and cytoplasmic domains in EAE. *Neurol. Neuroimmunol. Neuroinflam.* **2014**, *1*, e22. [CrossRef] [PubMed]

63. Kitley, J.; Waters, P.; Woodhall, M.; Leite, M.I.; Murchison, A.; George, J.; Küker, W.; Chandratre, S.; Vincent, A.; Palace, J. Neuromyelitis optica spectrum disorders with aquaporin-4 and myelin-oligodendrocyte glycoprotein antibodies: A comparative study. *JAMA Neurol.* **2014**, *71*, 276–283. [CrossRef] [PubMed]

64. Ikeda, K.; Kiyota, N.; Kuroda, H.; Sato, D.K.; Nishiyama, S.; Takahashi, T.; Misu, T.; Nakashima, I.; Fujihara, K.; Aoki, M. Severe demyelination but no astrocytopathy in clinically definite neuromyelitis optica with anti-myelin-oligodendrocyte glycoprotein antibody. *Mult. Scler.* **2015**, *21*, 656–659. [CrossRef] [PubMed]

65. Kaneko, K.; Sato, D.K.; Nakashima, I.; Nishiyama, S.; Tanaka, S.; Marignier, R.; Hyun, J.W.; Oliveira, L.M.; Reindl, M.; Seifert-Held, T.; et al. Myelin injury without astrocytopathy in neuroinflammatory disorders with MOG antibodies. *J. Neurol. Neurosurg. Psychiatry* **2016**, *87*, 1257–1259. [CrossRef] [PubMed]

66. Spadaro, M.; Gerdes, L.A.; Krumbholz, M.; Ertl-Wagner, B.; Thaler, F.S.; Schuh, E.; Metz, I.; Blaschek, A.; Dick, A.; Brück, W.; et al. Autoantibodies to MOG in a distinct subgroup of adult multiple sclerosis. *Neurol. Neuroimmunol. Neuroinflam.* **2016**, *3*, e257. [CrossRef] [PubMed]

67. Zamvil, S.S.; Slavin, A.J. Does MOG Ig-positive AQP4-seronegative opticospinal inflammatory disease justify a diagnosis of NMO spectrum disorder? *Neurol. Neuroimmunol. Neuroinflam.* **2015**, *2*, e62. [CrossRef] [PubMed]

68. Louveau, A.; Smirnov, I.; Keyes, T.J.; Eccles, J.D.; Rouhani, S.J.; Peske, J.D.; Derecki, N.C.; Castle, D.; Mandell, J.W.; Lee, K.S.; et al. Structural and functional features of central nervous system lymphatic vessels. *Nature* **2015**, *523*, 337–341. [CrossRef] [PubMed]

69. Aspelund, A.; Antila, S.; Proulx, S.T.; Karlsen, T.V.; Karaman, S.; Detmar, M.; Wiig, H.; Alitalo, K. A dural lymphatic vascular system that drains brain interstitial fluid and macromolecules. *J. Exp. Med.* **2015**, *212*, 991–999. [CrossRef] [PubMed]

70. Fabriek, B.O.; Zwemmer, J.N.; Teunissen, C.E.; Dijkstra, C.D.; Polman, C.H.; Laman, J.D.; Castelijns, J.A. In vivo detection of myelin proteins in cervical lymph nodes of MS patients using ultrasound-guided fine-needle aspiration cytology. *J. Neuroimmunol.* **2005**, *161*, 190–194. [CrossRef] [PubMed]

71. De Vos, A.F.; van Meurs, M.; Brok, H.P.; Boven, L.A.; Hintzen, R.Q.; van der Valk, P.; Ravid, R.; Rensing, S.; Boon, L.; Laman, J.D.; et al. Transfer of Central Nervous System Autoantigens and Presentation in Secondary Lymphoid Organs. *J. Immunol.* **2002**, *169*, 5415–5423. [CrossRef] [PubMed]
72. Van Zwam, M.; Huizinga, R.; Heijmans, N.; van Meurs, M.; Wierenga-Wolf, A.F.; Melief, M.J.; Hintzen, R.Q.; 't Hart, B.A.; Amor, S.; Boven, L.A.; et al. Surgical excision of CNS-draining lymph nodes reduces relapse severity in chronic-relapsing experimental autoimmune encephalomyelitis. *J. Pathol.* **2009**, *217*, 543–551. [CrossRef] [PubMed]

brain sciences

MDPI

Review

Multiple Sclerosis: Immunopathology and Treatment Update

Narges Dargahi [1], Maria Katsara [2], Theodore Tselios [3], Maria-Eleni Androutsou [4], Maximilian de Courten [1], **John Matsoukas [5] and Vasso Apostolopoulos [1,*]**

[1] Centre for Chronic Disease, College of Health and Biomedicine, Victoria University, Melbourne VIC 3030, Australia; narges.dargahi@live.vu.edu.au (N.D.); Maximilian.deCourten@vu.edu.au (M.d.C.)

[2] Medical Department, Novartis (Hellas) SACI, Metamorphosis, Athens 14452, Greece; maria.katsara@novartis.com

[3] Department of Chemistry, University of Patras, Rio, Patras 26500, Greece; ttselios@upatras.gr

[4] Vianex S.A., Metamorphosis, Attikis, Athens 14451, Greece; AndroutsouM@vianex.gr

[5] ELDrug S.A., Patras Science Park, Platani, Patras 26504, Greece; imats1953@gmail.com

* Correspondence: vasso.apostolopoulos@vu.edu.au; Tel.: +61-3-9919-2025

Academic Editor: Evanthia Bernitsas
Received: 25 June 2017; Accepted: 3 July 2017; Published: 7 July 2017

Abstract: The treatment of multiple sclerosis (MS) has changed over the last 20 years. All immunotherapeutic drugs target relapsing remitting MS (RRMS) and it still remains a medical challenge in MS to develop a treatment for progressive forms. The most common injectable disease-modifying therapies in RRMS include β-interferons 1a or 1b and glatiramer acetate. However, one of the major challenges of injectable disease-modifying therapies has been poor treatment adherence with approximately 50% of patients discontinuing the therapy within the first year. Herein, we go back to the basics to understand the immunopathophysiology of MS to gain insights in the development of new improved drug treatments. We present current disease-modifying therapies (interferons, glatiramer acetate, dimethyl fumarate, teriflunomide, fingolimod, mitoxantrone), humanized monoclonal antibodies (natalizumab, ofatumumab, ocrelizumab, alemtuzumab, daclizumab) and emerging immune modulating approaches (stem cells, DNA vaccines, nanoparticles, altered peptide ligands) for the treatment of MS.

Keywords: multiple sclerosis; immunotherapy; drug delivery; vaccine

1. Introduction

In the early 1900s, only a few cases of multiple sclerosis (MS) were reported, which quickly became a common occurrence for admission to neurological wards. Today, MS accounts over 2.5 million affected individuals with an estimated cost of US$2–3 billion per annum [1]. The distribution of MS varies according to geographic location. For example, the further north or south from the equator the higher the prevalence of MS; countries that lie on the equator have extremely low prevalence compared to Scotland, Norway, and Canada. The prevalence of MS has increased progressively over time with 30/100,000 diagnosed in 2008 to 33/100,000 diagnosed in 2013 globally. In fact, in a Norwegian cohort over 53 years (1961–2014), the prevalence increased from 20 to 203/100,000 and the incidence increased from 1.9 to 8/100,000 [2]. It is possible that the increase in prevalence is due to improved diagnostic procedures and reporting and changes in lifestyle (lack of vitamin D and increased smoking) [1]. MS is commonly diagnosed between 20 years and 40 years of age although it can affect younger and older individuals [3], and most commonly affects those with a genetic predisposition (major histocompatibility complex (MHC) class II phenotype, human leukocyte antigen (HLA)-DR2 and HLA-DR4 most commonly affected). In fact, the incidence of MS is increased 10-fold in monozygotic twins as compared to siblings of patients with MS [4–6]. In addition, viral infections can trigger disease where parts of the virus mimics that of the myelin sheath [7]. Although usually not life-shortening, MS is a chronic neurological disease often interfering with life and career plans of an individual [8].

MS is categorized into 4 distinct types, primarily based on its clinical course, which are characterized by increasing severity: (a) Relapsing/remitting MS (RRMS), the most common form, affecting 85% of all MS patients which involves relapses followed by remission; (b) secondary progressive MS (SPMS), which develops over time following diagnosis of RRMS; (c) primary progressive MS (PPMS) affecting 8–10% of patients, noted as gradual continuous neurologic deterioration; and (d) progressive relapsing MS (PRMS) the least common form (<5%), which is similar to PPMS but with overlapping relapses [9–11]. MS leads to a wide range of symptoms with various severity involving different parts of the body. MS diagnosis is mainly clinically based however, magnetic resonance imaging (MRI) assists in diagnosis [12]. As such, examination of the cerebrospinal fluid (CSF) and visual induced potentials with MRI can assist in confirming the clinical suspicion of MS [12,13]. MS symptoms and disease progression are varied, with some individuals experiencing little disability while most (up to 60%) require a wheelchair 20 years from diagnosis [9].

Although treatments against MS are able to decrease the relapse rate in RRMS, the prevention of long-term effects remains a problem; medications for progressive forms of MS are also limited in their efficacy. Hence, new improved drugs are required to effectively treat MS. One of the major pathophysiological mechanisms of MS involves autoreactive T cells, primarily T helper (Th)-1 CD4$^+$ T cells and Th17 cells leading to cytokine secretion and activation of an inflammatory cascade resulting in demyelination within the brain and spinal cord and axonal damage; autoreactive antibodies cannot be discounted. Indeed, MS is generally known as a chronic autoimmune disorder of the central nervous system (CNS) [14,15]. MS causes breakdown of the blood brain barrier (BBB) leading to migration of immune cells (macrophages, T cells, B cells) and secretion of pro-inflammatory cytokines and chemokines [16] which induces inflammation, formation of sclerotic plaques (lesions), demyelination and neurodegeneration [17]. MS lesions may form in any location of the CNS white matter or in grey matter, often leading to physical disability and sometimes, decline in cognitive ability [16,18]. It is therefore, conceivable to target immune cells and their products in order to prevent tissue damage by modulating inflammation [9,19] while reducing potential side effects such as global immunosuppression [6,19,20]. The major constituents of the myelin sheath in which autoreactive T cells and antibodies recognize, include, myelin basic protein (MBP), myelin oligodendrocyte glycoprotein (MOG) and proteolipid protein (PLP).

2. Immunopathophysiology of MS

The brain has primarily been considered to be an organ which is highly immune-advantaged, although a number of studies have challenged this [6]. In the last 10 years an important shift has surfaced in MS research, suggesting that MS is not just a disease of the immune system, but equally involves factors contributed by the CNS [21,22]. Immune cells residing in the CNS get activated following damage to CNS tissue; notably microglial cells whereby they upregulate MHC class I and II molecules and cell surface co-stimulatory molecules and secrete cytokines and chemokines, paving entry for T (CD4 and CD8) cells, B cells, monocytes, macrophages and dendritic (DC)-like cells into CNS lesions [6]. Infiltrating immune cells secrete pro-inflammatory cytokines, nitric oxide, and matrix metalloproteinases [23,24], leading to destruction of the myelin sheath.

It has been generally accepted that chronic inflammation is the hallmark of neurodegenerative diseases, such as MS, Alzheimer's disease and Parkinson's disease [6,7]. Myelin-reactive auto-T cells cross the BBB [19] and their migration into the CNS consequently initiates an inflammatory cascade followed by demyelination of the CNS and axonal damage. These cells reside in the perivenous demyelinating lesions which generate distinct inflammatory demyelinated plaques situated within the white matter [25]. MS lesions appear in the white matter inside the visual neuron, basal ganglia, brain stem and spinal cord [26]. White matter cells transmit neural signals from grey matter, where information is gathered, and transferred to the rest of the body [25,27]. MS involves 2 main steps, (i) myelin sheath damage resulting in formation of lesions in the CNS and (ii) inflammation, which together destroy the neuron tissue [25,28]. In MS, damage of oligodendrocytes and destruction of myelin sheath leads to breakdown of the nerve axon and loss of neuronal function [28]. Demyelination increases the inflammatory activation processes leading to damage of BBB and stimulation of macrophage activation and oxidative stress pathways [29]. The white matter lesions include myelin breakdown together with infiltration of monocytes, B cells, T cells and DC [30]. Microglia and macrophages are the main innate immune cells present in MS lesions where they either act together with T and B cells, or directly cause neuroinflammatory tissue damage [31]. Cells involved in the inflammatory process include those that are both in the innate and adaptive immune systems and are described below (Figure 1).

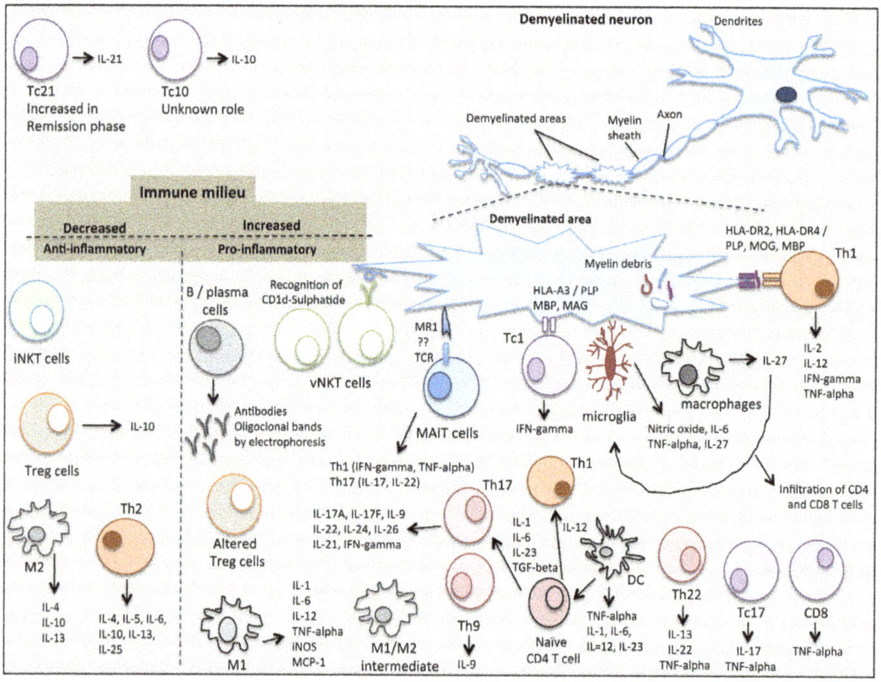

Figure 1. The immunological complexity of the immune/cytokine network in multiple sclerosis.

2.1. Natural Killer T (NKT) Cells

NKT cells share properties of both T cells and NK cells and recognize glycolipid antigens presented in complex with the MHC class I-like molecule, CD1d. Two subsets of NKT cells have been identified (type I, invariant NKT (iNKT) cells and type II, variant NKT (vNKT) cells) and are implicated in the pathogenesis of MS in humans and in the murine model of MS, experimental autoimmune encephalomyelitis (EME). iNKT cells express cell surface markers characteristic of activated or memory T cells (CD25, CD44, CD69) with the majority being CD4$^+$ as well as markers characteristic of NK cells (NK1.1 or CD161, Ly49). Following activation of iNKT cells (via binding to α-GalCer-CD1d complex) an array of cytokines is secreted that are associated with both pro- and anti-inflammatory immune responses and play a role in both innate and acquired immunity. As such, iNKT cells, (i) secrete interleukin (IL)-4 and IL-13 which stimulate CD4$^+$ T cells to differentiate into anti-inflammatory Th2 cells (IL-4, IL-10 producers) which inhibit Th17, Th1, CD8$^+$ T cells in the CNS; (ii) secrete IL-2 and tumor growth factor (TGF)-beta which stimulate the production of T regulatory (Treg) cells (IL-10, TGF-beta producers) which inhibit Th17, Th1 and CD8$^+$ T cells in the CNS; and (iii) secrete IL-4, IL-10, IL-13, interferon (IFN)-gamma and GM-CSF which activate suppressive myeloid derived suppressor cells (MDCs), DC and macrophages which in turn secrete IL-10 to activate Treg cells and suppress Th17, Th1 and CD8$^+$ T cells in the CNS [32]. Due to the pleiotropic properties of iNKT cells, they play a role in protecting the host against pathogens, tumors, autoimmunity and are involved in tissue rejection, ischemia reperfusion injury and obesity related diabetes [32]; deficiency or dysfunction of iNKT cells has been shown to be linked to the development of autoimmune diseases. Indeed, iNKT cell numbers are decreased in patients with MS [32] and are restored in patients in remission [33]. Analysis of iNKT cells in MS patients in remission showed a Th2 cytokine profile, suggesting an immunoregulatory effect of iNKT cells in MS [34]. Similarly, in the EAE mouse model, protection of EAE development is associated with high levels of iNKT cells and suppression of Th1 and Th17 cells [35]. Interestingly, injections of α-Galactosylceramide (α-GalCer), and analogues thereof, have potent activities in protecting mice against, cancer, infections, inflammatory conditions and autoimmune disorders. Hence, it is possible to develop iNKT cell based modulating therapies against MS [36,37]. Like iNKT cells, variant NKT (vNKT) cells also share properties of both T cells (CD4$^+$) and NK cells (NK1.1) and recognize β-linked glycolipid antigens in complex with CD1d. They are less common in mice compared to iNKT cells but are more abundant in humans. Of interest, vNKT cells recognize the self-glycolipid, sulphatide, which is abundantly expressed within the myelin sheath suggesting a role in MS although not yet established [38]. Likewise, vNKT cells recognizing sulphatide self-myelin ligand are present in high levels in mice with EAE suggesting their role in disease progression [38].

2.2. Mucosal-Associated Invariant T (MAIT) Cells

MAIT cells are a subset of T cells of the innate immune system to defend against microbial infections. They are present in the liver, lungs, mucosa and blood and make up to 25% of CD8 T cells in healthy individuals; they also support adaptive immune responses in that they have a memory like phenotype [39]. The MHC class I-like molecule, MR1, presents microbial antigens and vitamin B metabolites to MAIT cells, leading to their activation [39,40]. However, MAIT cells have also been implicated in autoimmune diseases such as MS, inflammatory bowel disease and rheumatoid arthritis where they are often noted at the site of autoimmune attack. Recently, it was reported that in MS, MAIT cells are highly present at the sites of demyelination and secrete pro-inflammatory Th1 cytokines (IFN-gamma and TNF-alpha) and activate Th17 cells (IL-17 and IL-22 cytokines) [22]; the major cytokines in the pathogenesis of chronic inflammatory and autoimmune diseases. In addition, MAIT cell have been noted in white matter inflammatory lesions [41] as well as transcription over expression of MR1 in MS lesions. Conversely, it has been reported that MAIT cells are decreased in blood of patients with RRMS [42]. It is not clear whether MAIT cells exert a protective or a non-protective role, thus a better understanding of how MAIT cells are involved in MS and of their interactions would aid in a better understanding of the pathogenesis of MS and development of therapeutic strategies.

2.3. Regulatory T Cells (Tregs)

Regulatory T cells (Tregs; originally known as suppressor T cells) are a subset of CD4$^+$ T cells that modulate immunity, maintain tolerance against self-antigens and prevent autoimmunity. Tregs are primarily characterized as Foxp3$^+$CD25$^+$CD4$^+$ and are anti-inflammatory (secrete IL-10). One of the first evidence of the role of Treg cells in MS was in mouse EAE models, where adoptive transfer of Treg cells from control mice into MOG or PLP induced EAE mice prevented the onset and progression of EAE [43,44]. Adoptive transfer of Treg cells recovering from EAE into MOG-induced active EAE mice resulted in resolution of EAE [45]. In addition, induction of Treg cells by estradiol or by monocytes under glatiramer acetate treatment reduced clinical signs of MOG-EAE [46,47]. Furthermore, injection anti-CD28 monoclonal antibody in Lewis rats results in Treg cell expansion and reduction in EAE disease severity [48]. Interestingly, injection of anti-CD25 monoclonal antibody, which blocks the effects of Treg cells into C57BL/6 mice increased susceptibility to EAE induction [45]. In patients with MS however, the frequency of Foxp3$^+$CD25$^+$CD4$^+$ Treg cells does not differ to those in healthy individuals, although the function of such cells are impaired (maturation and migration) [49]. In addition, mRNA and protein levels of Foxp3 are impaired in Treg cells of patients with MS especially in RRMS and are normalized during SPMS [49]. Hence, impaired functionality of Treg cells is primarily observed in the early stages of MS but not in their chronic stage, suggesting a causative role [50]. Further studies of Treg cells in MS may aid in the understanding for why tolerance against self-antigens is broken, leading to disease. However, it is not clear whether the impaired function of Treg cells is a direct cause of MS or whether such impairment is a general outcome for all autoimmune disorders.

2.4. Macrophages and Microglia

Macrophages are divided into M1 or M2 based on their pro- or anti-inflammatory cytokine secretion phenotype [51]. M1 macrophage phenotype of mice (F4/80$^+$CD11b$^+$CD11c$^+$iNOS$^+$) and human (CD40$^+$CD86$^+$CD64$^+$CD32$^+$) is induced in the presence of interferon (IFN)-gamma and/or toll-like receptor (TLR) ligands such as lipopolysaccharide (LPS). M1 macrophages are pro-inflammatory and primarily secrete IL-1, IL-6, IL-12, TNF-alpha, iNOS and MCP-1 [51]. In general, they stimulate adaptive immune responses. The M2 macrophage phenotype of mice (F4/80$^+$CD11c$^-$CD301$^+$Arg1$^+$CD206$^+$) and humans (CD163$^+$CD206$^+$) is induced in the presence of IL-4, IL-10, IL-13 and Arg1 that blocks iNOS activity [51]. M2 macrophages are anti-inflammatory and primarily secrete IL-1 receptor antagonist, IL-4, IL-10, transforming growth factor (TGF)-beta1. Macrophages play a crucial role in the pathogenesis of MS. In fact, in active demyelinating and early re-myelinating lesions, macrophages are highly present compared to inactive, demyelinated or late re-myelinated lesions [52]. However, a distinction of M1 vs M2 macrophages in human brain tissues is not so clear, with both M1 macrophages and an intermediate subtype (M1/M2, CD40$^+$CD206$^+$) being present [53]. Like macrophages, microglia cells are divided into M1- and M2-polarized microglia cells. M1 microglia cells are pro-inflammatory and express CD40, CD74, CD86 and CCR7, whereas, M2 microglia cells are anti-inflammatory and express mannose receptor (CD206) and CCL22. In MS brain lesions however, like macrophages, an intermediate microglia phenotype is present expressing CD40, CD74, CD86 and CCL22 but not CD206 markers [54]. Interestingly, in an EAE model it was shown that suppression of CCL22 decreased M1 macrophage accumulation in the CNS, thus therapies designed to suppress CCL22 have the potential to decrease demyelination and progression of disease. In addition, in mice M1 microglia cells have been found to switch to M2 microglia cells during remyelination, hence M2 polarization is necessary for efficient remyelination [55]. Indeed, fasudil (a selective Rho kinase inhibitor), injected into EAE bearing mice shifted M1 to M2 macrophages and ameliorated the clinical severity of EAE [56].

2.5. T Helper Cells

CD4 T cells or T helper (Th) cells, recognize short 9–17 amino acid peptides presented on the surface of antigen presenting cells (APC) in complex with MHC class II. CD4 T cells differentiate into distinct Th cells depending on the cytokine secretion profiles [57]. (i) Th1 cells are pro-inflammatory and produce high levels of IL-2, IL-12, TNF-alpha and IFN-gamma; (ii) Th2 cells are anti-inflammatory and secrete IL-4, IL-5, IL-6, IL-10, IL-13, IL-25; (iii) Th17 cells are pro-inflammatory and secrete high levels of IL-17A, IL-17F, IL-21, IL-22, IL-24, IL-26 and low levels of IL-9 and IFN-gamma; (iv) Th22 cells which are a combination of Th1, Th2, Th17 phenotype and secrete IL-13, IL-22 and TNF-alpha and (v) the newest addition to the Th subset, Th9, was identified for its potent secretion of IL-9. Th1, Th9, Th17 cells are key contributors to MS by increasing inflammation within the milieu of the myelin site.

Th1 cells and their pro-inflammatory cytokine products are present in high levels within the demyelinating axon and CNS lesions of humans and in MOG, PLP or MBP induced EAE in mice.

Th1 cells recognize MOG, PLP and MBP peptide epitopes presented in the context of MHC class II, HLA-DRB1*1501 (HLA-DR2, HLA-DR15) and HLA-DRB1*04 (HLA-DR4) alleles. As a result CD4 T cells become activated, cross the blood brain barrier and induce CNS autoimmunity. Some drug therapeutics target the MHC class II-peptide-T cell receptor (TCR) complex in an attempt to modulate or divert Th1 responses to therapeutic Th2 responses. Indeed, it was recently shown that dimethyl fumarate (DMF) injection in RRMS patients reduced Th1, Th17 and CD8 T cells and increased Th2 cells; this resulted in high levels of IL-4 and decreased levels of IFN-gamma and IL-17 [58]. In addition, we have shown that mannan conjugation of self-MBP, PLP or MOG native peptides or altered peptide ligands, are able to divert Th1 responses to Th2 responses in human PBMC from MS patients, in immunized mouse spleen cells and are able to ameliorate EAE in mice [59–73]. The role of Th9 cells in MS is not as clear although in mice, IL-9 and Th9 cells induce EAE and inflammation and IL-9 knockout mice are protected from developing EAE [74]. Th17 cells play a crucial role in the pathogenesis of MS in both mice and humans by inducing an inflammatory milieu. In fact, IL-17A is present at high levels in CNS lesions, cerebrospinal fluid and in the serum of patients with MS [75]. Th17 cells express high levels of CCR6 which binds to the ligand CCL20 on vascular endothelial cells, enabling their entry through the blood brain barrier where they secrete pro-inflammatory cytokines including IL-17A. In addition, IL-17 interferes with the remyelination process. Of interest, anti-IL-17A humanized neutralizing monoclonal antibody (AIN457 or Secukinumab) injected in patients with MS showed reduction of lesions compared to placebo-treated control subjects [75]. In addition, Th22 cells are highly present in the peripheral blood and cerebral spinal fluid of patients with active RRMS [76], and IL-22 mRNA and Th22 cells are increased in relapsing MS compared to remitting MS patients [77]. Furthermore, Th22 cells specifically recognize MBP and are resistant to IFN-beta therapy [76].

IL-27, a member of the IL-6/IL-12 cytokine family, is secreted by macrophages, dendritic cells and microglia cells, with pleiotropic roles in immunomodulation being either pro- or anti-inflammatory. IL-27 also stimulates or inhibits T cell differentiation. Th1 cells are induced by IL-27 whereas Th2, Th17 and Treg cells are inhibited by IL-27. In addition, Tr1 cells a specialized subset of T cells which secrete IL-10 are induced in the presence of IL-27 [78]. In 40 patients with RRMS, circulating plasma IL-27 levels were significantly higher compared to healthy control subjects [79]. Likewise, IL-27 and IL-27R are elevated in post-mortem MS brain lesions compared to non-MS control brains. Macrophages and microglia were identified to be the source of IL-27 and triggering infiltration of CD4 and CD8 T cells [80]. In addition, the effects of IL-27 on microglia cells showed that nitric oxide, TNF-alpha and IL-6 were secreted, promoting Th1 polarization, suggestive that IL-27 enhances microglia neuroinflammation [81]. Hence, suppressing IL-27 may be a strategy to modulate inflammatory responses in patients with MS.

2.6. CD8 T Cells

Classical CD8 T cells or cytotoxic T cells (Tc1 cells), recognize short antigenic 7-9-mer peptide epitopes presented on the surface of APC in complex with MHC class I. In MS there is a genetic association with HLA-A3 [82]; HLA-A2 has been shown to reduce the risk of MS in individuals that also express MHC class II, HLA-DRB1*1501. The antigen specificity of CD8 Tc1 cells isolated from patients with MS, has been suggested to be against MOG, MBP and PLP with cytolytic activity against neuronal cells in vitro [83] although their pathogenic role in MS is still not clear. More recently other subsets of CD8 T cells have been identified and are grouped into different subsets based on their cytokine profile. In as such, classical Tc1 cells secrete IFN-gamma, Tc2 secrete IL-4, Tc10 secrete IL-10, Tc17 secrete IL-17, Tc21 secrete IL-21, Tc22 secrete IL-22 and another subset is characterized by secreting TNF-alpha. In MS, regardless of the stage and activity of disease CD8 T cells are noted in high numbers, much higher than CD4 T cells at a ratio of 10:1 CD8:CD4 T cells. MHC class I is highly expressed within MS lesions and astrocytes, oligodendrocytes, neurons in addition to the classical APC, DCs and macrophages. In fact, CD8 T cells are found in great abundance within CNS tissues and cerebrospinal fluid of patients with MS. CD8 T cells present in both acute and chronic

MS lesions secrete high levels of IL-17 (classed as, Tc17 CD8 T cells) [84]. Tc17 cells secrete IL-17 and TNF-alpha and low IFN-gamma and are negative for granzyme B, perforin and cytolytic activity unlike the classical CD8 Tc1 cells. In peripheral blood of patients with SPMS and RRMS elevated levels of Tc1 and Tc17 cells are noted as well as a high percentage of TNF-alpha secreting CD8 T cells [85]; Tc21 cells are increased in the remission phase of RRMS compared to SPMS. In addition, higher levels of $CD8^+IFN\text{-}gamma^+TNF\text{-}alpha^+IL\text{-}17^+$ T cells in the relapsing phase of RRMS compared to remission phase, SPMS and controls [85]. It is clear that CD8 T cells contribute to the pathogenesis of MS, and it is important to understand how such cells escape T cell tolerance and induce CNS autoimmunity in order to design and develop new therapeutics against MS.

2.7. B Cells

Although there is a presence of T cells in MS plaques, B cells also contribute to the pathogenesis of MS where they secrete autoantibodies and cytokines and being APC they activate T cells. In patients with MS the presence of oligoclonal bands (OCB) in cerebrospinal fluid and brain parenchyma is a consistent finding in over 95% of patients. OCB is a product of clonally expanded B cells and IgG synthesis. In MS plaques plasma cells are noted in large numbers where antigen uptake, processing and presentation takes place as well as synthesis of IgG. Interestingly, over 50 antibodies isolated from cerebrospinal fluid from patients with MS did not react to MBP, PLP or MOG [86] but some groups reporting that they bind to intracellular proteins such as, MKNK1/2, FAM84A, AKAP12A and glial potassium channel KIR4.1, or, against intracellular lipid determinants [87,88]. Moreover, anti-MOG autoantibodies is a hallmark of childhood MS as well as in some patients with neuromyelitis optical spectrum disorder. It is clear, that abnormal activation of B cells within the CNS of patients with MS, suggests that B cells play a role in the pathophysiology of the disease. Further studies are required to ascertain whether B cell depletion is able to restore immune function and hence, be used as a therapeutic target against MS.

2.8. Dendritic Cells

DC are professional APC which process and present antigenic peptide epitopes on their surface in complex with MHC class I or class II, resulting in CD4 or CD8 T cell stimulation respectively. Even though MS is generally associated with predominant auto-reactive T cells, emerging evidence indicates that DCs play an important role in the pathophysiology of MS, primarily due to their T cell activating and cytokine secreting properties. Following activation of DCs in the periphery, T cells specific to myelin epitopes are activated inducing pro-inflammatory cytokines aiding their entry through the BBB into the CNS. In the CNS resident APC and T cells are further activated leading to demyelination and motor deficits. In patients with MS, DCs are abundantly present within inflamed lesions, cerebrospinal fluid and in the circulation and produce high levels of TNF-alpha, IFN-gamma and IL-6 [89]. In addition, the expression of co-stimulatory molecules, CD40 and CD80 on DCs are increased in RRMS and SPMS patients, suggesting an activated pro-inflammatory state of DCs, hence their contributing role in the pathogenesis of MS.

2.9. Myeloid Derived Suppressor Cells

Myeloid-derived suppressor cells (MDSC) are myeloid progenitors, the same lineage to that of macrophages, DC and neutrophils. However, MDSC have strong immunosuppressive properties rather than immune-stimulatory properties as noted with macrophages, DC and neutrophils [90]. Their major role is in tumor development and chronic inflammation having immune suppressive effects [90]. As such, it was recently shown following MBP_{1-11} peptide immunization in mice, that MDSCs were increased adopting a suppressive phenotype, inhibiting the activation of $CD4^+$ T cells via arginase-1 and inducible nitric oxide synthase; such approach inhibited the development of EAE in mice [91]. In addition, MDSC secrete inhibitory enzyme indoleamine 2,3-dioxygenase and Th2 cytokine, IL-10 [92]. It is not clear whether the number of MDSCs are reduced or whether their functionality is altered in patients with MS, leading to the failure of MDSCs to suppress autoimmune T cells, as a result of disease progression. The use of ex vivo cultured MDSCs could be a viable strategy to develop new improved treatments against MS.

3. Current Drug Therapies for Multiple Sclerosis

The majority of the treatments for MS are long term mainly suppressing the immune system however, such immune-suppressants pose increased risks for infections and cancer [27]. Alternative treatment options involve disease-modifying therapies such as, interferons, glatiramer acetate, monoclonal antibodies and sphingosine-1-phosphate receptor modulators (Table 1, Figure 2). These therapies have dramatically reduced the number of attacks and decreased disease progression. In fact, interferons are effective in the early relapsing phases of MS but not in the advanced phases of the disease [27]. Ultimately, induction of tolerance against self-antigens and re-establishing immune homeostasis can effectively "cure" the disease; such strategies have been the focus of recent research.

Table 1. Disease-modifying drugs available to patients with RRMS.

Drug	Brand	Dose	Number of of Injections, Route	Actions
IFN-β1a	Avonex®	7.5 mg 1st dose 15 mg 2nd dose 22.5 mg 3rd dose 30 mg all subsequent doses	1/week, i.m	Balances pro- and anti-inflammatory cytokines Decreases Th17 cells Decreases IL-17
	Rebif®	22 mg or 44 mg	3/week, s.c	
IFN-β1b	Betaseron®	62.5 mg and increase over 6 weeks to 250 mg	1/2 days, s.c	
	Extavia®	62.5 mg and increase over 6 weeks to 250 mg	1/2 days, s.c	
pegIFN-β1a	Plegridy®	63 mg 1st dose 95 mg 2nd dose 125 mg all subsequent doses	1/2 weeks, s.c	
Glatiramer acetate, EKAY	Copaxone®	20 mg or 40 mg	1/day, s.c 3/week, s.c	Blocks pMHC
Dimethyl fumarate	Tecfidera®	240 mg	2–3/day, oral	Anti-inflammatory Anti-oxidative stress
Teriflunomide	Aubagio®	7 or 14 mg	1/day, oral	Inhibits dihydroorotate dehydrogenase, T, B cells and IFN-γ secreting T cells
Fingolimod	Glenya®	0.5 mg	1/day, oral	Antagonist of SIP receptor Decrease T, B cells activates SIP signaling in CNS
Mitoxantrone	Novatrone®	12 mg/m²	1/3 months up to 2 years	Suppresses T, B cells and macrophages. Reduces Th1 cytokines
Dalfampridine	Ampyra®	10 mg	2/day, oral	Potassium channel blocker Improves motor symptoms, i.e., walking
Humanized Monoclonal Antibody Treatments				
Natalizumab	Tysabr®	300 mg	1/28 days, i.v	Humanized anti-α4-integrin Mab. Affects cell migration, division, growth and survival
Ofatumumab	Arzerra®	3–700 mg	1/2 weeks, i.v	Humanized anti-CD20 Mab. Cytotoxic to CD20+ cells via CDC and ADCC
Ocrelizumab	Ocrevus®	300–600 mg	300 mg weeks 1 and 3, then 600 mg 1/6 months, i.v	Humanized anti-CD20 Mab
Alemtuzumab	Lemtrada®	12 mg	5 days in a row; after 1 year, 3 days	Humanized anti-CD52 Mab. Depletes T, B cells, increases Treg, Th2, decrease Th1 cells
Daclizumab	Zinbryta®	150 mg	1/month, s.c	Humanized anti-CD25 Mab.Blocks IL-2R, decreases T cells, increases NK cells

ADCC, antibody-dependent cellular cytotoxicity; CDC, complement-dependent cytotoxicity; DC, dendritic cells; EKAY, single amino acid code for L-glutamic acid, lysine, alanine, tyrosine; IFN, interferon; IL-2R, interleukin-2 receptor; i.m, intramuscular; i.v, intravenous; Mab, monoclonal antibodies; NK, natural killer cells; pegIFN, polyethylene glycol linked to IFN; pMHC, peptide-major histocompatibility complex; RRMS, relapsing remitting multiple sclerosis; s.c, subcutaneous; SIP, sphingosine-1-phosphate; Th, helper T cells; Treg, regulatory T cells (CD4$^+$CD25$^+$FoxP3$^+$).

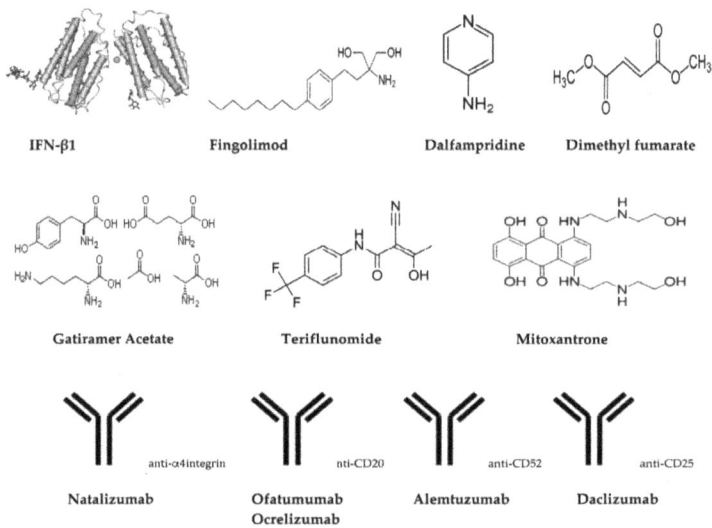

Figure 2. Chemical/schematic structures of treatments/drugs for MS.

3.1. Treatment of MS Relapses

Patients with MS who present with a relapse are generally treated with corticosteroids intravenously, plasma exchange or adrenocorticotropic hormone injections [50,93]. Although effective in reducing the duration of the relapse and patients recovery faster there are no long-term neuroprotective benefits [27,94–97].

3.2. Long-Term Treatment of MS with Disease-Modifying Agents

The treatment of MS has been a challenge with treatment options being limited mainly to corticosteroids, the potent alkylating agent cyclophosphamide and potent immunosuppressant methotrexate (Table 1, Figure 2). However, with the advent of immunomodulatory drugs in mid-1990s, a big shift was carried to treatment options for the first time [50]. The first disease-modifying drug for RRMS, interferon beta-1(IFNβ-1) was the primary key breakthrough for the treatment of MS [98,99]. Disease-modifying agents intend to modify the course of the disease rather than improving symptoms.

Until the approval of the first oral treatment in 2010 [11], all MS treatments consisted of either intramuscular or subcutaneous injectable drugs. To date, 13 FDA approved disease-modifying drugs are available for RRMS, and several more agents are in different developmental stages [9,11,65,66,69]. In the last 20 years there has been an evolving trend in novel treatments for MS and the global progress of therapies for MS has been quite promising. In general treatments consist of Ampyra®, Aubagio®, Avonex®, Betaseron®, Copaxone®, Extavia®, Gilenya®, Lemtrada®, Novantrone®, Plegridy®, Rebif®, Tecfidera® and Tysabri® [100]. Such treatment options consist of alemtuzumab (depletes lymphocytes), daclizumab (blocks the cytokine receptor IL-2), dimethylfumarate (combines features of immunomodulatory and immunosuppressive actions), fingolimod (modulates the sphingosine-receptor system), natalizumab (inhibits the migration of lymphocytes) and teriflunomide (inhibits activated T and B cells) [9,27,50]. Examples of current interferons include, Schering AG's Betaferon/Betaseron (IFNβ-1b), Biogen's Avonex (IFNβ-1a) and Serono/Pfizer's Rebif (IFNβ-1a). In addition, immune modulating agents include, Teva's Copaxone® (copolymer glatiramer acetate), Amgen/Serono's (Novantrone®; mitoxantrone), azathioprine, cyclophosphamide (Endoxan®) and Natalizumab® an a_4-integrin antagonist [101–103]. Disease-modifying agents have commonly been shown to reduce the rate of relapses, reduce MRI lesions and stabilize or delay MS disability.

The key therapeutic features of disease-modifying drugs are their anti-inflammatory effects in the relapsing phase of MS, although demyelination leading to chronic disability still remains a major hurdle [27,104–106]. Some studies, however, have shown that early intervention of disease-modifying drugs to patients with RRMS can reduce acute disability or death [27,107–110].

In general, disease-modifying drugs main action is by suppressing or altering the immune system. Hence, based on this theory that MS is, at least in part, a result of altered or abnormal immune response that results in attack of the myelin sheath. Current available drugs and their actions are described below (Table 1, Figure 2).

3.2.1. Interferons (Avonex®, Biogen, Cambridge, MA, USA; Betaseron®, Bayer, Leverkusen, Germany; Extavia®, Novartis Pharma AG, Basel, Switzerland; Rebif®, EMD Serono Inc., Darmstadt, Germany; Plegridy®, Biogen, Cambridge, MA, USA)

Interferon (IFN) type 1 consist of a group of IFNs (IFN-α, -β, -ε, -κ, -τ, -δ, -ζ, -ω, -ν) which help regulate the immune system. IFN-β is primarily produced by fibroblasts but other cells such as NK cells, B cells, T cells, macrophages also secrete IFN-β. IFN-β has anti-viral and anti-tumor activity as well as being effective in reducing the relapse rate in patients with MS [106]. The mechanism by which IFN-β acts, is that it balances the expression of pro- and anti-inflammatory cytokines in the brain and decreases the number of inflammatory cells crossing the blood brain barrier. As a consequence, there is decreased inflammation of neurons, increases nerve growth factors and improves neuronal survival. Moreover, IFN-β reduces Th17 population and IL-17 cytokine which are known to be involved in the immunopathophysiology of MS [111]. IFN-β injection subcutaneously or intramuscularly to patients with RRMS aims to decrease the relapse rate, duration and severity, however, there is lack of efficacy to long-term disability. Avonex was approved in 1996, the first FDA approved treatment for RRMS. To date there are 3 approaches using IFN-β; IFN-β1a low dosage (Avonex®), IFN-β1a (Rebif®) high dosage, and, IFN-β1b (Betaseron®, Extavia®) high dosage. Furthermore, pegIFN-β-1a (Plegridy®) has polyethylene glycol linked to IFN-β-1a allowing it to be active for longer in the body, hence fewer injections are required compared to Avonex®, Rebif®, Betaseron® and Extavia®. The first large scale human clinical trial in patients with RRMS using IFN-β was published in 1993 and showed that relapse rates were reduced by 34% in high dose IFN-β1b and by 8% in lower dose compared to placebo group and severity of relapses were also reduced [112]. Subsequent 5 year follow-up data showed that IFN-β1a and IFN-β1b decreased lesions up to 30% and reduced the formation of new lesions up to 50%, however, the study failed to show any reduction in disability progression in patients [113]. IFNs have no direct neuroprotective effects, however, through their direct effect on CD4$^+$Th1 cells and altering their profile results in decreased demyelination of neurons, which prevents further neuronal damage [114]. Despite the impact of IFN-β in disease progression in patients with RRMS there are limitations in their use, with side effects ranging from local body aches, skin reactions (swelling, redness), fever, myalgia, flu-like symptoms to more serious side effects such as suicidal thoughts, hallucinations, seizures and heart and liver problems [9]. As a result, many patients have stopped treatment and overall the benefit of using IFNs is relatively small.

3.2.2. Glatiramer Acetate (Copaxone®, Inc., Petah Tikva, Israel)

Glatiramer acetate (GA) is a synthetic 4-mer peptide (L-glutamic acid, lysine, alanine, and tyrosine) mimic of MBP, which competes with short antigenic MBP peptides in complex with MHC class II. Initially, GA was designed to induce EAE but instead it suppressed EAE, which was quickly translated into human trials with MS in order to prevent disease progression, as it bound to MHC class II and inhibited the activation of encephalitogenic T cells [115–118]. GA diverts Th1 cells to Th2 cells that suppress inflammatory responses and activate Tregs in the periphery [119]. In patients, GA significantly reduced disease symptoms and development of new lesions by up to 30% in RRMS, although it showed no improvement in long-term efficacy on progression of disability [120]. GA injection in patients results in side effects ranging from minor (fever, chills) to more serious (cardiovascular, digestive, muscular, respiratory issues).

3.2.3. Dimethyl Fumarate (Tecfidera®, Biogen, Cambridge, MA, USA)

Dimethyl fumarate (BG-12) is a methyl ester of fumaric acid that modulates immune responses and was approved by the FDA in 2013. BG-12 was shown in phase III clinical trials to reduce relapse rate and increase the time to disability progression in patients with RRMS [121]. BG-12 reduces the migration of inflammatory cells through the blood brain barrier and activates nuclear factor erythroid 2-related factor (Nrf2) [122]. Nrf2 regulates anti-oxidative proteins that protect cells against oxidative damage and inflammation. In fact, BG-12 protects neuronal cells from oxidative stress by increasing glutathione levels and suppressing pro-inflammatory cytokines from splenocytes in vitro [123]. Side effects of BG-12 include diarrhea, abdominal pain, nausea, abnormal liver enzymes and decreased lymphocyte counts.

3.2.4. Teriflunomide (Aubagio®, Sanofi Genzyme, Cambridge, MA, USA)

Teriflunomide is an active metabolite of leflunomide (an immunosuppressive disease-modifying drug used for rheumatoid arthritis) which inhibits the enzyme dihydroorotate dehydrogenase [124] and inhibits the proliferation of B and T cells. In addition, teriflunomide exerts anti-inflammatory properties by inhibiting IFN-gamma producing T cells while IL-4 and IL-10 producing T cells are unaffected [125]. In MS, oral administration of teriflunomide reduced relapse rates, MS lesions and decreased disability progression [126–131]. Moreover, permanent discontinuation due to side effects was substantially less common in MS patients who received teriflunomide compared to IFN-β-1a. Side effects include, reduced white blood cell count, alopecia, hepatic effects, nausea, diarrhea, numbness in hand and feet, allergic reactions, breathing issues and increased blood pressure. Teriflunomide was approved by the FDA in 2012 and by EMA in 2013 for use in patients with RRMS.

3.2.5. Fingolimod (Gilenya®, FTY720, Novartis Pharma AG, Basel, Switzerland)

Fingolimod was granted FDA approval in 2010 and was the first oral therapy (0.5 mg once daily) available for patients with relapsing forms of MS. Fingolimod is a sphingosine 1-phosphate (S1P) receptor modulator, which acts as a super agonist of S1P receptor causing receptor internalization and leading to reduced infiltration of potentially auto-reactive lymphocytes into the CNS, and as such, they remain localized in the lymph nodes [132–134]. In addition, a secondary beneficial effects of fingolimod is that it targets S1P receptors on glia cells in the CNS, activating signaling pathways within the CNS [132,135]. Based on Phase III human clinical trials in patients with RRMS (TRANSFORMS, FREEDOMS and FREEDOMS II), fingolimod was more effective compared to first line treatment IFNβ-1a and placebo, in reducing the frequency of flare-ups (clinical exacerbations), disability progression, MRI outcome measures, including brain volume loss and was associated with clearly identified adverse events [103,136,137]. More than 180,000 patients have been treated with fingolimod in clinical trials and post-marketing settings globally, and the total patient exposure now exceeds 395,000 patient-years. Side effects include bradycardia (within 6 h after treatment initiation), blurred vision, diarrhea, back pain, headache, cough and vomiting. With reasonable data showing its long-term safety and disease improvement, fingolimod is a great alternative choice for patients with highly active RRMS and who prefer the oral treatment option.

3.2.6. Mitoxantrone (Novantrone®, Immunex/Amgen, Thousand Oaks, CA, USA)

Mitoxantrone is primarily used to treat certain types of cancers, in particular, non-Hodgkin's lymphoma, acute myeloid leukemia, breast and prostate cancer. Mitoxantrone is a type-II topoisomerase inhibitor, which disrupts DNA synthesis and DNA repair of cancer cells, however, normal cells are also affected. It is a potent immune suppressant, suppressing T cells, B cells and macrophages and reduces pro-inflammatory cytokines (IFN-γ, TNF-α, and IL-2) [138,139]. In patients with SPMS, intravenous injection of 12 mg/m^2 mitoxantrone every 3 months up to 2 years resulted in reduced disability progression by 84% [140,141]. However, several side effects are associated with mitoxantrone which range from nausea, vomiting, hair loss, to, cardiotoxicity, leukemia, infertility, infection, leukopenia and thrombocytopenia [11]. As a result, its use has significantly been reduced over time. Furthermore, due to the risk of cardiotoxicity and leukemia, there is a limit on the cumulative lifetime dose to be administered to patients [11,142].

3.3. Treatment Using Humanized Monoclonal Antibodies

3.3.1. Natalizumab (Tysabr®, Biogen, Cambridge, MA, USA)

Natalizumab is a humanized monoclonal antibody against the cellular adhesion molecule α4- integrin. Integrins are transmembrane receptors that enable cell-extracellular matrix adhesion activating cell signaling which regulate cell growth, division, survival, differentiation and migration. Integrins are expressed on T cells, B cells, monocytes, macrophages, NK cells, DC, neutrophils and eosinophils. Interfering or blocking α4-integrin affects immune cell migration across the blood brain barrier, thus, by blocking the interaction between α4-integrin and vascular endothelial adhesion molecule-1, inhibits transendothelial migration to the CNS [143]. Natalizumab is administered intravenously once a month [144] which reduces activated T cells within the CNS, resulting in anti-inflammatory responses and hence, neuroprotective effects [114]. In a phase III clinical trial natalizumab reduced brain lesions and the rate of disability progression up to 24 months [12,145]. In addition, natalizumab decreased by 92% of contrast-enhancing lesions, by 83% of new or expanding T2-weighted lesions, and by 76% in new T1-weighted hypointense lesions [146,147]. Natalizumab, was approved by the FDA in 2004, but was withdrawn due to 3 cases of rare brain infection, progressive multifocal leukoencephalopathy (PML; that usually leads to death or severe disability), but was re-introduced in 2006 under a special prescription program. However, by 2012 a further 212 cases (or 2.1/1000) of PML were reported to be attributed to natalizumab [148]. Despite these reports the FDA has not withdrawn natalizumab from the market as the clinical benefits outweigh the risks involved. Other side effects include, hepatotoxicity, allergic reactions and increased risks of infection. Due to the risks involved with natalizumab, there are reservations over its use as a preferred treatment option.

3.3.2. Ofatumumab (Arzerra®, Novartis Pharma AG, Basel, Switzerland)

Ofatumumab (OMB157) is the first fully human type 1 IgG1 kappa (IgG1κ) monoclonal antibody and is currently licensed for the treatment (of patients with chronic lymphocytic leukemia (intravenously (iv), Arzerra®). It has also been shown to be beneficial to patients with rheumatoid arthritis, follicular non-Hodgkin's lymphoma, diffuse B cell lymphoma and MS. B cells play a role in the pathogenesis of MS. B cells have essential functions in regulating immune response, by activating CD4$^+$ T-cells and regulating T-cell responses via the secretion of cytokines and antibodies. B cells are present at demyelinating areas and in cerebrospinal fluid of patients with MS [149]. CD20 is a marker and present on the cell surface of all B cells. In an attempt to reduce the number of B cells including autoreactive B cells, the use of anti-CD20 antibodies would conceivably improve MS relapses and progression. In fact, there are several humanized anti-CD20 antibodies, such as rituximab [150], ocrelizumab [151] and ofatumumab [152], which have shown high efficacy in patients with RRMS. In 2015, Novartis acquired the rights from GlaxoSmithKline for the development of ofatumumab in oncology and other autoimmune indications. Ofatumumab binds to 2 unique novel epitopes on the CD20 molecule, induces B-cell depletion via complement dependent cytotoxicity and antibody-dependent cell-mediated cytotoxicity causing B cell apoptosis [153]. Ofatumumab has demonstrated high efficacy in hematologic malignancies and in rheumatoid arthritis. Based on 2 Phase II dosing human clinical studies, ofatumumab demonstrated high efficacy in reducing new MRI lesion activity more than 90% and was well tolerated in patients with MS [152]. Currently, ofatumumab is being further investigated in 2 Phase III trials (ASCLEPIOS I AND ASCLEPIOS II) and are recruiting patients with relapsing forms of MS (ofatumumab versus teriflunomide). The adaptive study design of both trials was recently presented by Hauser SL and colleagues at the American Academy of Neurology April 2017 in Boston, USA and results are highly anticipated [154].

3.3.3. Ocrelizumab (Ocrevus®, Genentech Inc., San Fransisco, CA, USA)

A few months ago (March 2017), the FDA approved ocrelizumab to be used in PPMS, the first drug approved by the FDA for this form of MS and phase IV clinical trials were a requirement of the FDA to be conducted in order to determine the safety of ocrelizumab in younger patients with MS, ie, risk of cancer and effects on pregnancy (study outcomes due by 2024); although clinical trials in patients with lupus and rheumatoid arthritis were halted due to high rates of infections and increased risk of progressive multifocal leukoencephalopathy [155]. In addition, in patients with MS, there was an increased risk of breast cancer (6/781 females with MS on ocrelizumab compared to 0/668 females with MS in other trials) [155].

3.3.4. Alemtuzumab (Lemtrada®, Sanofi Genzyme, Cambridge, MA, USA)

Alemtuzumab is a humanized monoclonal antibody against CD52, a cell surface molecule expressed on B and T cells; mature NK cells, plasma cells, neutrophils and importantly, hematological stem cells do not express CD52. In phase III clinical trials in patients with RRMS, alemtuzumab showed significantly lower annualized relapse rates and MRI measures (gadolinium-enhancing lesions, new or enlarging T2 lesions and brain atrophy) and were free of clinical disease longer, compared to IFNβ-1a [156,157]. Alemtuzumab can cause serious side effects including, immune thrombocytopenia, kidney problems, serious infusion problems (trouble breathing, swelling, chest pain, irregular heart beat), certain cancers (blood cancers, thyroid cancer), cytopenia and serious infections. It was approved by the FDA in 2014 to be used in RRMS patients, but due to the frequent and significant adverse events of alemtuzumab, it is generally used in patients with RRMS who have used 2 or more MS drugs and have failed to work.

3.3.5. Daclizumab (Zinbryta®, Biogen, Cambridge, MA, USA)

Daclizumab is a humanized monoclonal antibody against CD25, the IL-2 receptor expressed on the surface of T cells. The mechanism by which daclizumab works is that it blocks the IL-2 receptor on T cells, preventing the activation of T cells. It was originally approved by the FDA in 1997 to prevent acute kidney transplants (together with corticosteroids and cyclosporine) however its use was halted due to low market demand. In recent years its use has re-emerged to treat patients with RRMS, it is injected subcutaneously, once a month [158]. In human clinical trials, daclizumab showed 45% reduced annualized relapse rates and 54% lower in the number of new lesions [158]. The side effects associated with daclizumab are relatively minor compared to other MS drugs, and include infections, skin rashes and liver complications.

4. New and Emerging Immunotherapeutic Strategies against MS

Antigen/peptide specific immunotherapy or using immune cells (i.e., stems cells), aim to restore tolerance while avoiding the use of non-specific immunosuppressive drugs as describe in Section 3, is a promising approach to fight autoimmune diseases including MS. As such, a number of approaches have been utilized.

4.1. Stem Cells

Multipotent hematopoietic stem cells (HSC) are cells isolated either from the bone marrow, umbilical cord blood or peripheral blood and are transplanted into the recipient. More commonly used for hematological malignancies (leukemia, multiple myeloma) its application has also expanded into autoimmune diseases. The first report of a bone marrow transplant in 1997 in a chronic myelogenous leukemia patient with MS which showed marked improvements in MS brain lesions [159] quickly led to the use of HSC transplantation (HSCT) in MS patients. HSCT in patients with active RRMS, reduce progression in about 70% of patients, decrease relapses dramatically and suppresses inflammatory MRI activity [160]. MS patients who have not responded to conventional therapy, who's disease is aggressive with relapsing-remitting course and who are not presenting with high level of disability, are considered appropriate candidates for such treatment [161]. Although the clinical efficacy of HSCT long term has not been established. The mechanism by which HSCT works is that HSCT "reboots" the immune system and thus, prevents inflammation associated with the disease.

Mesenchymal stem cells (MSC) are isolated from an adult's bone marrow, are differentiated in vitro for 2–3 weeks and re-injected back into the patient. In recent years a vast amount of research has been conducted in MSCs to treat MS with most studies being in mice and EAE models, and more recently in human clinical trials. In fact, in a pilot study in advanced MS patients, MSC transplantation improved expanded disability scale score with stabilization in 1/7 and disease progression in 1/7 patients and vision and low contrast sensitivity test showed improvement in 5/6 patients with 1/6 showing worsening effects [162]. In a phase II randomized double-blind, placebo-controlled crossover clinical trial showed lower mean cumulative number of lesions in patients receiving MSCs compared to placebo [163]. No serious adverse events were reported. The mechanism of action of MSC includes immunomodulation, neuroprotection and neuroregeneration [162]. The use of MSCs that reduce MRI parameters is a new and emerging research focus to develop new improved treatments for MS.

4.2. DNA Vaccine Studies

BHT-3009, a DNA vaccine that encodes the full-length human MBP, was developed with the aim to tolerize patients with MS against MBP [9,164,165]. In fact, in 30 patients with RRMS or SPMS who received 4 injections of BHT-3009 on weeks 1, 3, 5, 9 with escalating doses of 0.5 mg, 1.5 mg or 3 mg was reported to be safe and conferred positive changes on brain MRI and reduced the number of $CD4^+$ T cells [9,164,166]. In addition, in a retrospective, randomized double blind, phase II study in 155 MS patients, BHT-3009 had no impact on the risk for persistent black holes (axonal loss and disability progression). However, there was a correlation to those who had generated high anti-IgM MBP antibodies to reduced risk of persistent black holes [167].

4.3. Nanoparticles

Nanoparticles have extensively been characterized and used as vaccine formulations in pre-clinical models of cancer and infectious diseases [168,169]. Polymeric biodegradable lactic-glycolic acid (PLGA) nanoparticles loaded with MOG_{35-55} peptide together with recombinant IL-10, were partially endocytozed by dendritic cells, secreted both MOG_{35-55} peptide and IL-10 in culture media for several weeks in vitro [170]. In mice, PLGA nanoparticles (MOG_{35-55} + IL-10) showed significant amelioration of EAE and reduction of IL-17 and IFN-gamma secretion by splenic T cells in vitro [170]. Recently, poly(ε-caprolactone) nanoparticles loaded with recombinant human MBP reduced IFN-gamma cytokines, reduced the clinical score and showed only mild histological changes of the myelin sheath [171]. Hence, nanoparticles as a delivery method of self-antigens are a promising tool to treat MS.

4.4. Altered Peptide Ligands

Altered peptide ligands (APL) are peptides closely related to the native (agonist) peptide with defined 1–2 substituted amino acid residues which interact with the T cell receptor (TCR) yet retains its binding ability to the MHC [65]. In phase I/II clinical trial by Neurocrine Biosciences Inc, used an APL of MBP_{83-99}, where L-amino acids were changed to D-amino acids at positions 83, 84, 89, 91 (NBI-5788) [172]. However, this mode of APL induced T cell cross reactivity between the APL and the wild-type/agonist MBP_{83-99} peptide and adverse events in some patients resulted [173]. A subsequent multi-center double-blinded phase II clinical trial with NBI-5788 was suspended—Th2 responses were induced (IL-5, IL-13), however, 13/142 patients developed immediate-type hypersensitivity, who also generated anti-NBI-5788 antibodies which cross-reacted with native agonist MBP$_{83-99}$ peptide [172,174]. National Institute of Neurological Disorders and Stroke sponsored trial, CGP77116, was used in a MRI-controlled phase II clinical trial. CGP77116, has Ala D-amino acids of MBP$_{83-99}$ peptide at positions 83, 84, 89, 91 (CGP77116) of MBP$_{83-99}$ peptide, in order to enhance stability [174]. However, this peptide was poorly tolerated at the dose tested, and the trial had to be discontinued. Three patients showed exacerbations to disease of which two were linked to CGP77116 injection with high IFN-gamma and low IL-4 (Th1-skewing) were secreted by activated CD4$^+$ T cells. These CD4$^+$ T cells also cross reacted with the native agonist MBP$_{83-99}$ peptide [175]. Accordingly, the problems noted with both NBI-5788 and CGP77116 were likely due to inadequate pre-screening of APL effects on the many clonotypes against the targeted epitopes. Thus, although the APL was highly effective at blocking or switching some clones, it activated others. Thus, further pre-clinical testing is required and new modified peptides need to be designed, or a carrier needs to be used which further changes the resulting immune response.

4.4.1. Cyclic Peptides

Cyclization of peptides increases the stability, since linear peptides are sensitive to proteolytic enzymes. In addition, cyclic peptides are an important intermediate step and a useful template towards the rational design and development of non-peptide mimetics. While mimetic strategy is a challenging perspective it is worth pursuing in particular for MBP epitope-based MS therapy as it is still in its infancy. Efforts to design semi-mimetics of MBP$_{72-85}$ epitope by combining non-natural amino acids as spacers and MBP epitope immunophores (Ser, Arg, Glu, Ala, Gln), led to substances that were effective to some extent in inducing the onset of EAE. Cyclic peptides are not only as a step towards non-peptide mimetics but also as putative therapeutics in MS [66].

Structure activity studies of the immunodominant agonist peptide MBP$_{87-99}$, have shown that K^{91}, P^{96} are important T cell receptor contact residues. Double mutation of K^{91}, P^{96} to R^{91},A^{96} or single mutation of P^{96} to A^{96} (APL) of either in their linear or cyclic forms, results in suppression of EAE and decreased inflammation in the spinal cord of Lewis rats [71]. Single and double cyclic[A^{91}]MBP$_{83-99}$ peptide and cyclic[A^{91}A^{96}]MBP$_{83-99}$ peptides emulsified in CFA induced IL-4 cytokines in SJL/J mice [62] however conjugation to reduced mannan further enhanced IL-4 cytokines with no IFN-gamma responses [63]. In guinea pigs and Lewis rats, cyclic[A^{91}A^{96}]MBP$_{83-99}$ showed significantly reduced mechanical pain hypersensitivity compared to cyclic MBP$_{83-99}$ peptide. This was associated with reduced T cell and macrophage infiltration to injured nerves of the spinal cord of animals [176–178]. In addition, these APL decreased CD4$^+$ T cell line proliferation raised from a patient with MS, increased IL-10 cytokine secretion, bound to HLA-DR4 and were more stable to lysosomal enzymes (cathepsin B, D, H) compared to their linear counterparts [70]. Double mutation of K^{91}, P^{96} to A^{91}, A^{96} in either linear or cyclic forms were also shown to be active, with suppression of EAE in SJL/J mice, higher Th2 over Th1 cytokines produced, bound to HLA-DR4, the cyclic forms were more stable to lysosomal enzymes and induced high levels of IL-10 of peripheral blood mononuclear cells from patients with MS [61]. Recently, cyclic native agonist MOG$_{35-55}$ peptide was shown to ameliorate clinical and neuropathological features of EAE in mice compared to its linear counterpart [179]. Thus, cyclic peptides, which offer greater stability and are able to modulate immune responses, are novel leads for the immunotherapy of many diseases, such as MS [66].

4.4.2. Mannan as a Carrier to Modulate Immune Responses

Mannan, a polymannose, isolated from the wall of yeast cells has been shown to bind to the mannose receptor on dendritic cells as well as being a ligand for toll-like receptor 4 [180,181]. Mannan conjugated to MUC1 cancer protein induces immune responses in mice and protects mice against tumor challenge. This work was translated into human phase I, II and pilot III clinical trials; mannan-MUC1 induces protection against cancer recurrence at 18 years follow-up [182–185]. Furthermore, ex vivo cultured dendritic cells pulsed with mannan-MUC1 (CVac^{TM}) and re-injection into patients induces strong cellular and clinical responses in ovarian cancer patients [186,187]. Due to the immunomodulatory properties of mannan, its effects as a carrier to MS peptides were determined.

Mutations of MBP$_{83-99}$ agonist native peptide to result in mutant peptides (APL)—linear [A^{91}]MBP$_{83-99}$, [E^{91}]MBP$_{83-99}$, [F^{91}]MBP$_{83-99}$, [Y^{91}]MBP$_{83-99}$ and [R^{91}, A^{96}]MBP$_{83-99}$, induced IFN-gamma albeit reduced compared to the native agonist peptide, however, only the double APL [R^{91}, A^{96}]MBP$_{83-99}$ induced IL-4 secretion by T cells and antagonized IFN-gamma production in vitro by T cells against the native MBP$_{83-99}$ peptide [67]. In addition, T cells against the native MBP$_{83-99}$ peptide cross-reacted with all peptides except [Y^{91}]MBP$_{83-99}$ and [R^{91},A^{96}]MBP$_{83-99}$ [68]. Conjugation of [R^{91},A^{96}]MBP$_{83-99}$, [A^{91},A^{96}]MBP$_{83-99}$, F^{91}]MBP$_{83-99}$, [Y^{91}]MBP$_{83-99}$ peptides to mannan, completed abrogated IFN-gamma responses and elicited high IL-4 (i.e., Th1 to Th2 switch) [63,69,188]. Likewise, linear double-mutant APL [L^{144}R^{147}]PLP$_{139-151}$ induces high levels of IL-4, and cyclization of this analog elicited low levels of IFN-gamma. When conjugated to mannan, [L^{144}R^{147}]PLP$_{139-151}$ peptide completely abrogated IFN-gamma, while both linear and cyclic native agonist PLP$_{139-151}$ peptides stimulated IFN-gamma secreting T cells [64]. Furthermore, mannan conjugated to the immunodominant agonist MOG$_{35-55}$ peptide primes non-pathogenic Th1 and Th17 cells and ameliorates EAE in mice [73]; a phase I human clinical trial is planned using mannan conjugated to MOG$_{35-55}$ peptide later this year. It is clear that, mannan is able to divert immune responses from Th1 to Th2 and is a promising carrier for further studies for the development of immunotherapeutics against MS.

5. Symptomatic Medication

Dalfampridine (Ampyra/Fampyra®, Acorda Therapeutics)

Dalfampridine is not intended to delay symptoms or change the course of disease, but rather, to improve motor symptoms such as walking. Dalfampridine, is a potassium channel blocker, resulting in improved potassium currents and nerve conductance. Dalfampridine is used in patients who have had MS for more than 3 years and it was approved by the FDA in 2010. Common side effects include nausea, nervousness and dizziness, which are relatively minor compared to other MS drugs.

6. Conclusions and Future Prospects

MS is an autoimmune disorder of the CNS with an array of immune cells being either activated or suppressed leading to demyelination and disease progression. In addition, genetic predisposition, viral mimicry, vitamin and mineral deficiency, geographical location are also etiological factors that contribute to disease. More recently, citrulination of myelin peptides have been shown to contribute to disease activation [59,60]. A number of treatment options are available to patients with MS, in particular those with active disease, however due to side effects, limited long term effectiveness and inability to reverse disease, new improved treatment options are required. As described here a number of new and upcoming promising therapeutic candidates are becoming available, although their effectiveness in human clinical trials remains to be determined. Recently, it was reported that non-peptide mimetics mapping the MBP$_{83-96}$ T cell epitope can function as T cell receptor antagonists, hence such an approach may pave the way to developing alternative and improved immunotherapeutics against MS [189]. With the plethora of information regarding the immunopathophysiology of MS and availability of treatment options and new upcoming treatments, the future holds promise for managing and treating the disease.

Acknowledgments: V.A. would like to thank Vianex S.A. Greece for support (Specific task agreement MS immunotherapeutics). V.A. and J.M. would like to thank Vianex S.A. Greece for their enthusiasm, support and helpful discussions regarding drug development and immunotherapeutics against MS.

Author Contributions: N.D. and V.A. wrote the article and all authors reviewed and edited the article.

Conflicts of Interest: V.A. is supported by Vianex S.A. Greece in developing immunotherapeutics against MS; N.D. is supported under VU-Vianex contract 2 (specific task agreement MS immunotherapeutics) in developing immunotherapeutics against MS; J.M. is head of the scientific advisory board of ELDrug a spin off company of Vianex S.A.; M.-E.A. works for Vianex S.A. Greece; T.T. has an association with Vianex S.A. Greece in relation to supporting his research; M.K. is an employee of Novartis (Hellas) Greece; M.d.C. declares no conflicts of interest. The review represents a detailed literature search in the areas of drugs and treatments against MS with no bias towards immunotherapeutics developed by Vianex S.A.

References

1. Compston, A.; Coles, A. Multiple sclerosis. *Lancet* **2002**, *359*, 1221–1231. [CrossRef]
2. Grytten, N.; Torkildsen, O.; Myhr, K.M. Time trends in the incidence and prevalence of multiple sclerosis in norway during eight decades. *Acta Neurol. Scand.* **2015**, *132*, 29–36. [CrossRef] [PubMed]
3. Antel, J.; Antel, S.; Caramanos, Z.; Arnold, D.L.; Kuhlmann, T. Primary progressive multiple sclerosis: Part of the ms disease spectrum or separate disease entity? *Acta Neuropathol.* **2012**, *123*, 627–638. [CrossRef] [PubMed]
4. Sadovnick, A.D.; Ebers, G.C.; Dyment, D.A.; Risch, N.J. Evidence for genetic basis of multiple sclerosis. *Lancet* **1996**, *347*, 1728–1730. [CrossRef]
5. Dai, H.; Ciric, B.; Zhang, G.X.; Rostami, A. Interleukin-10 plays a crucial role in suppression of experimental autoimmune encephalomyelitis by bowman-birk inhibitor. *J. Neuroimmunol.* **2012**, *245*, 1–7. [CrossRef] [PubMed]
6. Hemmer, B.; Nessler, S.; Zhou, D.; Kieseier, B.; Hartung, H.P. Immunopathogenesis and immunotherapy of multiple sclerosis. *Nat. Clin. Pract. Neurol.* **2006**, *2*, 201–211. [CrossRef] [PubMed]
7. Sospedra, M.; Martin, R. Immunology of multiple sclerosis. *Annu. Rev. Immunol.* **2005**, *23*, 683–747. [CrossRef] [PubMed]
8. Rieckmann, P. Improving ms patient care. *J. Neurol. Suppl.* **2004**, *251*, v69–v73. [CrossRef] [PubMed]
9. Katsara, M.; Matsoukas, J.; Deraos, G.; Apostolopoulos, V. Towards immunotherapeutic drugs and vaccines against multiple sclerosis. *Acta Biochim. Biophys. Sin.* **2008**, *40*, 636–642. [CrossRef] [PubMed]
10. Lublin, F.D.; Reingold, S.C. Defining the clinical course of multiple sclerosis: Results of an international survey. *Neurology* **1996**, *46*, 907–911. [CrossRef] [PubMed]
11. Eckstein, C.; Bhatti, M.T. Currently approved and emerging oral therapies in multiple sclerosis: An update for the ophthalmologist. *Surv. Ophthalmol.* **2016**, *61*, 318–332. [CrossRef] [PubMed]
12. Polman, C.H.; Reingold, S.C.; Banwell, B.; Clanet, M.; Cohen, J.A.; Filippi, M.; Fujihara, K.; Havrdova, E.; Hutchinson, M.; Kappos, L.; et al. Diagnostic criteria for multiple sclerosis: 2010 revisions to the mcdonald criteria. *Ann. Neurol.* **2011**, *69*, 292–302. [CrossRef] [PubMed]
13. Lunde Larsen, L.S.; Larsson, H.B.W.; Frederiksen, J.L. The value of conventional high-field mri in ms in the light of the mcdonald criteria: A literature review. *Acta Neurol. Scand.* **2010**, *122*, 149–158. [CrossRef] [PubMed]
14. Gafson, A.; Giovannoni, G.; Hawkes, C.H. The diagnostic criteria for multiple sclerosis: From charcot to mcdonald. *Mult. Scler. Relat. Disord.* **2012**, *1*, 9–14. [CrossRef] [PubMed]
15. Mahad, D.H.; Trapp, B.D.; Lassmann, H. Pathological mechanisms in progressive multiple sclerosis. *Lancet Neurol.* **2015**, *14*, 183–193. [CrossRef]
16. Minagar, A.; Alexander, J.S. Blood-brain barrier disruption in multiple sclerosis. *Mult. Scler.* **2003**, *9*, 540–549. [CrossRef] [PubMed]
17. Steinman, L. Multiple sclerosis: A coordinated immunological attack against myelin in the central nervous system. *Cell* **1996**, *85*, 299–302. [CrossRef]
18. Bennett, J.; Basivireddy, J.; Kollar, A.; Biron, K.E.; Reickmann, P.; Jefferies, W.A.; McQuaid, S. Blood–brain barrier disruption and enhanced vascular permeability in the multiple sclerosis model eae. *J. Neuroimmunol.* **2010**, *229*, 180–191. [CrossRef] [PubMed]
19. Farjam, M.; Zhang, G.X.; Ciric, B.; Rostami, A. Emerging immunopharmacological targets in multiple sclerosis. *J. Neurol. Sci.* **2015**, *358*, 22–30. [CrossRef] [PubMed]
20. Dandekar, A.A.; Wu, G.F.; Pewe, L.; Perlman, S. Axonal damage is t cell mediated and occurs concomitantly with demyelination in mice infected with a neurotropic coronavirus. *J. Virol.* **2001**, *75*, 6115–6120. [CrossRef] [PubMed]

21. Jiang, J.; Kelly, K.A. Phenotype and function of regulatory t cells in the genital tract. *Curr. Trends Immunol.* **2011**, *12*, 89–94. [PubMed]
22. Bianchini, E.; De Biasi, S.; Simone, A.M.; Ferraro, D.; Sola, P.; Cossarizza, A.; Pinti, M. Invariant natural killer T cells and mucosal-associated invariant T cells in multiple sclerosis. *Immunol. Lett.* **2017**, *183*, 1–7. [CrossRef] [PubMed]
23. Tabarkiewicz, J.; Pogoda, K.; Karczmarczyk, A.; Pozarowski, P.; Giannopoulos, K. The role of il-17 and th17 lymphocytes in autoimmune diseases. *Arch. Immunol. Ther. Exp.* **2015**, *63*, 435–449. [CrossRef] [PubMed]
24. Van Hamburg, J.P.; Asmawidjaja, P.S.; Davelaar, N.; Mus, A.M.C.; Colin, E.M.; Hazes, J.M.W.; Dolhain, R.J.E.M.; Lubberts, E. Th17 cells, but not th1 cells, from patients with early rheumatoid arthritis are potent inducers of matrix metalloproteinases and proinflammatory cytokines upon synovial fibroblast interaction, including autocrine interleukin-17a production. *Arthritis Rheum.* **2011**, *63*, 73–83. [CrossRef] [PubMed]
25. Dolati, S.; Babaloo, Z.; Jadidi-Niaragh, F.; Ayromlou, H.; Sadreddini, S.; Yousefi, M. Multiple sclerosis: Therapeutic applications of advancing drug delivery systems. *Biomed. Pharmacother.* **2017**, *86*, 343–353. [CrossRef] [PubMed]
26. Münzel, E.J.; Williams, A. Promoting remyelination in multiple sclerosis-recent advances. *Drugs* **2013**, *73*, 2017–2029. [CrossRef] [PubMed]
27. Inglese, M.; Petracca, M. Therapeutic strategies in multiple sclerosis: A focus on neuroprotection and repair and relevance to schizophrenia. *Schizophr. Res.* **2015**, *161*, 94–101. [CrossRef] [PubMed]
28. Koriem, K.M.M. Multiple sclerosis: New insights and trends. *Asian Pac. J. Trop. Biomed.* **2016**, *6*, 429–440. [CrossRef]
29. Kallaur, A.P.; Lopes, J.; Oliveira, S.R.; Simão, A.N.; Reiche, E.M.; de Almeida, E.R.D.; Morimoto, H.K.; de Pereira, W.L.; Alfieri, D.F.; Borelli, S.D.; et al. Immune-inflammatory and oxidative and nitrosative stress biomarkers of depression symptoms in subjects with multiple sclerosis: Increased peripheral inflammation but less acute neuroinflammation. *Mol. Neurobiol.* **2016**, *53*, 5191–5202. [CrossRef] [PubMed]
30. Mirshafiey, A.; Jadidi-Niaragh, F. Prostaglandins in pathogenesis and treatment of multiple sclerosis. *Immunopharmacol. Immunotoxicol.* **2010**, *32*, 543–554. [CrossRef] [PubMed]
31. Fischer, M.T.; Sharma, R.; Lim, J.L.; Haider, L.; Frischer, J.M.; Drexhage, J.; Mahad, D.; Bradl, M.; Van Horssen, J.; Lassmann, H. Nadph oxidase expression in active multiple sclerosis lesions in relation to oxidative tissue damage and mitochondrial injury. *Brain* **2012**, *135*, 886–899. [CrossRef] [PubMed]
32. Van Kaer, L.; Wu, L.; Parekh, V.V. Natural killer t cells in multiple sclerosis and its animal model, experimental autoimmune encephalomyelitis. *Immunology* **2015**, *146*, 1–10. [CrossRef] [PubMed]
33. Gigli, G.; Caielli, S.; Cutuli, D.; Falcone, M. Innate immunity modulates autoimmunity: Type 1 interferon-beta treatment in multiple sclerosis promotes growth and function of regulatory invariant natural killer t cells through dendritic cell maturation. *Immunology* **2007**, *122*, 409–417. [CrossRef] [PubMed]
34. Araki, M.; Kondo, T.; Gumperz, J.E.; Brenner, M.B.; Miyake, S.; Yamamura, T. Th2 bias of cd4+ nkt cells derived from multiple sclerosis in remission. *Int. Immunol.* **2003**, *15*, 279–288. [CrossRef] [PubMed]
35. Mars, L.T.; Laloux, V.; Goude, K.; Desbois, S.; Saoudi, A.; Van Kaer, L.; Lassmann, H.; Herbelin, A.; Lehuen, A.; Liblau, R.S. Cutting edge: V alpha 14-j alpha 281 nkt cells naturally regulate experimental autoimmune encephalomyelitis in nonobese diabetic mice. *J. Immunol.* **2002**, *168*, 6007–6011. [CrossRef] [PubMed]
36. Van Kaer, L. Alpha-galactosylceramide therapy for autoimmune diseases: Prospects and obstacles. *Nat. Rev. Immunol.* **2005**, *5*, 31–42. [CrossRef] [PubMed]
37. Van Kaer, L.; Parekh, V.V.; Wu, L. Invariant nk t cells: Potential for immunotherapeutic targeting with glycolipid antigens. *Immunotherapy* **2011**, *3*, 59–75. [CrossRef] [PubMed]
38. Jahng, A.; Maricic, I.; Aguilera, C.; Cardell, S.; Halder, R.C.; Kumar, V. Prevention of autoimmunity by targeting a distinct, noninvariant cd1d-reactive t cell population reactive to sulfatide. *J. Exp. Med.* **2004**, *199*, 947–957. [CrossRef] [PubMed]
39. Napier, R.J.; Adams, E.J.; Gold, M.C.; Lewinsohn, D.M. The role of mucosal associated invariant t cells in antimicrobial immunity. *Front. Immunol.* **2015**, *6*, 344. [CrossRef] [PubMed]
40. Kjer-Nielsen, L.; Patel, O.; Corbett, A.J.; Le Nours, J.; Meehan, B.; Liu, L.; Bhati, M.; Chen, Z.; Kostenko, L.; Reantragoon, R.; et al. Mr1 presents microbial vitamin b metabolites to mait cells. *Nature* **2012**, *491*, 717–723. [CrossRef] [PubMed]

41. Abrahamsson, S.V.; Angelini, D.F.; Dubinsky, A.N.; Morel, E.; Oh, U.; Jones, J.L.; Carassiti, D.; Reynolds, R.; Salvetti, M.; Calabresi, P.A.; et al. Non-myeloablative autologous haematopoietic stem cell transplantation expands regulatory cells and depletes il-17 producing mucosal-associated invariant t cells in multiple sclerosis. *Brain* **2013**, *136*, 2888–2903. [CrossRef] [PubMed]

42. Miyazaki, Y.; Miyake, S.; Chiba, A.; Lantz, O.; Yamamura, T. Mucosal-associated invariant t cells regulate th1 response in multiple sclerosis. *Int. Immunol.* **2011**, *23*, 529–535. [CrossRef] [PubMed]

43. Kohm, A.P.; Carpentier, P.A.; Anger, H.A.; Miller, S.D. Cutting edge: CD4+CD25+ regulatory t cells suppress antigen-specific autoreactive immune responses and central nervous system inflammation during active experimental autoimmune encephalomyelitis. *J. Immunol.* **2002**, *169*, 4712–4716. [CrossRef] [PubMed]

44. Zhang, X.; Koldzic, D.N.; Izikson, L.; Reddy, J.; Nazareno, R.F.; Sakaguchi, S.; Kuchroo, V.K.; Weiner, H.L. Il-10 is involved in the suppression of experimental autoimmune encephalomyelitis by CD25+CD4+ regulatory t cells. *Int. Immunol.* **2004**, *16*, 249–256. [CrossRef] [PubMed]

45. McGeachy, M.J.; Stephens, L.A.; Anderton, S.M. Natural recovery and protection from autoimmune encephalomyelitis: Contribution of CD4+CD25+ regulatory cells within the central nervous system. *J. Immunol.* **2005**, *175*, 3025–3032. [CrossRef] [PubMed]

46. Matejuk, A.; Bakke, A.C.; Hopke, C.; Dwyer, J.; Vandenbark, A.A.; Offner, H. Estrogen treatment induces a novel population of regulatory cells, which suppresses experimental autoimmune encephalomyelitis. *J. Neurosci. Res.* **2004**, *77*, 119–126. [CrossRef] [PubMed]

47. Weber, M.S.; Prod'homme, T.; Youssef, S.; Dunn, S.E.; Rundle, C.D.; Lee, L.; Patarroyo, J.C.; Stuve, O.; Sobel, R.A.; Steinman, L.; et al. Type ii monocytes modulate t cell-mediated central nervous system autoimmune disease. *Nat. Med.* **2007**, *13*, 935–943. [CrossRef] [PubMed]

48. Beyersdorf, N.; Gaupp, S.; Balbach, K.; Schmidt, J.; Toyka, K.V.; Lin, C.H.; Hanke, T.; Hunig, T.; Kerkau, T.; Gold, R. Selective targeting of regulatory t cells with cd28 superagonists allows effective therapy of experimental autoimmune encephalomyelitis. *J. Exp. Med.* **2005**, *202*, 445–455. [CrossRef] [PubMed]

49. Zozulya, A.L.; Wiendl, H. The role of regulatory t cells in multiple sclerosis. *Nat. Clin. Pract. Neurol.* **2008**, *4*, 384–398. [CrossRef] [PubMed]

50. Diebold, M.; Derfuss, T. Immunological treatment of multiple sclerosis. *Semin. Hematol.* **2016**, *53* (Suppl. 1), S54–S57. [CrossRef] [PubMed]

51. Mosser, D.M.; Edwards, J.P. Exploring the full spectrum of macrophage activation. *Nat. Rev. Immunol.* **2008**, *8*, 958–969. [CrossRef] [PubMed]

52. Bruck, W.; Sommermeier, N.; Bergmann, M.; Zettl, U.; Goebel, H.H.; Kretzschmar, H.A.; Lassmann, H. Macrophages in multiple sclerosis. *Immunobiology* **1996**, *195*, 588–600. [CrossRef]

53. Vogel, D.Y.; Vereyken, E.J.; Glim, J.E.; Heijnen, P.D.; Moeton, M.; van der Valk, P.; Amor, S.; Teunissen, C.E.; van Horssen, J.; Dijkstra, C.D. Macrophages in inflammatory multiple sclerosis lesions have an intermediate activation status. *J. Neuroinflamm.* **2013**, *10*, 35. [CrossRef] [PubMed]

54. Peferoen, L.A.; Vogel, D.Y.; Ummenthum, K.; Breur, M.; Heijnen, P.D.; Gerritsen, W.H.; Peferoen-Baert, R.M.; van der Valk, P.; Dijkstra, C.D.; Amor, S. Activation status of human microglia is dependent on lesion formation stage and remyelination in multiple sclerosis. *J. Neuropathol. Exp. Neurol.* **2015**, *74*, 48–63. [CrossRef] [PubMed]

55. Miron, V.E.; Boyd, A.; Zhao, J.W.; Yuen, T.J.; Ruckh, J.M.; Shadrach, J.L.; van Wijngaarden, P.; Wagers, A.J.; Williams, A.; Franklin, R.J.; et al. M2 microglia and macrophages drive oligodendrocyte differentiation during cns remyelination. *Nat. Neurosci.* **2013**, *16*, 1211–1218. [CrossRef] [PubMed]

56. Liu, C.; Li, Y.; Yu, J.; Feng, L.; Hou, S.; Liu, Y.; Guo, M.; Xie, Y.; Meng, J.; Zhang, H.; et al. Targeting the shift from m1 to m2 macrophages in experimental autoimmune encephalomyelitis mice treated with fasudil. *PLoS ONE* **2013**, *8*, e54841. [CrossRef] [PubMed]

57. Apostolopoulos, V.; de Courten, M.P.; Stojanovska, L.; Blatch, G.L.; Tangalakis, K.; de Courten, B. The complex immunological and inflammatory network of adipose tissue in obesity. *Mol. Nutr. Food Res.* **2016**, *60*, 43–57. [CrossRef] [PubMed]

58. Wu, Q.; Wang, Q.; Mao, G.; Dowling, C.A.; Lundy, S.K.; Mao-Draayer, Y. Dimethyl fumarate selectively reduces memory t cells and shifts the balance between th1/th17 and th2 in multiple sclerosis patients. *J. Immunol.* **2017**, *198*, 3069–3080. [CrossRef] [PubMed]

59. Apostolopoulos, V.; Deraos, G.; Matsoukas, M.T.; Day, S.; Stojanovska, L.; Tselios, T.; Androutsou, M.E.; Matsoukas, J. Cyclic citrullinated mbp87–99 peptide stimulates t cell responses: Implications in triggering disease. *Bioorg. Med. Chem.* **2017**, *25*, 528–538. [CrossRef] [PubMed]

60. Deraos, G.; Chatzantoni, K.; Matsoukas, M.T.; Tselios, T.; Deraos, S.; Katsara, M.; Papathanasopoulos, P.; Vynios, D.; Apostolopoulos, V.; Mouzaki, A.; et al. Citrullination of linear and cyclic altered peptide ligands from myelin basic protein (mbp(87–99)) epitope elicits a th1 polarized response by t cells isolated from multiple sclerosis patients: Implications in triggering disease. *J. Med. Chem.* **2008**, *51*, 7834–7842. [CrossRef] [PubMed]

61. Deraos, G.; Rodi, M.; Kalbacher, H.; Chatzantoni, K.; Karagiannis, F.; Synodinos, L.; Plotas, P.; Papalois, A.; Dimisianos, N.; Papathanasopoulos, P.; et al. Properties of myelin altered peptide ligand cyclo(87–99) (ala91,ala96)mbp87–99 render it a promising drug lead for immunotherapy of multiple sclerosis. *Eur. J. Med. Chem.* **2015**, *101*, 13–23. [CrossRef] [PubMed]

62. Katsara, M.; Deraos, G.; Tselios, T.; Matsoukas, J.; Apostolopoulos, V. Design of novel cyclic altered peptide ligands of myelin basic protein mbp83–99 that modulate immune responses in sjl/j mice. *J. Med. Chem.* **2008**, *51*, 3971–3978. [CrossRef] [PubMed]

63. Katsara, M.; Deraos, G.; Tselios, T.; Matsoukas, M.T.; Friligou, I.; Matsoukas, J.; Apostolopoulos, V. Design and synthesis of a cyclic double mutant peptide (cyclo(87–99)[a91,a96]mbp87–99) induces altered responses in mice after conjugation to mannan: Implications in the immunotherapy of multiple sclerosis. *J. Med. Chem.* **2009**, *52*, 214–218. [CrossRef] [PubMed]

64. Katsara, M.; Deraos, S.; Tselios, T.V.; Pietersz, G.; Matsoukas, J.; Apostolopoulos, V. Immune responses of linear and cyclic plp139–151 mutant peptides in sjl/j mice: Peptides in their free state versus mannan conjugation. *Immunotherapy* **2014**, *6*, 709–724. [CrossRef] [PubMed]

65. Katsara, M.; Minigo, G.; Plebanski, M.; Apostolopoulos, V. The good, the bad and the ugly: How altered peptide ligands modulate immunity. *Expert Opin. Biol. Ther.* **2008**, *8*, 1873–1884. [CrossRef] [PubMed]

66. Katsara, M.; Tselios, T.; Deraos, S.; Deraos, G.; Matsoukas, M.T.; Lazoura, E.; Matsoukas, J.; Apostolopoulos, V. Round and round we go: Cyclic peptides in disease. *Curr. Med. Chem.* **2006**, *13*, 2221–2232. [PubMed]

67. Katsara, M.; Yuriev, E.; Ramsland, P.A.; Deraos, G.; Tselios, T.; Matsoukas, J.; Apostolopoulos, V. A double mutation of mbp83–99 peptide induces il-4 responses and antagonizes ifn-γ responses. *J. Neuroimmunol.* **2008**, *200*, 77–89. [CrossRef] [PubMed]

68. Katsara, M.; Yuriev, E.; Ramsland, P.A.; Deraos, G.; Tselios, T.; Matsoukas, J.; Apostolopoulos, V. Mannosylation of mutated mbp83–99 peptides diverts immune responses from th1 to th2. *Mol. Immunol.* **2008**, *45*, 3661–3670. [CrossRef] [PubMed]

69. Katsara, M.; Yuriev, E.; Ramsland, P.A.; Tselios, T.; Deraos, G.; Lourbopoulos, A.; Grigoriadis, N.; Matsoukas, J.; Apostolopoulos, V. Altered peptide ligands of myelin basic protein (mbp87–99) conjugated to reduced mannan modulate immune responses in mice. *Immunology* **2009**, *128*, 521–533. [CrossRef] [PubMed]

70. Matsoukas, J.; Apostolopoulos, V.; Kalbacher, H.; Papini, A.M.; Tselios, T.; Chatzantoni, K.; Biagioli, T.; Lolli, F.; Deraos, S.; Papathanassopoulos, P.; et al. Design and synthesis of a novel potent myelin basic protein epitope 87–99 cyclic analogue: Enhanced stability and biological properties of mimics render them a potentially new class of immunomodulators. *J. Med. Chem.* **2005**, *48*, 1470–1480. [CrossRef] [PubMed]

71. Tselios, T.; Apostolopoulos, V.; Daliani, I.; Deraos, S.; Grdadolnik, S.; Mavromoustakos, T.; Melachrinou, M.; Thymianou, S.; Probert, L.; Mouzaki, A.; et al. Antagonistic effects of human cyclic mbp(87–99) altered peptide ligands in experimental allergic encephalomyelitis and human t-cell proliferation. *J. Med. Chem.* **2002**, *45*, 275–283. [CrossRef] [PubMed]

72. Tselios, T.V.; Lamari, F.N.; Karathanasopoulou, I.; Katsara, M.; Apostolopoulos, V.; Pietersz, G.A.; Matsoukas, J.M.; Karamanos, N.K. Synthesis and study of the electrophoretic behavior of mannan conjugates with cyclic peptide analogue of myelin basic protein using lysine-glycine linker. *Anal. Biochem.* **2005**, *347*, 121–128. [CrossRef] [PubMed]

73. Tseveleki, V.; Tselios, T.; Kanistras, I.; Koutsoni, O.; Karamita, M.; Vamvakas, S.S.; Apostolopoulos, V.; Dotsika, E.; Matsoukas, J.; Lassmann, H.; et al. Mannan-conjugated myelin peptides prime non-pathogenic th1 and th17 cells and ameliorate experimental autoimmune encephalomyelitis. *Exp. Neurol.* **2015**, *267*, 254–267. [CrossRef] [PubMed]

74. Deng, Y.; Wang, Z.; Chang, C.; Lu, L.; Lau, C.S.; Lu, Q. Th9 cells and il-9 in autoimmune disorders: Pathogenesis and therapeutic potentials. *Hum. Immunol.* **2017**, *78*, 120–128. [CrossRef] [PubMed]

75. Volpe, E.; Batistini, L.; Borsellino, G. Advances in t helper 17 cell biology: Pathogenic role and potential therapy in multiple sclerosis. *Mediat. Inflamm.* **2015**, 475158. [CrossRef] [PubMed]

76. Rolla, S.; Bardina, V.; De Mercanti, S.; Quaglino, P.; De Palma, R.; Gned, D.; Brusa, D.; Durelli, L.; Novelli, F.; Clerico, M. Th22 cells are expanded in multiple sclerosis and are resistant to ifn-beta. *J. Leukoc. Biol.* **2014**, *96*, 1155–1164. [CrossRef] [PubMed]

77. Muls, N.; Nasr, Z.; Dang, H.A.; Sindic, C.; van Pesch, V. Il-22, gm-csf and il-17 in peripheral CD4+ t cell subpopulations during multiple sclerosis relapses and remission. Impact of corticosteroid therapy. *PLoS ONE* **2017**, *12*, e0173780. [CrossRef] [PubMed]

78. Iwasaki, Y.; Fujio, K.; Okamura, T.; Yamamoto, K. Interleukin-27 in t cell immunity. *Int. J. Mol. Sci.* **2015**, *16*, 2851–2863. [CrossRef] [PubMed]

79. Naderi, S.; Hejazi, Z.; Shajarian, M.; Alsahebfosoul, F.; Etemadifar, M.; Sedaghat, N. Il-27 plasma level in relapsing remitting multiple sclerosis subjects: The double-faced cytokine. *J. Immunoass. Immunochem.* **2016**, *37*, 659–670. [CrossRef] [PubMed]

80. Senecal, V.; Deblois, G.; Beauseigle, D.; Schneider, R.; Brandenburg, J.; Newcombe, J.; Moore, C.S.; Prat, A.; Antel, J.; Arbour, N. Production of il-27 in multiple sclerosis lesions by astrocytes and myeloid cells: Modulation of local immune responses. *Glia* **2016**, *64*, 553–569. [CrossRef] [PubMed]

81. Kawanokuchi, J.; Takeuchi, H.; Sonobe, Y.; Mizuno, T.; Suzumura, A. Interleukin-27 promotes inflammatory and neuroprotective responses in microglia. *Clin. Exp. Neuroimmunol.* **2013**, *4*, 36–45. [CrossRef]

82. Sawcer, S.; Hellenthal, G. The major histocompatibility complex and multiple sclerosis: A smoking gun? *Brain* **2011**, *134*, 638–640. [CrossRef] [PubMed]

83. Dressel, A.; Chin, J.L.; Sette, A.; Gausling, R.; Hollsberg, P.; Hafler, D.A. Autoantigen recognition by human cd8 t cell clones: Enhanced agonist response induced by altered peptide ligands. *J. Immunol.* **1997**, *159*, 4943–4951. [PubMed]

84. Tzartos, J.S.; Friese, M.A.; Craner, M.J.; Palace, J.; Newcombe, J.; Esiri, M.M.; Fugger, L. Interleukin-17 production in central nervous system-infiltrating t cells and glial cells is associated with active disease in multiple sclerosis. *Am. J. Pathol.* **2008**, *172*, 146–155. [CrossRef] [PubMed]

85. Salehi, Z.; Doosti, R.; Beheshti, M.; Janzamin, E.; Sahraian, M.A.; Izad, M. Differential frequency of CD8+ T cell subsets in multiple sclerosis patients with various clinical patterns. *PLoS ONE* **2016**, *11*, e0159565. [CrossRef] [PubMed]

86. Disanto, G.; Morahan, J.M.; Barnett, M.H.; Giovannoni, G.; Ramagopalan, S.V. The evidence for a role of b cells in multiple sclerosis. *Neurology* **2012**, *78*, 823–832. [CrossRef] [PubMed]

87. Wekerle, H. B cells in multiple sclerosis. *Autoimmunity* **2017**, *50*, 57–60. [CrossRef] [PubMed]

88. Winger, R.C.; Zamvil, S.S. Antibodies in multiple sclerosis oligoclonal bands target debris. *Proc. Natl. Acad. Sci. USA* **2016**, *113*, 7696–7698. [CrossRef] [PubMed]

89. Huang, Y.M.; Xiao, B.G.; Ozenci, V.; Kouwenhoven, M.; Teleshova, N.; Fredrikson, S.; Link, H. Multiple sclerosis is associated with high levels of circulating dendritic cells secreting pro-inflammatory cytokines. *J. Neuroimmunol.* **1999**, *99*, 82–90. [CrossRef]

90. Kong, Y.Y.; Fuchsberger, M.; Xiang, S.D.; Apostolopoulos, V.; Plebanski, M. Myeloid derived suppressor cells and their role in diseases. *Curr. Med. Chem.* **2013**, *20*, 1437–1444. [CrossRef] [PubMed]

91. Wegner, A.; Verhagen, J.; Wraith, D.C. Myeloid-derived suppressor cells mediate tolerance induction in autoimmune disease. *Immunology* **2017**, *151*, 26–42. [CrossRef] [PubMed]

92. Yu, J.; Du, W.; Yan, F.; Wang, Y.; Li, H.; Cao, S.; Yu, W.; Shen, C.; Liu, J.; Ren, X. Myeloid-derived suppressor cells suppress antitumor immune responses through ido expression and correlate with lymph node metastasis in patients with breast cancer. *J. Immunol.* **2013**, *190*, 3783–3797. [CrossRef] [PubMed]

93. Filippini, G.; Brusaferri, F.; Sibley, W.A.; Citterio, A.; Ciucci, G.; Midgard, R.; Candelise, L. Corticosteroids or acth for acute exacerbations in multiple sclerosis. *Cochrane Database Syst. Rev.* **2000**. [CrossRef]

94. Havrdova, E.; Zivadinov, R.; Krasensky, J.; Dwyer, M.G.; Novakova, I.; Dolezal, O.; Ticha, V.; Dusek, L.; Houzvickova, E.; Cox, J.L.; et al. Randomized study of interferon beta-1a, low-dose azathioprine, and low-dose corticosteroids in multiple sclerosis. *Mult. Scler.* **2009**, *15*, 965–976. [CrossRef] [PubMed]

95. Morrow, S.A.; Metz, L.M.; Kremenchutzky, M. High dose oral steroids commonly used to treat relapses in canadian ms clinics. *Can. J. Neurol. Sci.* **2009**, *36*, 213–215. [CrossRef] [PubMed]

96. Myhr, K.M.; Mellgren, S.I. Corticosteroids in the treatment of multiple sclerosis. *Acta Neurol. Scand.* **2009**, *120*, 73–80. [CrossRef] [PubMed]

97. Van Der Voort, L.F.; Visser, A.; Knol, D.L.; Oudejans, C.B.M.; Polman, C.H.; Killestein, J. Lack of interferon-beta bioactivity is associated with the occurrence of relapses in multiple sclerosis. *Eur. J. Neurol.* **2009**, *16*, 1049–1052. [CrossRef] [PubMed]

98. Kappos, L.; Freedman, M.S.; Polman, C.H.; Edan, G.; Hartung, H.-P.; Miller, D.H.; Montalbán, X.; Barkhof, F.; Radü, E.-W.; Bauer, L.; et al. Effect of early versus delayed interferon beta-1b treatment on disability after a first clinical event suggestive of multiple sclerosis: A 3-year follow-up analysis of the benefit study. *Lancet* **2007**, *370*, 389–397. [CrossRef]

99. Kappos, L.; Polman, C.H.; Freedman, M.S.; Edan, G.; Hartung, H.P.; Miller, D.H.; Montalban, X.; Barkhof, F.; Bauer, L.; Jakobs, P.; et al. Treatment with interferon beta-1b delays conversion to clinically definite and mcdonald ms in patients with clinically isolated syndromes. *Neurology* **2006**, *67*, 1242–1249. [CrossRef] [PubMed]

100. Huang, D.R. Challenges in randomized controlled trials and emerging multiple sclerosis therapeutics. *Neurosci. Bull.* **2015**, *31*, 745–754. [CrossRef] [PubMed]

101. Fenu, G.; Lorefice, L.; Frau, F.; Coghe, G.C.; Marrosu, M.G.; Cocco, E. Induction and escalation therapies in multiple sclerosis. *Anti-Inflamm. Anti-Allergy Agents Med. Chem.* **2015**, *14*, 26–34. [CrossRef]

102. Kipp, M.; Wagenknecht, N.; Beyer, C.; Samer, S.; Wuerfel, J.; Nikoubashman, O. Thalamus pathology in multiple sclerosis: From biology to clinical application. *Cell. Mol. Life Sci.* **2015**, *72*, 1127–1147. [CrossRef] [PubMed]

103. Calabresi, P.A.; Radue, E.W.; Goodin, D.; Jeffery, D.; Rammohan, K.W.; Reder, A.T.; Vollmer, T.; Agius, M.A.; Kappos, L.; Stites, T.; et al. Safety and efficacy of fingolimod in patients with relapsing-remitting multiple sclerosis (freedoms ii): A double-blind, randomised, placebo-controlled, phase 3 trial. *Lancet Neurol.* **2014**, *13*, 545–556. [CrossRef]

104. Greenberg, B.M.; Balcer, L.; Calabresi, P.A.; Cree, B.; Cross, A.; Frohman, T.; Gold, R.; Havrdova, E.; Hemmer, B.; Kieseier, B.C.; et al. Interferon beta use and disability prevention in relapsing-remitting multiple sclerosis. *JAMA Neurol.* **2013**, *70*, 248–251. [CrossRef] [PubMed]

105. Noyes, K.; Weinstock-Guttman, B. Impact of diagnosis and early treatment on the course of multiple sclerosis. *Am. J. Manag. Care* **2013**, *19*, s321–s331.

106. Shirani, A.; Zhao, Y.; Karim, M.E.; Evans, C.; Kingwell, E.; Van Der Kop, M.L.; Oger, J.; Gustafson, P.; Petkau, J.; Tremlett, H. Association between use of interferon beta and progression of disability in patients with relapsing-remitting multiple sclerosis. *JAMA* **2012**, *308*, 247–256. [CrossRef] [PubMed]

107. Rommer, P.S.; Stüve, O. Management of secondary progressive multiple sclerosis: Prophylactic treatment - past, present, and future aspects. *Curr. Treat. Options Neurol.* **2013**, *15*, 241–258. [CrossRef] [PubMed]

108. Trojano, M.; Pellegrini, F.; Paolicelli, D.; Fuiani, A.; Zimatore, G.B.; Tortorella, C.; Simone, I.L.; Patti, F.; Ghezzi, A.; Zipoli, V.; et al. Real-life impact of early interferonβ therapy in relapsing multiple sclerosis. *Ann. Neurol.* **2009**, *66*, 513–520. [CrossRef] [PubMed]

109. Goodin, D.S.; Ebers, G.C.; Cutter, G.; Cook, S.D.; O'Donnell, T.; Reder, A.T.; Kremenchutzky, M.; Oger, J.; Rametta, M.; Beckmann, K.; et al. Cause of death in ms: Long-term follow-up of a randomised cohort, 21 years after the start of the pivotal ifnβ-1b study. *BMJ Open* **2012**, *2*. [CrossRef] [PubMed]

110. Goodin, D.S.; Reder, A.T.; Ebers, G.C.; Cutter, G.; Kremenchutzky, M.; Oger, J.; Langdon, D.; Rametta, M.; Beckmann, K.; DeSimone, T.M.; et al. Survival in ms a randomized cohort study 21 years after the start of the pivotal ifnβ-1b trial. *Neurology* **2012**, *78*, 1315–1322. [CrossRef] [PubMed]

111. Mitsdoerffer, M.; Kuchroo, V. New pieces in the puzzle: How does interferon-beta really work in multiple sclerosis? *Ann. Neurol.* **2009**, *65*, 487–488. [CrossRef] [PubMed]

112. The IFNB Multiple Sclerosis Study Group. Interferon beta-1b is effective in relapsing-remitting multiple sclerosis. I. Clinical results of a multicenter, randomized, double-blind, placebo-controlled trial. *Neurology* **1993**, *43*, 655–661.

113. The IFNB Multiple Sclerosis Study Group; The University of British Columbia MS/MRI Analysis Group. Interferon beta-1b in the treatment of multiple sclerosis: Final outcome of the randomized controlled trial. *Neurology* **1995**, *45*, 1277–1285.

114. Yong, V.W.; Giuliani, F.; Xue, M.; Bar-Or, A.; Metz, L.M. Experimental models of neuroprotection relevant to multiple sclerosis. *Neurology* **2007**, *68*, S32–S37. [CrossRef] [PubMed]

115. Wolinsky, J.S.; Narayana, P.A.; O'Connor, P.; Coyle, P.K.; Ford, C.; Johnson, K.; Miller, A.; Pardo, L.; Kadosh, S.; Ladkani, D.; et al. Glatiramer acetate in primary progressive multiple sclerosis: Results of a multinational, multicenter, double-blind, placebo-controlled trial. *Ann. Neurol.* **2007**, *61*, 14–24. [CrossRef] [PubMed]

116. Wolinsky, J.S. Copolymer 1: A most reasonable alternative therapy for early relapsing–remitting multiple sclerosis with mild disability. *Neurology* **1995**, *45*, 1245–1247. [CrossRef] [PubMed]

117. Neuhaus, O.; Farina, C.; Wekerle, H.; Hohlfeld, R. Mechanisms of action of glatiramer acetate in multiple sclerosis. *Neurology* **2001**, *56*, 702–708. [CrossRef] [PubMed]

118. Ragheb, S.; Abramczyk, S.; Lisak, D.; Lisak, R. Long-term therapy with glatiramer acetate in multiple sclerosis: Effect on t-cells. *Mult. Scler.* **2001**, *7*, 43–47. [CrossRef] [PubMed]

119. Haas, J.; Korporal, M.; Balint, B.; Fritzsching, B.; Schwarz, A.; Wildemann, B. Glatiramer acetate improves regulatory t-cell function by expansion of naive CD4(+)CD25(+)Foxp3(+)CD31(+) t-cells in patients with multiple sclerosis. *J. Neuroimmunol.* **2009**, *216*, 113–117. [CrossRef] [PubMed]

120. Johnson, K.P.; Brooks, B.R.; Cohen, J.A.; Ford, C.C.; Goldstein, J.; Lisak, R.P.; Myers, L.W.; Panitch, H.S.; Rose, J.W.; Schiffer, R.B.; et al. Copolymer 1 reduces relapse rate and improves disability in relapsing-remitting multiple sclerosis: Results of a phase iii multicenter, double-blind placebo-controlled trial. *Neurology* **1995**, *45*, 1268–1276. [CrossRef] [PubMed]

121. Gold, R.; Kappos, L.; Arnold, D.L.; Bar-Or, A.; Giovannoni, G.; Selmaj, K.; Tornatore, C.; Sweetser, M.T.; Yang, M.; Sheikh, S.I.; et al. Placebo-controlled phase 3 study of oral bg-12 for relapsing multiple sclerosis. *N. Engl. J. Med.* **2012**, *367*, 1098–1107. [CrossRef] [PubMed]

122. Moharregh-Khiabani, D.; Linker, R.A.; Gold, R.; Stangel, M. Fumaric acid and its esters: An emerging treatment for multiple sclerosis. *Curr. Neuropharmacol.* **2009**, *7*, 60–64. [CrossRef] [PubMed]

123. Albrecht, P.; Bouchachia, I.; Goebels, N.; Henke, N.; Hofstetter, H.H.; Issberner, A.; Kovacs, Z.; Lewerenz, J.; Lisak, D.; Maher, P.; et al. Effects of dimethyl fumarate on neuroprotection and immunomodulation. *J. Neuroinflamm.* **2012**, *9*. [CrossRef] [PubMed]

124. Palmer, A.M. Teriflunomide, an inhibitor of dihydroorotate dehydrogenase for the potential oral treatment of multiple sclerosis. *Curr. Opin. Investig. Drugs* **2010**, *11*, 1313–1323. [PubMed]

125. Korn, T.; Magnus, T.; Toyka, K.; Jung, S. Modulation of effector cell functions in experimental autoimmune encephalomyelitis by leflunomide–mechanisms independent of pyrimidine depletion. *J. Leukoc. Biol.* **2004**, *76*, 950–960. [CrossRef] [PubMed]

126. O'Connor, P.; Wolinsky, J.S.; Confavreux, C.; Comi, G.; Kappos, L.; Olsson, T.P.; Benzerdjeb, H.; Truffinet, P.; Wang, L.; Miller, A.; et al. Randomized trial of oral teriflunomide for relapsing multiple sclerosis. *N. Engl. J. Med.* **2011**, *365*, 1293–1303. [CrossRef] [PubMed]

127. Sanvito, L.; Constantinescu, C.S.; Gran, B. Novel therapeutic approaches to autoimmune demyelinating disorders. *Curr. Pharm. Des.* **2011**, *17*, 3191–3201. [CrossRef] [PubMed]

128. Yeh, E.A. Current therapeutic options in pediatric multiple sclerosis. *Curr. Treat. Options Neurol.* **2011**, *13*, 544–559. [CrossRef] [PubMed]

129. O'Connor, P.W.; Li, D.; Freedman, M.S.; Bar-Or, A.; Rice, G.P.A.; Confavreux, C.; Paty, D.W.; Stewart, J.A.; Scheyer, R. A phase ii study of the safety and efficacy of teriflunomide in multiple sclerosis with relapses. *Neurology* **2006**, *66*, 894–900. [CrossRef] [PubMed]

130. Confavreux, C.; O'Connor, P.; Comi, G.; Freedman, M.S.; Miller, A.E.; Olsson, T.P.; Wolinsky, J.S.; Bagulho, T.; Delhay, J.-L.; Dukovic, D.; et al. Oral teriflunomide for patients with relapsing multiple sclerosis (tower): A randomised, double-blind, placebo-controlled, phase 3 trial. *Lancet Neurol.* **2014**, *13*, 247–256. [CrossRef]

131. Vermersch, P.; Czlonkowska, A.; Grimaldi, L.M.; Confavreux, C.; Comi, G.; Kappos, L.; Olsson, T.P.; Benamor, M.; Bauer, D.; Truffinet, P.; et al. Teriflunomide versus subcutaneous interferon beta-1a in patients with relapsing multiple sclerosis: A randomised, controlled phase 3 trial. *Mult. Scler. J.* **2014**, *20*, 705–716. [CrossRef] [PubMed]

132. Brinkmann, V.; Davis, M.D.; Heise, C.E.; Albert, R.; Cottens, S.; Hof, R.; Bruns, C.; Prieschl, E.; Baumruker, T.; Hiestand, P.; et al. The immune modulator fty720 targets sphingosine 1-phosphate receptors. *J. Biol. Chem.* **2002**, *277*, 21453–21457. [CrossRef] [PubMed]

133. Mandala, S.; Hajdu, R.; Bergstrom, J.; Quackenbush, E.; Xie, J.; Milligan, J.; Thornton, R.; Shei, G.J.; Card, D.; Keohane, C.; et al. Alteration of lymphocyte trafficking by sphingosine-1-phosphate receptor agonists. *Science* **2002**, *296*, 346–349. [CrossRef] [PubMed]

134. Matloubian, M.; Lo, C.G.; Cinamon, G.; Lesneski, M.J.; Xu, Y.; Brinkmann, V.; Allende, M.L.; Proia, R.L.; Cyster, J.G. Lymphocyte egress from thymus and peripheral lymphoid organs is dependent on s1p receptor 1. *Nature* **2004**, *427*, 355–360. [CrossRef] [PubMed]

135. Choi, J.W.; Gardell, S.E.; Herr, D.R.; Rivera, R.; Lee, C.W.; Noguchi, K.; Teo, S.T.; Yung, Y.C.; Lu, M.; Kennedy, G.; et al. Fty720 (fingolimod) efficacy in an animal model of multiple sclerosis requires astrocyte sphingosine 1-phosphate receptor 1 (s1p1) modulation. *Proc. Natl. Acad. Sci. USA* **2011**, *108*, 751–756. [CrossRef] [PubMed]

136. Cohen, J.A.; Barkhof, F.; Comi, G.; Hartung, H.P.; Khatri, B.O.; Montalban, X.; Pelletier, J.; Capra, R.; Gallo, P.; Izquierdo, G.; et al. Oral fingolimod or intramuscular interferon for relapsing multiple sclerosis. *N. Engl. J. Med.* **2010**, *362*, 402–415. [CrossRef] [PubMed]

137. Kappos, L.; Radue, E.W.; O'Connor, P.; Polman, C.; Hohlfeld, R.; Calabresi, P.; Selmaj, K.; Agoropoulou, C.; Leyk, M.; Zhang-Auberson, L.; et al. A placebo-controlled trial of oral fingolimod in relapsing multiple sclerosis. *N. Engl. J. Med.* **2010**, *362*, 387–401. [CrossRef] [PubMed]

138. Huang, B.; Wang, Q.T.; Song, S.S.; Wu, Y.J.; Ma, Y.K.; Zhang, L.L.; Chen, J.Y.; Wu, H.X.; Jiang, L.; Wei, W. Combined use of etanercept and mtx restores $CD4^+/CD8^+$ ratio and tregs in spleen and thymus in collagen-induced arthritis. *Inflamm. Res.* **2012**, *61*, 1229–1239. [CrossRef] [PubMed]

139. Lenk, H.; Muller, U.; Tanneberger, S. Mitoxantrone: Mechanism of action, antitumor activity, pharmacokinetics, efficacy in the treatment of solid tumors and lymphomas, and toxicity. *Anticancer Res.* **1987**, *7*, 1257–1264. [PubMed]

140. Hartung, H.-P.; Gonsette, R.; Konig, N.; Kwiecinski, H.; Guseo, A.; Morrissey, S.P.; Krapf, H.; Zwingers, T. Mitoxantrone in progressive multiple sclerosis: A placebo-controlled, double-blind, randomised, multicentre trial. *Lancet* **2002**, *360*, 2018–2025. [CrossRef]

141. Edan, G.; Miller, D.; Clanet, M.; Confavreux, C.; Lyon-Caen, O.; Lubetzki, C.; Brochet, B.; Berry, I.; Rolland, Y.; Froment, J.C.; et al. Therapeutic effect of mitoxantrone combined with methylprednisolone in multiple sclerosis: A randomised multicentre study of active disease using mri and clinical criteria. *J. Neurol. Neurosurg. Psychiatry* **1997**, *62*, 112–118. [CrossRef] [PubMed]

142. Martinelli, V.; Radaelli, M.; Straffi, L.; Rodegher, M.; Comi, G. Mitoxantrone: Benefits and risks in multiple sclerosis patients. *Neurol. Sci.* **2009**, *30*, S167–S170. [CrossRef] [PubMed]

143. Sheremata, W.A.; Minagar, A.; Alexander, J.S.; Vollmer, T. The role of alpha-4 integrin in the aetiology of multiple sclerosis: Current knowledge and therapeutic implications. *CNS Drugs* **2005**, *19*, 909–922. [CrossRef] [PubMed]

144. Rice, G.P.A.; Hartung, H.P.; Calabresi, P.A. Anti-α4 integrin therapy for multiple sclerosis: Mechanisms and rationale. *Neurology* **2005**, *64*, 1336–1342. [CrossRef] [PubMed]

145. Klotz, L.; Gold, R.; Hemmer, B.; Korn, T.; Zipp, F.; Hohlfeld, R.; Kieseier, B.C.; Wiendl, H. Diagnosis of multiple sclerosis 2010 revision of the mcdonald criteria. *Nervenarzt* **2011**, *82*, 1302–1309. [CrossRef] [PubMed]

146. Jarius, S.; Hohlfeld, R.; Voltz, R. Diagnosis and therapy of multiple sclerosis—Update 2003. *MMW Fortschr. Med.* **2003**, *145*, 88–95. [PubMed]

147. Miller, D.H.; Khan, O.A.; Sheremata, W.A.; Blumhardt, L.D.; Rice, G.P.A.; Libonati, M.A.; Willmer-Hulme, A.J.; Dalton, C.M.; Miszkiel, K.A.; O'Connor, P.W. A controlled trial of natalizumab for relapsing multiple sclerosis. *N. Engl. J. Med.* **2003**, *348*, 15–23. [CrossRef] [PubMed]

148. Natalizumab: New drug. Multiple sclerosis: Risky market approval. *Prescrire Int.* **2008**, *17*, 7–10.

149. Frohman, E.M.; Racke, M.K.; Raine, C.S. Multiple sclerosis–the plaque and its pathogenesis. *N. Engl. J. Med.* **2006**, *354*, 942–955. [CrossRef] [PubMed]

150. Hauser, S.L.; Waubant, E.; Arnold, D.L.; Vollmer, T.; Antel, J.; Fox, R.J.; Bar-Or, A.; Panzara, M.; Sarkar, N.; Agarwal, S.; et al. B-cell depletion with rituximab in relapsing-remitting multiple sclerosis. *N. Engl. J. Med.* **2008**, *358*, 676–688. [CrossRef] [PubMed]

151. Kappos, L.; Li, D.; Calabresi, P.A.; O'Connor, P.; Bar-Or, A.; Barkhof, F.; Yin, M.; Leppert, D.; Glanzman, R.; Tinbergen, J.; et al. Ocrelizumab in relapsing-remitting multiple sclerosis: A phase 2, randomised, placebo-controlled, multicentre trial. *Lancet* **2011**, *378*, 1779–1787. [CrossRef]

152. Sorensen, P.S.; Lisby, S.; Grove, R.; Derosier, F.; Shackelford, S.; Havrdova, E.; Drulovic, J.; Filippi, M. Safety and efficacy of ofatumumab in relapsing-remitting multiple sclerosis: A phase 2 study. *Neurology* **2014**, *82*, 573–581. [CrossRef] [PubMed]

153. Bleeker, W.K.; Munk, M.E.; Mackus, W.J.; van den Brakel, J.H.; Pluyter, M.; Glennie, M.J.; van de Winkel, J.G.; Parren, P.W. Estimation of dose requirements for sustained in vivo activity of a therapeutic human anti-cd20 antibody. *Br. J. Haematol.* **2008**, *140*, 303–312. [CrossRef] [PubMed]

154. ClinicalTrials.gov. Identifier: NCT02792218 and NCT02792231. Available online: https://www.mdanderson. org/patients-family/diagnosis-treatment/clinical-trials.html (accessed on 20 June 2017).

155. U.S. Food & Drug Administration (FDA). Drugs@fda: FDA Approved Drug Products. Available online: http://www.accessdata.fda.gov/scripts/cder/daf/index.cfm?event=overview.process&varApplNo=761053 (accessed on 20 June 2017).

156. Cohen, J.A.; Coles, A.J.; Arnold, D.L.; Confavreux, C.; Fox, E.J.; Hartung, H.P.; Havrdova, E.; Selmaj, K.W.; Weiner, H.L.; Fisher, E.; et al. Alemtuzumab versus interferon beta 1a as first-line treatment for patients with relapsing-remitting multiple sclerosis: A randomised controlled phase 3 trial. *Lancet* **2012**, *380*, 1819–1828. [CrossRef]

157. Coles, A.J.; Twyman, C.L.; Arnold, D.L.; Cohen, J.A.; Confavreux, C.; Fox, E.J.; Hartung, H.P.; Havrdova, E.; Selmaj, K.W.; Weiner, H.L.; et al. Alemtuzumab for patients with relapsing multiple sclerosis after disease-modifying therapy: A randomised controlled phase 3 trial. *Lancet* **2012**, *380*, 1829–1839. [CrossRef]

158. Lycke, J. Monoclonal antibody therapies for the treatment of relapsing-remitting multiple sclerosis: Differentiating mechanisms and clinical outcomes. *Ther. Adv. Neurol. Disord.* **2015**, *8*, 274–293. [CrossRef] [PubMed]

159. McAllister, L.D.; Beatty, P.G.; Rose, J. Allogeneic bone marrow transplant for chronic myelogenous leukemia in a patient with multiple sclerosis. *Bone Marrow Transplant.* **1997**, *19*, 395–397. [CrossRef] [PubMed]

160. Mancardi, G.; Saccardi, R. Autologous haematopoietic stem-cell transplantation in multiple sclerosis. *Lancet Neurol.* **2008**, *7*, 626–636. [CrossRef]

161. Sormani, M.P.; Muraro, P.A.; Schiavetti, I.; Signori, A.; Laroni, A.; Saccardi, R.; Mancardi, G.L. Autologous hematopoietic stem cell transplantation in multiple sclerosis: A meta-analysis. *Neurology* **2017**, *88*, 2115–2122. [CrossRef] [PubMed]

162. Yamout, B.; Hourani, R.; Salti, H.; Barada, W.; El-Hajj, T.; Al-Kutoubi, A.; Herlopian, A.; Baz, E.K.; Mahfouz, R.; Khalil-Hamdan, R.; et al. Bone marrow mesenchymal stem cell transplantation in patients with multiple sclerosis: A pilot study. *J. Neuroimmunol.* **2010**, *227*, 185–189. [CrossRef] [PubMed]

163. Llufriu, S.; Sepulveda, M.; Blanco, Y.; Marin, P.; Moreno, B.; Berenguer, J.; Gabilondo, I.; Martinez-Heras, E.; Sola-Valls, N.; Arnaiz, J.A.; et al. Randomized placebo-controlled phase ii trial of autologous mesenchymal stem cells in multiple sclerosis. *PLoS ONE* **2014**, *9*, e113936. [CrossRef] [PubMed]

164. Correale, J.; Fiol, M. Bht-3009, a myelin basic protein-encoding plasmid for the treatment of multiple sclerosis. *Curr. Opin. Mol. Ther.* **2009**, *11*, 463–470. [PubMed]

165. Kang, Y.; Sun, Y.; Zhang, J.; Gao, W.; Kang, J.; Wang, Y.; Wang, B.; Xia, G. Treg cell resistance to apoptosis in DNA vaccination for experimental autoimmune encephalomyelitis treatment. *PLoS ONE* **2012**, *7*, e49994. [CrossRef] [PubMed]

166. Bar-Or, A.; Vollmer, T.; Antel, J.; Arnold, D.L.; Bodner, C.A.; Campagnolo, D.; Gianettoni, J.; Jalili, F.; Kachuck, N.; Lapierre, Y.; et al. Induction of antigen-specific tolerance in multiple sclerosis after immunization with DNA encoding myelin basic protein in a randomized, placebo-controlled phase 1/2 trial. *Arch. Neurol.* **2007**, *64*, 1407–1415. [CrossRef] [PubMed]

167. Papadopoulou, A.; Von Felten, S.; Traud, S.; Rahman, A.; Quan, J.; King, R.; Garren, H.; Steinman, L.; Cutter, G.; Kappos, L.; et al. Evolution of ms lesions to black holes under DNA vaccine treatment. *J. Neurol.* **2012**, *259*, 1375–1382. [CrossRef] [PubMed]

168. Xiang, S.D.; Scholzen, A.; Minigo, G.; David, C.; Apostolopoulos, V.; Mottram, P.L.; Plebanski, M. Pathogen recognition and development of particulate vaccines: Does size matter? *Methods* **2006**, *40*, 1–9. [CrossRef] [PubMed]

169. Xiang, S.D.; Selomulya, C.; Ho, J.; Apostolopoulos, V.; Plebanski, M. Delivery of DNA vaccines: An overview on the use of biodegradable polymeric and magnetic nanoparticles. *Wiley Interdiscip. Rev. Nanomed. Nanobiotechnol.* **2010**, *2*, 205–218. [CrossRef] [PubMed]

170. Cappellano, G.; Woldetsadik, A.D.; Orilieri, E.; Shivakumar, Y.; Rizzi, M.; Carniato, F.; Gigliotti, C.L.; Boggio, E.; Clemente, N.; Comi, C.; et al. Subcutaneous inverse vaccination with plga particles loaded with a mog peptide and il-10 decreases the severity of experimental autoimmune encephalomyelitis. *Vaccine* **2014**, *32*, 5681–5689. [CrossRef] [PubMed]

171. Al-Ghobashy, M.A.; ElMeshad, A.N.; Abdelsalam, R.M.; Nooh, M.M.; Al-Shorbagy, M.; Laible, G. Development and pre-clinical evaluation of recombinant human myelin basic protein nano therapeutic vaccine in experimental autoimmune encephalomyelitis mice animal model. *Sci. Rep.* **2017**, *7*, 46468. [CrossRef] [PubMed]

172. Crowe, P.D.; Qin, Y.; Conlon, P.J.; Antel, J.P. Nbi-5788, an altered mbp83–99 peptide, induces a t-helper 2-like immune response in multiple sclerosis patients. *Ann. Neurol.* **2000**, *48*, 758–765. [CrossRef]

173. Hartung, H.P.; Kieseier, B.C.; Hemmer, B. Purely systemically active anti-inflammatory treatments are adequate to control multiple sclerosis. *J. Neurol.* **2005**, *252*, v30–v37. [CrossRef] [PubMed]

174. Kappos, L.; Comi, G.; Panitch, H.; Oger, J.; Antel, J.; Conlon, P.; Steinman, L.; Comi, G.; Kappos, L.; Oger, J.; et al. Induction of a non-encephalitogenic type 2 t helper-cell autoimmune response in multiple sclerosis after administration of an altered peptide ligand in a placebo-controlled, randomized phase ii trial. *Nat. Med.* **2000**, *6*, 1176–1182. [PubMed]

175. Bielekova, B.; Goodwin, B.; Richert, N.; Cortese, I.; Kondo, T.; Afshar, G.; Gran, B.; Eaton, J.; Antel, J.; Frank, J.A.; et al. Encephalitogenic potential of the myelin basic protein peptide (amino acids 83–99) in multiple sclerosis: Results of a phase ii clinical trial with an altered peptide ligand. *Nat. Med.* **2000**, *6*, 1167–1175. [PubMed]

176. Perera, C.J.; Duffy, S.S.; Lees, J.G.; Kim, C.F.; Cameron, B.; Apostolopoulos, V.; Moalem-Taylor, G. Active immunization with myelin-derived altered peptide ligand reduces mechanical pain hypersensitivity following peripheral nerve injury. *J. Neuroinflamm.* **2015**, *12*, 28. [CrossRef] [PubMed]

177. Perera, C.J.; Lees, J.G.; Duffy, S.S.; Makker, P.G.; Fivelman, B.; Apostolopoulos, V.; Moalem-Taylor, G. Effects of active immunisation with myelin basic protein and myelin-derived altered peptide ligand on pain hypersensitivity and neuroinflammation. *J. Neuroimmunol.* **2015**, *286*, 59–70. [CrossRef] [PubMed]

178. Tian, D.H.; Perera, C.J.; Apostolopoulos, V.; Moalem-Taylor, G. Effects of vaccination with altered peptide ligand on chronic pain in experimental autoimmune encephalomyelitis, an animal model of multiple sclerosis. *Front. Neurol.* **2013**, *4*, 168. [CrossRef] [PubMed]

179. Lourbopoulos, A.; Deraos, G.; Matsoukas, M.; Touloumi, O.; Giannakopoulou, A.; Kalbacher, H.; Grigoriadis, N.; Apostolopoulos, V.; Matsoukas, J. Cyclic mog 35–55 ameliorates clinical and neuropathological features of experimental autoimmune encephalomyelitis. *Bioorg. Med. Chem.* **2017**. [CrossRef] [PubMed]

180. Apostolopoulos, V.; Pietersz, G.A.; Loveland, B.E.; Sandrin, M.S.; McKenzie, I.F. Oxidative/reductive conjugation of mannan to antigen selects for t1 or t2 immune responses. *Proc. Natl. Acad. Sci. USA* **1995**, *92*, 10128–10132. [CrossRef] [PubMed]

181. Apostolopoulos, V.; Pietersz, G.A.; McKenzie, I.F. Cell-mediated immune responses to muc1 fusion protein coupled to mannan. *Vaccine* **1996**, *14*, 930–938. [CrossRef]

182. Apostolopoulos, V.; Pietersz, G.A.; Tsibanis, A.; Tsikkinis, A.; Drakaki, H.; Loveland, B.E.; Piddlesden, S.J.; Plebanski, M.; Pouniotis, D.S.; Alexis, M.N.; et al. Pilot phase iii immunotherapy study in early-stage breast cancer patients using oxidized mannan-muc1 [isrctn71711835]. *Breast Cancer Res.* **2006**, *8*, R27. [CrossRef] [PubMed]

183. Apostolopoulos, V.; Pietersz, G.A.; Tsibanis, A.; Tsikkinis, A.; Stojanovska, L.; McKenzie, I.F.; Vassilaros, S. Dendritic cell immunotherapy: Clinical outcomes. *Clin. Transl. Immunol.* **2014**, *3*, e21. [CrossRef] [PubMed]

184. Karanikas, V.; Hwang, L.A.; Pearson, J.; Ong, C.S.; Apostolopoulos, V.; Vaughan, H.; Xing, P.X.; Jamieson, G.; Pietersz, G.; Tait, B.; et al. Antibody and t cell responses of patients with adenocarcinoma immunized with mannan-muc1 fusion protein. *J. Clin. Investig.* **1997**, *100*, 2783–2792. [CrossRef] [PubMed]

185. Vassilaros, S.; Tsibanis, A.; Tsikkinis, A.; Pietersz, G.A.; McKenzie, I.F.; Apostolopoulos, V. Up to 15-year clinical follow-up of a pilot phase iii immunotherapy study in stage ii breast cancer patients using oxidized mannan-muc1. *Immunotherapy* **2013**, *5*, 1177–1182. [CrossRef] [PubMed]

186. Loveland, B.E.; Zhao, A.; White, S.; Gan, H.; Hamilton, K.; Xing, P.X.; Pietersz, G.A.; Apostolopoulos, V.; Vaughan, H.; Karanikas, V.; et al. Mannan-muc1-pulsed dendritic cell immunotherapy: A phase i trial in patients with adenocarcinoma. *Clin. Cancer Res.* **2006**, *12*, 869–877. [CrossRef] [PubMed]

187. Mitchell, P.L.; Quinn, M.A.; Grant, P.T.; Allen, D.G.; Jobling, T.W.; White, S.C.; Zhao, A.; Karanikas, V.; Vaughan, H.; Pietersz, G.; et al. A phase 2, single-arm study of an autologous dendritic cell treatment against mucin 1 in patients with advanced epithelial ovarian cancer. *J. Immunother. Cancer* **2014**, *2*, 16. [CrossRef] [PubMed]

188. Day, S.; Tselios, T.; Androutsou, M.E.; Tapeinou, A.; Frilligou, I.; Stojanovska, L.; Matsoukas, J.; Apostolopoulos, V. Mannosylated linear and cyclic single amino acid mutant peptides using a small 10 amino acid linker constitute promising candidates against multiple sclerosis. *Front. Immunol.* **2015**, *6*, 136. [CrossRef] [PubMed]

189. Yannakakis, M.P.; Simal, C.; Tzoupis, H.; Rodi, M.; Dargahi, N.; Prakash, M.; Mouzaki, A.; Platts, J.A.; Apostolopoulos, V.; Tselios, T.V. Design and synthesis of non-peptide mimetics mapping the immunodominant myelin basic protein (MBP$_{83-96}$) epitope to function as t-cell receptor antagonists. *Int. J. Mol. Sci.* **2017**, *18*. [CrossRef] [PubMed]

brain
sciences

MDPI

Article

Structural and Neuronal Integrity Measures of Fatigue Severity in Multiple Sclerosis

Evanthia Bernitsas [1,*], Kalyan Yarraguntla [2], Fen Bao [2], Rishi Sood [1], Carla Santiago-Martinez [1], Rajkumar Govindan [3], Omar Khan [1,2] and Navid Seraji-Bozorgzad [2]

[1]   Multiple Sclerosis Center, Department of Neurology, Detroit Medical Center, 4201 St Antoine, 8C-UHC Detroit 48201, MI, USA; rsood@med.wayne.edu (R.S.); du2067@wayne.edu (C.S.-M.); okhan@med.wayne.edu (O.K.)

[2]   The Sastry Foundation Advanced Imaging Laboratory, Wayne State School of Medicine, 4201 St Antoine, Detroit 48201, MI, USA; kalyancy@wayne.edu (K.Y.); fbao@med.wayne.edu (F.B.); nseraji@wayne.edu (N.S.-B.)

[3]   Department of Pediatric Neurology/PET Center, Children's Hospital of Michigan, 4201 St Antoine, Detroit 48201, MI, USA; mgrajkumar45@gmail.com

*   Correspondence: ebernits@med.wayne.edu; Tel.: +1-313-745-4994

Received: 24 June 2017; Accepted: 7 August 2017; Published: 12 August 2017

Abstract: Fatigue is a common and disabling symptom in Multiple Sclerosis (MS). However, consistent neuroimaging correlates of its severity are not fully elucidated. In this article, we study the neuronal correlates of fatigue severity in MS. Forty-three Relapsing Remitting MS (RRMS) patients with MS-related fatigue (Fatigue Severity Scale (FSS) range: 1–7) and Expanded Disability Status Scale (EDSS) $\leq$ 4, were divided into high fatigue (HF, FSS $\geq$ 5.1) and low fatigue groups (LF, FSS $\leq$ 3). We measured T2 lesion load using a semi-automated technique. Cortical thickness, volume of sub-cortical nuclei, and brainstem structures were measured using Freesurfer. Cortical Diffusion Tensor Imaging (DTI) parameters were extracted using a cross modality technique. A correlation analysis was performed between FSS, volumetric, and DTI indices across all patients. HF patients showed significantly lower volume of thalamus, ($p = 0.02$), pallidum ($p = 0.01$), and superior cerebellar peduncle ((SCP), $p = 0.002$). The inverse correlation between the FSS score and the above volumes was significant in the total study population. In the right temporal cortex (RTC), the Radial Diffusivity ((RD), $p = 0.01$) and Fractional Anisotropy ((FA), $p = 0.01$) was significantly higher and lower, respectively, in the HF group. After Bonferroni correction, thalamic volume, FA-RTC, and RD-RTC remained statistically significant. Multivariate regression analysis identified FA-RTC as the best predictor of fatigue severity. Our data suggest an association between fatigue severity and volumetric changes of thalamus, pallidum, and SCP. Early neuronal injury in the RTC is implicated in the pathogenesis of MS-related fatigue.

Keywords: fatigue; multiple sclerosis; diffuse tensor imaging; fatigue severity scale; deep gray matter nuclei volume; cortical thickness

1. Introduction

Multiple Sclerosis (MS) is a chronic immune-mediated central nervous system disease and a leading cause of non-traumatic disability in the young adult population [1]. Up to 80% of patients with MS report fatigue that severely impacts their daily activities, quality of life, and employment status, frequently leading to part-time employment or early retirement [2–4]. Furthermore, the impairment resulting from MS-related fatigue is recognized by the United States Social Security Administration as a criterion for disability [2,3].

MS-related fatigue is characterized by the constant feeling of exhaustion and limited endurance of sustained physical and mental activities [4–6]. Since fatigue is a subjective symptom with a physical and mental component, the objective assessment of its severity is often challenging. The Fatigue Severity Scale (FSS) and Modified Fatigue Impact Scale (MFIS) are the most commonly used measures of fatigue severity. FSS is reported to have higher test-retest consistency compared to MFIS [7–9]. It seems that MS-related fatigue is multidimensional and cannot merely be explained by the degree of clinical disease activity, neurological disability, or the extent of Magnetic Resonance Imaging (MRI) abnormalities [10]. Although the pathology of MS-related fatigue is not clear, previous studies have reported atrophy of gray (GM) and white matter (WM), disruptions in cortico-subcortical connections involving the fronto-parietal cortex, and reduction in thalamus and basal ganglia nuclei volume in MS patients compared to healthy controls [11–13]. In contrast, other studies reported no correlation between fatigue severity and white matter disease, and the role of cortical atrophy in MS-related fatigue remains controversial [14,15]. In this retrospective cross-sectional study, we used volumetric and diffusion metrics to investigate the anatomical and neuronal integrity of specific brain structures in relation to fatigue severity in MS patients with low disability.

2. Methods

2.1. Participant Recruitment and Selection Criteria

Forty-three relapsing remitting multiple sclerosis patients (RRMS) patients were enrolled in this retrospective cross-sectional study from the MS Center, Division of Neurology, Detroit Medical Center, Michigan. We included patients from 18 to 65 years of age, diagnosed with MS per the revised McDonald criteria. Patients had a relapsing remitting course and were relapse and steroid treatment-free for at least one month prior to MRI scan. We excluded patients who were pregnant or had other neurological or psychiatric disorders, such as depression or anxiety, because of their established association with fatigue [16]. Patient on antidepressants, psychoactive medications, stimulants, or medications for symptomatic treatment of fatigue were excluded. All included patients denied sleep disorders and other causes of fatigue such as active infection, malignancy, anemia, thyroid or adrenal disease. On the same day of MRI acquisition, MS patients underwent a neurological evaluation, including the Expanded Disability Status Scale (EDSS). Patients with an EDSS > 4 were excluded in order to minimize the effect of physical disability on fatigue [17]. Fatigue severity was assessed using FSS, given the higher Cronbach's alpha value (0.89) of FSS compared to MFIS (0.81) [7,8]. FSS is a self-report questionnaire consisting of nine statements with a seven-point scale response per statement, with lower scores indicating less fatigue [18]. The RRMS patients with a mean FSS score ≥ 5.1 were categorized as high fatigued (HF), those with a mean FSS score ≤ 3 as low fatigued (LF), and those between FSS score 3.1–5 were classified as moderately fatigued (MF). The effect of various MS medications was minimal, as all the patients included in this study were on fingolimod.

The study was approved by the Wayne State University Institutional Review Board. A signed informed consent was obtained from all enrolled participants.

2.2. MRI Image Acquisition

Whole-brain MRI scan was performed using a 3-Tesla Siemens Verio System (Siemens Medical Systems, Erlangen, Germany). The following protocols were used for this analysis: (1) localizer sequence, (2) 3-D T1 weighted Magnetization Prepared Rapid Acquisition Gradient Echo (MPRAGE) images [Repetition time/Echo time (TR/TE) = 1680/3.52 ms, flip angle 9°, acquisition matrix size 384 × 384, with 176 slices, giving a nominal voxel size of 0.7 × 0.7 × 1.3 mm], and (3) a DTI sequence using a single-shot spin-echo diffusion sensitized echo-planar imaging sequence with balanced Icosa21 tensor encoding scheme (TR/TE = 10,400/126 ms, flip angle 90°, acquisition matrix size 200 × 200, with 46 consecutive slices, giving a voxel size of 1.3 × 1.3 × 3 mm).

2.3. MRI Data Processing and Analysis

Cortical reconstruction and volumetric segmentation was obtained from the 3-D T1 images using Freesurfer image analysis version 5 [19] as described in prior publications [20–23]. We used the standard protocol including skull stripping and image registration to Talairach brain atlas, followed by segmentation, topology correction, and placement of gray matter-white matter (GM-WM), white matter- cerebrospinal fluid (WM-CSF) boundaries using surface normalization and intensity gradients. In addition to the cortical thickness and volume of subcortical structures that were obtained from Freesurfer, we used the coordinates of the cortical boundaries to extract the diffusion parameters from the DTI images, as follows.

The diffusivity maps—no diffusion (b0), Mean Diffusivity (ADC), Radial Diffusivity (RD) and Fractional Anisotropy (FA) maps—were generated from the DTI sequence using DTI studio 3.0 [24]. We used FSL software to spatially register the Freesurfer brain-extracted FA images of each subject onto the subject's 3-D T1 images using a simple rigid body (six degrees of freedom) registration, followed by a rigorous non-linear image registration (FNIRT) [23]. FA, ADC, and RD images were then resampled to the structural image space for further analysis. Subsequently, the reconstructed Gray-White and the Gray-Pial interface surfaces of the structural image from Freesurfer analysis were used as the boundaries of the gray and the white matter and to generate new surface images [25], which were then used to obtain the average diffusion value along the normal vector. Finally, the labeled masks of frontal, temporal, parietal, and occipital lobes were used to measure the FA, ADC, and RD values for each subject.

2.4. Statistical Methods

Population demographics, diffusivity parameters of cortices, and subcortical nuclei volume variation between HF and LF groups were analyzed using an independent *t*-test. Boot strapping was performed to avoid the assumption of normality. A Mann-Whitney test was used to verify the findings. We did not find any discrepancy between the results from the non-parametric and boot-strapped *t*-tests. We used the Bonferroni method to correct for multiple comparisons, given the large number of variables used to model the fatigue scores, and both uncorrected and corrected *p*-values are reported. Pearson correlation with bootstrapping was used to explore the relation between the fatigue score and volumetric or DTI measures. A binomial, multivariate regression analysis was used to evaluate the best predictors of fatigue score from among those independent variables that showed a significant difference between the high and low fatigue scores. Two-sided bootstrapped *p*-value < 0.05 was considered statistically significant. All results are expressed as means ± standard error of mean (SEM), and statistical analysis was performed using SPSS v23.(International Business Machines Corporation, SPSS statistics for Windows, version 23.0, Armonk, NY, United States)

3. Results

3.1. Demographics and Clinical Data

Overall, 43 patients with RRMS participated: 15 patients were classified as the HF group, 14 as the MF, and 14 as the LF group. The groups did not differ in age, disease duration, medication, EDSS, or T2 lesion volume. The sample was primarily female (*n* = 26) with a mean age of 41 (±2.4) years. Seventeen patients were men with a mean age of 39 (±2.3) years. Demographic and clinical characteristics are shown in Table 1.

Table 1. Demographic and clinical data.

RRMS Population	HF Group	MF Group	LF Group	Total	p-Value
Number of patients	15	14	14	43	
Ethnicity (Cau vs. AA)	9 vs. 6	6 vs. 8	9 vs. 5	24 vs. 19	
Age (years) Range	43 ± 2.9 (23–55)	39 ± 3 (26–45)	39 ± 1.7 (29–47)	41 ± 1.7 (23–55)	0.102
Mean FSS score Range	6 ± 0.12 (5.1–7)	4 ± 0.14 (3.1–5)	1.89 ± 0.2 (1–3)	4.35 ± 0.26 (1–7)	
Median EDSS score Range	2 (1–4)	2 (1–4)	1.5 (1–4)	2 (1–4)	0.754
T2 lesion volume (mL) Range	14 ± 2.5 (7.4–27.16)	18.8 ± 4.8 (2.6–40.5)	15.3 ± 5.9 (1.8–39.7)	15.6 ± 2.3 (1.8–40.5)	0.859
Disease period (years) Range	10 ± 1.7 (0.5–19.17)	9.2 ± 1.2 (0.67–14.4)	8.6 ± 1.9 (0.25–15)	9.3 ± 1 (0.25–19.17)	0.136

RRMS—Relapsing Remitting Multiple Sclerosis, FSS—Fatigue Severity Scale, EDSS—Expanded Disability Status Scale, HF—High Fatigue, MF—Moderate Fatigue, LF—Low Fatigue, M—Male, F—Female, Cau—Caucasian, AA—African American, ml—milliliter. The data represents average and standard error of mean.

3.2. Structural Imaging Findings

Given the number of variables used in this study, we first looked at those variables that were most strikingly different between the HF and LF groups. The subcortical nuclei volumes were lower in the high fatigue group compared to the low fatigue group, prior to correcting for multiple comparisons: thalamic (HF: 11.5 ± 0.3 mL vs. LF: 14 ± 0.6 mL; $p = 0.001$), pallidal (HF: 2.6 ± 0.07 mL vs. LF: 3 ± 0.13 mL; $p = 0.013$), and SCP (HF: 207.3 ± 7.1 mL vs. LF: 246.1 ± 9.6 mL; $p = 002$, Figure 1a–c). Of these structures, only the thalamic volume retained statistical significance after Bonferroni correction for multiple comparisons (corrected p-value = 0.007, Supplementary Table S1). In addition to the HF and LF group differences, we looked at the correlation between the fatigue score and the volume of thalamus, pallidi, and SCP across all patients. Figure 2a–c show the inverse correlation with the FSS score: thalamic ($r^2 = 0.416$, $p = 0.006$), pallidal ($r^2 = 0.399$, $p = 0.005$), and SCP ($r^2 = 0.293$, $p = 0.04$). The relationship between these brain structure volumes and EDSS was not significant.

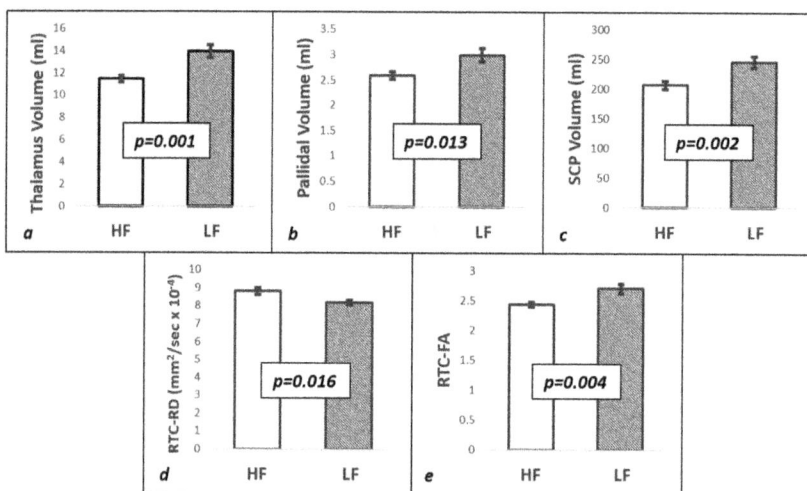

Figure 1. Volumetric and Diffusion measures showing lower Thalamic (a) Pallidal (b), Superior Cerebellar Peduncle (SCP, c) volume, and higher Radial Diffusivity (RD, d) and lower Fractional Anisotropy (FA, e) in Right Temporal Cortex (RTC) of High Fatigue group (HF) vs. Low Fatigue Group (LF). The error bars represent the standard error of mean.

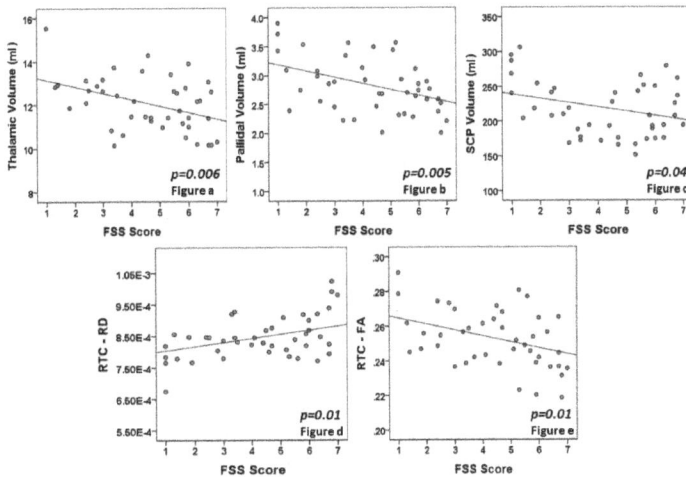

Figure 2. Correlation graphs showing inverse relation between Thalamic (**a**), Pallidal (**b**), Superior Cerebellar Peduncle (SCP, **c**) volume, Right Temporal Cortex-Fractional Anisotropy (RTC-FA, **e**) vs. FSS score, and positive relation between Right Temporal Cortex-Radial Diffusivity (RTC-RD, **d**) vs. Fatigue Severity Scale (FSS) score.

3.3. Diffuse Tensor Imaging Findings

A significant difference in diffusion tensor parameters was observed in the right temporal cortex (RTC) between the HF and LF groups. Radial Diffusivity (HF: $8.82 \times 10^{-4} \pm 0.2 \times 10^{-4}$ mm^2/s vs. LF: $8.18 \times 10^{-4} \pm 0.13 \times 10^{-4}$ mm^2/s; $p = 0.016$) was significantly higher in the HF group compared to the LF group (Figure 1d). In contrast, the HF group had significantly lower FA compared to the LF group (HF: $2.44 \times 10^{-1} \pm 0.04 \times 10^{-1}$ vs. LF: $2.71 \times 10^{-1} \pm 0.08 \times 10^{-1}$; $p = 0.004$) in RTC (Figure 1e). Both RD and FA retained statistical significance after correction for multiple comparisons (Bonferroni uncorrected p-values = 0.004 and 0.016, corrected p-values = 0.005 and 0.026 for FA and RD, respectively). Furthermore, the RD of RTC ($r^2 = 0.349$, $p = 0.01$) had a significant inverse correlation with the FSS score across all patients (Figure 2d), and the FA of RTC ($r^2 = 0.358$, $p = 0.01$) had a significant positive correlation with the FSS score (Figure 2e). The RD and FA of other cortices was not significantly different between the groups, and the variation in ADC of RTC did not reach statistical significance between the groups (Supplementary Table S1).

3.4. Multivariate Regression Analysis

We further explored the best MRI correlates of fatigue score in our patient group using a binomial regression analysis. The binomial defendant variable representing HF or LF was modeled using the independent variables that we found to be significantly different between the two groups in the univariate analysis, namely, subcortical volumes (thalamus, pallidum, and SCP), as well as FA and RD of the right temporal cortex. Since fatigue can be significantly affected by age, we included age in one model; however, since our sample did not have a large age difference between the groups, we also ran the model without age as a variable to avoid overestimation. Results are presented in Table 2, and demonstrate that in a multivariate regression model, right temporal cortex fractional anisotropy is the best correlate of fatigue status among the variables examined ($p = 0.11$ after correction for age, and $p = 0.023$ when age was excluded from the model).

Table 2. (a). Regression model to predict fatigue level based on neural structures of significance. (b). Regression model, corrected for age, to predict fatigue level based on neural structures of significance. FA—Fractional Anisotropy, RD—Radial Diffusivity, RTC—Right Temporal Cortex, SCP—Superior Cerebellar Peduncle.

(a)			
	Standarized Beta Coefficient	**Standard Error**	***p*-Value**
Thalamic volume	−1.042	0.514	0.043
Pallidal volume	0.068	1.435	0.962
SCP volume	−0.014	0.013	0.291
FA-RTC	−82.839	36.316	0.023
RD-RTC	−8567.04	10.782.51	0.427
(b)			
	Standarized Beta Coefficient	**Standard Error**	***p*-Value**
Age	0.119	0.061	0.058
Thalamic volume	−0.96	0.505	0.050
Pallidal volume	−0.14	1.474	0.924
SCP volume	−0.027	0.016	0.099
FA-RTC	−113.826	44.699	0.011
RD-RTC	−11070.1	10884.88	0.309

4. Discussion

The main focus of our study was to explore the structural and neuronal integrity measures of fatigue severity in RRMS patients with low disability, using the combination of structural and diffuse tensor imaging techniques. The correlation between fatigue and disability status is debated. Flachenecker et al. showed that fatigue is strongly related to physical disability [7]. Biberacher et al. reported a significant correlation between fatigue severity and EDSS in all three study cohorts (discovery, MRI, and CSF validation cohorts) [26]. In contrast, Krupp et al. found no significant relationship between fatigue and EDSS score [27]. In a large prospective study, Bakshi et al. reported no significant association between fatigue severity and EDSS [28]. In our study, we recruited MS patients with low disability in an attempt to study a homogenous population, by minimizing the potential impact of disability on fatigue severity and potentially on volumetric and DTI indices.

Given the complex nature of fatigue in MS, a wide variety of imaging techniques have been used in previous studies, and multi-regional damage rather than global brain damage was implicated. Research has identified a strong link between thalamic and basal ganglia nuclei abnormalities and the pathophysiology of fatigue [11,29–31]. A recent study reported lower thalamic volume of RRMS patients with fatigue compared to healthy controls [27,28]. Finke et al. described disrupted functional connectivity within the basal ganglia nuclei (including pallidi, putamen, and caudate nuclei), which correlates with fatigue severity in MS patients [29]. However, the literature on the anatomical variation of the aforementioned structures and their possible correlation with fatigue severity and lesion load in RRMS is limited.

In our study, we initially divided the MS patients into low and high fatigue groups, and eliminated patients with moderate fatigue, in order to augment potential groups' differences in cortical thicknesses, subcortical volumes, or diffusion indices. We observed significantly lower thalamic, pallidal, and superior cerebellar peduncles volumes in the HF group compared to the LF group. Furthermore, the magnitude of thalamic, pallidal, and SCP atrophy correlated positively with the FSS score in the total study population, which suggests that these structures are affected proportionately to fatigue severity in RRMS. Our observations support prior findings obtained by using different approaches, such as functional Magnetic Resonance Imaging (fMRI)and voxel based morphometry [29,32]. Nourbakhsh et al. [33], studied a cohort of early relapsing MS patients and reported an association between lower thalamic volume at baseline and increasing physical subscale of MFIS during the study. They also found a trend for baseline thalamic and cerebellar cortical volume to predict subsequent change in total MFIS in the same cohort of patients.

Previous studies have reported the involvement of cortical white matter in the fatigue pathology of RRMS patients. Notably, disruption of the fronto-parietal pathways was implicated based on increased FA in their respective white matter tract [6,13]. Advanced imaging methods, such as fluorodeoxyglucose Positron Emission Tomography (PET) have demonstrated dysfunction of the basal ganglia, fronto-parietal pathways, and cerebellar vermis [30]. Given the wide variability in imaging techniques used to study the cortical pathology of MS-related fatigue, we implemented a unique cross modality technique of registering DTI images to structurally sound MPRAGE data and obtained the DTI measures of their respective cortices [25]. The higher RD in parallel to lower FA indicated more severe neuronal injury in the right temporal cortex (RTC) of the HF group compared to the LF group. Furthermore, the RD and FA having direct and inverse correlation, respectively, with FSS score in RTC across all patients, suggest that the neuronal injury in RTC is proportionate to fatigue severity. Of note, RD and FA are not significantly affected in the left temporal cortex (LTC), which indicates neuronal integrity in the LTC. Additionally, the cortical thickness variation in RTC and in other cortices was not statistically significant between the study groups. These findings suggest that disruption of the neuronal integrity may occur prior to the evidence of cortical atrophy in the RTC. Moreover, they support and further expand previous evidence regarding the role of the right temporal lobe in the pathophysiology of fatigue.

Rocca et al. reported significant atrophy in the right inferior temporal gyrus in pure MS fatigue patients compared to a healthy control group, using voxel-based morphometry [31]. Hanken et al. described cortical thinning in the right middle temporal lobe in a subgroup of patients with both fatigue and depression [34]. Bisecco et al. reported decreased FA and increased RD in the left superior longitudinal fasciculus (SLF) in a fatigued MS group. However, correlation analysis showed a significant association between higher FSS scores and lower FA at the level of right SLF [35]. Future studies on arcuate fasciculus, a part of the SLF that links the temporal cortex with the frontal and parietal lobes, may confirm these findings and elucidate its role in fatigue pathology. Given the initial clinical presentation of RRMS in patients enrolled, the aforementioned findings indicate that the RTC is involved in early fatigue pathology and may occur prior to the involvement of other cortices in RRMS. Given that the majority of studies in MS-related fatigue focus on white matter pathology, in our study we provide evidence for the neuronal integrity measures in cortical areas of RRMS patients. Gray matter imaging studies in MS will eliminate the confounding effect of the variability of FA due to MS lesions, and can still be able to detect the disease burden on the cortex and deep nuclei. Longitudinal studies with advanced gray matter imaging techniques may elucidate the role of the RTC in fatigue and could explain the role of other cortices in the prognosis of MS-related fatigue.

Previous studies have examined the relationship between T2 lesion load and fatigue severity and yielded conflicting results [36,37]. In our study, no significant correlation was found between fatigue severity and T2 lesion load. Strengths of this study include a homogenous study population consisting of patients with RRMS and low EDSS, careful selection criteria, close timing of MRI and fatigue assessment, a blinded MRI reader to clinical characteristics, and the application of robust statistical analyses. Our study is not without limitations. First, the small sample size, mainly due to the strict exclusion criteria, and the cross-sectional design, which interferes with group comparison, limit the generalization of our findings. Second, the location of MS lesions and their effect on the integrity of structures located in proximity to them, was not taken into account. Third, this study provides limited information on the structures involved in the fatigue circuitry of moderately fatigued RRMS patients. Finally, the exclusion of patients who were on stimulants or other medications for symptomatic treatment of fatigue may lead to sample bias, as the enrolled patients may represent a subgroup with the worst fatigue, resistant to therapeutic modalities. However, the robust statistical methods implemented to analyze the brain structures strengthen the significance of our observations, and the correlation we observed is unlikely to be affected by these shortcomings.

5. Conclusions

In conclusion, our study shows that FA-RTC is the best predictor of fatigue severity in RRMS patients. Additionally, variation in thalamic volume may serve as a biomarker of fatigue in RRMS. Our study provides further evidence of an association between pallidal and SCP volume variation and fatigue severity, however, further investigation is required to reveal their role in MS-related fatigue. Moreover, higher RD in addition to reduced FA in the right temporal cortex indicates the early involvement of RD in fatigue pathology compared to other cortices in RRMS. Longitudinal integrated cortical DTI and volumetric studies in a large sample size are needed to confirm our findings and help in determining the prognosis of fatigue and its response to different treatment modalities.

Supplementary Materials: The supplementary file is available online at www.mdpi.com/2076-3425/7/8/102/s1.

Acknowledgments: This study was supported by a grant from the Sastry Foundation. We wish to thank Sara Razmjou and Samuel Mikol for their contribution to manuscript preparation.

Author Contributions: E.B. contributed to study design, study conceptualization and manuscript writing. K.Y. contributed to statistical analysis, imaging analysis and manuscript writing. F.B. contributed to imaging analysis. C.S.M. and R.S. contributed to data collection and data processing. R.G. contributed to statistical and imaging analysis. O.K. contributed to study conceptualization and study design. N.S.B. contributed to imaging and statistical analysis and manuscript writing.

Conflicts of Interest: K.Y., F.B., R.S., C.S.M., R.G., N.S.B. report no conflict of interest. O.K. received lecture fees from TEVA, Genzyme and research support from Chugai, Medimmune, and Biogen. EB received lecture fees from Biogen, TEVA, and EMD Serono and research support from Roche, Chugai, and Medimmune.

References

1. Noseworthy, J.H.; Lucchinetti, C.; Rodriguez, M.; Weinshenker, B.G. Multiple sclerosis. *N. Eng. J. Med.* **2000**, *343*, 938–952. [CrossRef] [PubMed]
2. Bakshi, R. Fatigue associated with multiple sclerosis: Diagnosis, impact and management. *Mult. Scler. J.* **2003**, *9*, 219–227. [CrossRef] [PubMed]
3. Goodin, D.S.; Frohman, E.M.; Garmany, G.P., Jr.; Halper, J.; Likosky, W.H.; Lublin, F.D.; Donald, M.; Silberberg, H.; William, M.; Stuart, H.; et al. Disease modifying therapies in multiple sclerosis: Report of the Therapeutics and Technology Assessment Subcommittee of the American Academy of Neurology and the MS Council for Clinical Practice Guidelines. *Neurology* **2002**, *58*, 169–278. [CrossRef] [PubMed]
4. Chaudhuri, A.; Behan, P.O. Fatigue in neurological disorders. *Lancet.* **2004**, *363*, 978–988. [CrossRef]
5. Mills, R.J.; Young, C.A. A medical definition of fatigue in multiple sclerosis. *QJM Int. J. Med.* **2008**, *101*, 49–60. [CrossRef] [PubMed]
6. Pardini, M.; Bonzano, L.; Mancardi, G.L.; Roccatagliata, L. Frontal networks play a role in fatigue perception in multiple sclerosis. *Behav. Neurosci.* **2010**, *124*, 329–336. [CrossRef] [PubMed]
7. Flachenecker, P.; Kumpfel, T.; Kallmann, B.; Gottschalk, M.; Grauer, O.; Rieckmann, P. Fatigue in multiple sclerosis: A comparison of different rating scales and correlation to clinical parameters. *Mult. Scler. J.* **2002**, *8*, 523–526. [CrossRef] [PubMed]
8. Krupp, L.B.; LaRocca, N.G.; Muir-Nash, J.; Steinberg, A.D. The fatigue severity scale. Application to patients with multiple sclerosis and systemic lupus erythematosus. *Arch. Neurol.* **1989**, *46*, 1121–1123. [CrossRef] [PubMed]
9. Tellez, N.; Rio, J.; Tintore, M.; Nos, C.; Galan, I.; Montalban, X. Does the Modified Fatigue Impact Scale offer a more comprehensive assessment of fatigue in MS? *Mult. Scler. J.* **2005**, *11*, 198–202. [CrossRef] [PubMed]
10. Van der Werf, S.P.; Jonge, J.; Lycklama a Nijeholt, G.J.; Barkhof, F.; Hommes, O.R.; Bleijenberg, G. Fatigue in multiple sclerosis: Interrelations between fatigue complaints, cerebral MRI abnormalities and neurological disability. *J. Neurol. Sci.* **1998**, *160*, 164–170. [CrossRef]
11. Calabrese, M.; Rinaldi, F.; Grossi, P.; Mattisi, I.; Bernardi, V.; Favaretto, A. Basal ganglia and frontal/parietal cortical atrophy is associated with fatigue in relapsing-remitting multiple sclerosis. *Mult. Scler. J.* **2010**, *16*, 1220–1228. [CrossRef] [PubMed]

12. Pellicano, C.; Gallo, A.; Li, X.; Ikonomidou, V.N.; Evangelou, I.E.; Ohayon, J.M.; Stern, S.K.; Ehrmantraunt, M.; Cantor, F.; McFarland, H.F.; et al. Relationship of cortical atrophy to fatigue in patients with multiple sclerosis. *Arch. Neurol.* **2010**, *67*, 447–453. [CrossRef] [PubMed]

13. Sepulcre, J.; Masdeu, J.C.; Goni, J.; Arrondo, G.; Velez de Mendizabal, N.; Bejarano, B. Fatigue in multiple sclerosis is associated with the disruption of frontal and parietal pathways. *Mult. Scler. J.* **2009**, *15*, 337–344. [CrossRef] [PubMed]

14. Bakshi, R.; Miletich, R.S.; Henschel, K.; Shaikh, Z.A.; Janardhan, V.; Wasay, M. Fatigue in multiple sclerosis: cross-sectional correlation with brain MRI findings in 71 patients. *Neurology* **1999**, *53*, 1151–1153. [CrossRef] [PubMed]

15. Mainero, C.; Faroni, J.; Gasperini, C.; Filippi, M.; Giugni, E.; Ciccarelli, O. Fatigue and magnetic resonance imaging activity in multiple sclerosis. *J. Neurol.* **1999**, *246*, 454–458. [CrossRef] [PubMed]

16. Simpson, S., Jr.; Tan, H.; Otahal, P.; Taylor, B.; Ponsonby, A.L.; Lucas, R.M.; Blizzard, L.; Valery, P.C.; Lechner-Scott, J.; Shaw, C.; et al. Anxiety, depression and fatigue at 5-year review following CNS demyelination. *Acta Neurol. Scand.* **2016**, *134*, 403–413. [CrossRef] [PubMed]

17. Hameau, S.; Zory, R.; Latrille, C.; Roche, N.; Bensmail, D. Relationship between Neuromuscular and Perceived Fatigue and Locomotor Performance in Patients with Multiple Sclerosis. Available online: https://www.ncbi.nlm.nih.gov/m/pubmed/27164538/ (accessed on 11 August 2017).

18. Clerx, L.; Jacobs, H.I.; Burgmans, S.; Gronenschild, E.H.; Uylings, H.B.; Echavarri, C. Sensitivity of different MRI-techniques to assess gray matter atrophy patterns in Alzheimer's disease is region-specific. *Curr. Alzheimer Res.* **2013**, *10*, 940–951. [CrossRef] [PubMed]

19. FreeSurfer. Available online: http://surfer.nmr.mgh.harvard.edu/ (accessed on 11 August 2017).

20. Dale, A.M.; Fischl, B.; Sereno, M.I. Cortical surface-based analysis. I. Segmentation and surface reconstruction. *Neuroimage* **1999**, *9*, 179–194. [CrossRef] [PubMed]

21. Fischl, B.; Dale, A.M. Measuring the thickness of the human cerebral cortex from magnetic resonance images. *Proc. Natl. Acad. Sci. USA* **2000**, *97*, 11050–11055. [CrossRef] [PubMed]

22. Fischl, B.; Salat, D.H.; Busa, E.; Albert, M.; Dieterich, M.; Haselgrove, C.; van der Kouwe, A.; Killiany, R.; Kennedy, D.; Klaveness, S.; et al. Whole brain segmentation: Automated labeling of neuroanatomical structures in the human brain. *Neuron* **2002**, *33*, 341–355. [CrossRef]

23. Fischl, B.; Sereno, M.I.; Dale, A.M. Cortical surface-based analysis. II: Inflation, flattening, and a surface-based coordinate system. *Neuroimage* **1999**, *9*, 195–207. [CrossRef] [PubMed]

24. Jiang, H.; van Zijl, P.C.; Kim, J.; Pearlson, G.D.; Mori, S. DtiStudio: Resource program for diffusion tensor computation and fiber bundle tracking. *Comput. Methods Progr. Biomed.* **2006**, *81*, 106–116. [CrossRef] [PubMed]

25. Govindan, R.M.; Asano, E.; Juhasz, C.; Jeong, J.W.; Chugani, H.T. Surface-based laminar analysis of diffusion abnormalities in cortical and white matter layers in neocortical epilepsy. *Epilepsia* **2013**, *54*, 667–677. [CrossRef] [PubMed]

26. Biberacher, V.; Schmidt, P.; Selter, R.; Pernpeinter, V.; Kowarik, M.; Knier, B.; Duck, D.; Hoshi, M.-M.; Korn, T.; Berthele, A.; Kirschke, J.S.; et al. Fatigue in multiple sclerosis: Association with clinical, MRI and CSF parameters. *Mult. Scler. J.* **2017**, 1–11. [CrossRef] [PubMed]

27. Krupp, L.B.; Alvarez, L.A.; LaRocca, N.G.; Scheinberg, L.C. Fatigue in multiple sclerosis. *Arch Neurol.* **1988**, *45*, 435–437. [CrossRef] [PubMed]

28. Bakshi, R.; Shaikh, Z.A.; Miletich, R.S. Fatigue in multiple sclerosis and its relationship to depression and neurological disability. *Mult. Scler. J.* **2000**, *6*, 181–185. [CrossRef] [PubMed]

29. Finke, C.; Schlichting, J.; Papazoglou, S.; Scheel, M.; Freing, A.; Soemmer, C.; Pech, L.M.; Pajkert, A.; Pfüller, C.; Wuerfel, J.T.; et al. Altered basal ganglia functional connectivity in multiple sclerosis patients with fatigue. *Mult. Scler. J.* **2015**, *21*, 925–934. [CrossRef] [PubMed]

30. Roelcke, U.; Kappos, L.; Lechner-Scott, J.; Brunnschweiler, H.; Huber, S.; Ammann, W.; Plohmann, A.; Dellas, S.; Maguire, R.P.; Missimer, J.; et al. Reduced glucose metabolism in the frontal cortex and basal ganglia of multiple sclerosis patients with fatigue: A 18F-fluorodeoxyglucose positron emission tomography study. *Neurology* **1997**, *48*, 1566–1571. [CrossRef] [PubMed]

31. Rocca, M.A.; Parisi, L.; Pagani, E.; Copetti, M.; Rodegher, M.; Colombo, B. Regional but not global brain damage contributes to fatigue in multiple sclerosis. *Radiology* **2014**, *273*, 511–520. [CrossRef] [PubMed]

32. Wilting, J.; Rolfsnes, H.O.; Zimmermann, H.; Behrens, M.; Fleischer, V.; Zipp, F. Structural correlates for fatigue in early relapsing remitting multiple sclerosis. *Eur. Radiol.* **2016**, *26*, 515–523. [CrossRef] [PubMed]

33. Nourbakhsh, B.; Azevedo, C.; Nunah-Saah, J.; Maghzi, A.; Spain, R.; Pelletier, D.; Waubant, E. Longitudinal associations between brain structural changes and fatigue in early MS. *Mult. Scler. Relat. Disord.* **2016**, *5*, 29–33. [CrossRef] [PubMed]

34. Hanken, K.; Eling, P.; Klein, J.; Klaene, E.; Hildebrandt, H. Different cortical underpinnings for fatigue and depression in MS? *Mult. Scler. Relat. Disord.* **2016**, *6*, 81–86. [CrossRef] [PubMed]

35. Bisecco, A.; Caiazzo, G.; Dambrosio, A.; Sacco, R.; Bonavita, S.; Docimo, R.; Cirillo, M.; Pagani, E.; Filippi, M.; Esposito, F.; et al. Faitgue in multiple sclerosis: The contribution of occult white mattre damage. *Mult. Scler. J.* **2016**, *22*, 1676–1684. [CrossRef] [PubMed]

36. Tedeschi, G.; Dinacci, D.; Lavorgna, L.; Prinster, A.; Savettieri, G.; Quattrone, A.; Livera, P.; Messina, C.; Reggio, A.; Servillo, G.; et al. Correlation between fatigue and brain atrophy and lesion load in multiple sclerosis patients independent of disability. *J. Neurol. Sci.* **2007**, *263*, 15–19. [CrossRef] [PubMed]

37. Bester, M.; Lazar, M.; Petracca, M.; Babb, J.S.; Herbert, J.; Grossman, R.I. Tract-specific white matter correlates of fatigue and cognitive impairment in benign multiple sclerosis. *J. Neurol. Sci.* **2013**, *330*, 61–66. [CrossRef] [PubMed]

Review

Melanocortins, Melanocortin Receptors and Multiple Sclerosis

Robert P. Lisak * and Joyce A. Benjamins

Department of Neurology, Wayne State University School of Medicine, Detroit, MI 48201, USA;
jbenjami@med.wayne.edu
* Correspondence: rlisak@med.wayne.edu; Tel.: +1-313-577-1249

Received: 30 June 2017; Accepted: 8 August 2017; Published: 14 August 2017

Abstract: The melanocortins and their receptors have been extensively investigated for their roles in the hypothalamo-pituitary-adrenal axis, but to a lesser extent in immune cells and in the nervous system outside the hypothalamic axis. This review discusses corticosteroid dependent and independent effects of melanocortins on the peripheral immune system, central nervous system (CNS) effects mediated through neuronal regulation of immune system function, and direct effects on endogenous cells in the CNS. We have focused on the expression and function of melanocortin receptors in oligodendroglia (OL), the myelin producing cells of the CNS, with the goal of identifying new therapeutic approaches to decrease CNS damage in multiple sclerosis as well as to promote repair. It is clear that melanocortin signaling through their receptors in the CNS has potential for neuroprotection and repair in diseases like MS. Effects of melanocortins on the immune system by direct effects on the circulating cells (lymphocytes and monocytes) and by signaling through CNS cells in regions lacking a mature blood brain barrier are clear. However, additional studies are needed to develop highly effective MCR targeted therapies that directly affect endogenous cells of the CNS, particularly OL, their progenitors and neurons.

Keywords: ACTH; melanocortins; melanocortin receptors; multiple sclerosis; neuroprotection; oligodendroglia; repair

1. Introduction

Melanocortins and melanocortin receptors (MCR) have been extensively investigated for their roles in the hypothalamo-pituitary-adrenal axis [1–5], and to a lesser extent in immune cells [6–10] and the nervous system outside the hypothalamic axis [3,11,12]. We have focused on the expression and function of MCR in oligodendroglia (OL), the myelin producing cells of the central nervous system (CNS), with the goal of identifying new therapeutic approaches to decrease CNS damage in multiple sclerosis (MS) as well as to promote repair.

2. Melanocortins

The melanocortins—adrenocorticotropic hormone (ACTH), α-MSH, β-MSH and γ-MSH—are polypeptides derived from a common precursor, pro-opiomelanocortin (POMC) (Figure 1). ACTH is 39 amino acids in length, and can be cleaved to the smaller 13 amino acid α-MSH. The 12 amino acid γ-MSH is cleaved from the amino terminus of POMC, while the 20 amino acid β-MSH is cleaved from POMC towards the carboxy terminus. ACTH has both steroidogenic and nonsteroidogenic actions, while the other three melanocortins are nonsteroidogenic.

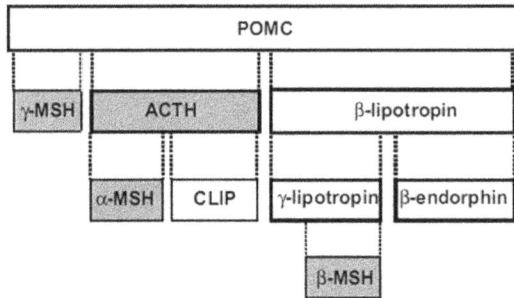

Figure 1. Melanocortin peptides (shaded boxes) derived from POMC. ACTH: adrenocorticotropic hormone; CLIP: corticotropin-like intermediate lobe peptide; MSH: melanocyte-stimulating hormone; POMC: proopiomelanocortin, Reprinted by permission from [8].

3. Melanocortin Receptors

Five subtypes of MCR have been identified and cloned; their distribution, function and pharmacology are characterized in part [1,2,4,13–17] (reviews). ACTH can activate all five receptor subtypes; of the melanocortins, only ACTH activates MC2R; ACTH as well as α-MSH, β-MSH and γ-MSH activate MC1R, MC3R, MC4R and MC5R, although with varying affinities. For example, α-MSH and β-MSH bind with higher affinity to MC1R, MC3R and MC4R than ACTH, while γ-MSH has relatively low binding affinities for all the MCR except MC3R [3,7].

All five MCR are G protein-coupled receptors of the Class A rhodopsin-like family and share 7 homologous membrane spanning domains but differ in their N-terminus and C-terminus sequences [13]. MC1R has been associated primarily with pigmentation, MC2R with glucocorticoid biosynthesis, MC3R and MC4R with energy homeostasis and MC5R with exocrine gland regulation [2,3,13]. However, each of the subtypes is widely distributed in various parts of the body, where they serve a variety of functions [2,3,6,7,12] (reviews).

Mutations in the genes of each of the MCR subtypes have been identified in humans and other mammals, as summarized by Switonski et al. [18]. MC1R is prominent in the synthesis of melanin in melanocytes, and mutations are associated with various skin phenotypes and diseases, including increased cancer risk, especially for melanomas [19–21]. Of interest, three reports indicate an association between disability in MS and MC1R gene single nucleotide polymorphisms leading to MC1R hyporesponsiveness [22–24]. Mutations in the human MC2R gene cause familial glucocorticoid deficiency; of the 25 missense mutations identified, most result in decreased trafficking of MC2R to the cell surface. A mutation in the MC2R gene with increased expression and stronger response to ACTH has been associated with increased responsiveness to ACTH treatment of infantile spasms [25,26]. MC3R, like MC4R, is involved in energy homeostasis; while several human mutations have been identified [20], there is a less clear association with obesity for these mutations than for MC4R mutations. MC4R has been extensively investigated in obesity research, and decreased activity of MC4R is the leading monogenic cause of severe early onset obesity [27,28]. Over 166 mutations in the human MC4R gene have been identified, many in obese individuals. MC5R is involved in lipid metabolism, exocrine function and inflammatory activity. Only a few polymorphisms have been identified in the human MC5R gene; associations with obesity, type 2 diabetes, schizophrenia and bipolar disorder have been reported [29,30].

4. Melanocortin Receptor Signaling

Melanocortin receptors are coupled to G proteins, and signal primarily via adenylyl cyclase and multiple down-stream pathways [31,32] (reviews). Other pathways independent of adenylyl cyclase and cAMP have also been identified. The signaling pathways are complex and pleiotropic, depending on the ligand, cell type, MCR surface expression, associated proteins, time of receptor occupancy and other factors. All 5 MCR are known to be coupled to Gs, but in some instances can be coupled to Gq or Gi [32] (review). MC1R activation through Gs stimulates adenylyl cyclase to increase cAMP with activation of protein kinase A (PKA), increases Ca^{++} levels, and can independently stimulate the ERK1/2 pathway, but appears to have little effect on the protein kinase C (PKC) pathway. In general, MC2R and MC3R signal in a similar fashion, but can also activate PKC. MC4R, in addition to coupling to Gs and activating adenylyl cyclase, can also couple to Gq or Gi to activate other signaling cascades, including PKC, PI3 kinase and ERK1/2 pathways. MC5R can independently activate the PKA pathway through Gs or the ERK1/2 pathway through Gi. G protein independent pathways involving a variety of kinases have been characterized for GPCRs; for example, MC1R activation via Src kinase has been reported [21]. Activation of MCR can also be regulated by receptor internalization; for all five MCR, binding of melanocortins or other agonists decreases association of β-arrestins, disrupting signaling via GPCRs and leading to internalization through clathrin-coated pits [32] (review). Regulation of MCR function occurs by multiple mechanisms, including transport to the plasma membrane by chaperones [33], attenuation and selectivity for signaling pathways via MCR membrane-associated proteins (MRAP) [34] and inhibition by the naturally occurring inhibitors agouti and agouti-related proteins.

In addition to melanocortins, the naturally occurring agonists of MCR, many pharmacologic agonists have been synthesized and tested for MCR subtype specificity, for longer half-lives than the rapidly degraded melanocortin peptides, and for reduction in steroidogenic or other side effects [4,11,35] (reviews). These include agents targeted to allosteric (extracellular), orthosteric (transmembrane, where ACTH and melanocortins bind) and signal transduction (intracellular) sites on MCR. The recent development of allosteric modulators [36,37] and biased agonists [4,38] present new approaches to activate MCR with potential therapeutic advantages. As one example, modulation of constitutively activated MC4R with inverse agonists has received recent attention as a promising area for drug development [39]. Conversely, naturally occurring antagonists include the proteins agouti and agouti-related protein. While development of specific antagonists for the ACTH-specific MC2R [40] (review) as well as for MC1R and MC5R has been problematic, synthetic antagonists are available for MC3R and MC4R [41]. The current development of cyclic peptides or site specific antibodies holds the promise of more subtype specific agents in the future.

5. Anti-Inflammatory Effects on the Peripheral Immune System

5.1. Corticosteroid Dependent Effects

Activation of MC2R expressed by cells of the adrenal cortex results in increase in circulating levels of corticosteroids. The control of corticosteroid production and secretion is under control of ACTH, which in turn is controlled by corticotropin releasing hormone (CRH) made by the hypothalamus. Levels of CRH and ACTH are controlled by levels of corticosteroids vain a feedback loop referred to as the pituitary-adrenal axis. Administration of ACTH, the only melanocortin with the ability to strongly bind to MC2R [2,4,6] and to initiate signaling of MC2R, results in an increase of corticosteroids. The corticosteroids have extensive effects on many body functions including exerting anti-inflammatory effects on many cells of the immune system. The effects of corticosteroids on the trafficking, number and function of the lymphocytes and monocytes have been assumed to be the mechanism of therapeutic efficacy of ACTH in MS and other immune/inflammatory mediated diseases. As noted in Section 6, there is evidence that effects of ACTH and other melanocortins on immune and other inflammatory processes may also involve direct effects on circulating cells of the immune system, effects on the immune system via the CNS and effects on cells within the CNS [4,6,8,9,42].

5.2. Corticosteroid Independent Effects

Inflammatory cells including lymphocytes, monocytes/macrophages, and neutrophils as well as tissue-based cells including mast cells express MCR. Monocyte/macrophages as well as microglia express MC1R, MC3R and MC5R and lymphocytes express MC1R, MC3R and MC5R as well, reviewed in [6]. Signaling through these receptors inhibits inflammatory processes and has been associated with shifts from proinflammatory to inhibitory effects of lymphocytes, perhaps in part through effects on antigen presenting cells such as monocytes/macrophages. While this is somewhat difficult to demonstrate with administration of ACTH which increases endogenous corticosteroids, studies with α-MSH, which does not signal through MC2R, show induction of anti-inflammatory effects [43–46]. Administration of intravenous ACTH is able to inhibit maturation of B cells obtained from those MS patients into immunoglobulin secreting cells in vitro [47]. α-MSH inhibits inflammation in experimental autoimmune uveitis (EAU), induces CD25+ regulatory CD4+ T-cells [45,46,48] and regulates ubiquitination in T cells as well [49]. MC5R appears to be the important MCR in these latter functions. Nonsteroidogenic effects of ACTH were demonstrated in a rat model of gouty arthritis; ACTH administered systemically did not reduce joint inflammation, whereas ACTH or the MC3R agonist γ-MSH injected locally reduced inflammation in both normal and adrenalectomized rats [50]. Additional studies are clearly needed to study longer term effects of ACTH treatment in patients as well as in animal models to determine if there are long-term effects that outlast any immediate effects of the increase in corticosteroids on the peripheral immune system.

6. Direct Effects in the CNS

6.1. Effects Mediated through Neuronal Regulation of Immune System Function

Endogenous ACTH and other melanocortins, as well as presumably exogenously administered ACTH, can access the CNS in the brain stem and the hypothalamus, bind to MCR, particularly MC4R, and initiate signaling [10,51–58]. These brain stem neurons trigger vagal activity with release of acetylcholine (ACh) in peripheral tissue, with binding and activation of acetylcholine receptors (AChR). An important receptor is α7-AChR, which triggers anti-inflammatory processes and inhibition of excitatory and other damaging processes [59]. It has been suggested that some of this anti-inflammatory activity may then feed back to the CNS and further provide neuroprotective effects within the CNS [6,9,60] (reviews). Bilateral vagotomy interferes with melanocortin protective effects in the CNS [58], supporting the importance of the vagal pathway in melanocortin neuroprotection and reparative processes [51,57]. Activation of hypothalamic neurons via MCR seems to be important in hormonal and metabolic processes as noted in Section 3 [11,13,61]. These important pathways are covered in greater detail in several review articles [4,6,8,9,11].

6.2. Effects on CNS Neurons

Within the CNS, neuronal expression of MCR has been characterized most extensively in the central hypothalamic melanocortin pathway, where MC4Rs are the subtype involved in regulation of metabolism [5] (review). For example, genetic regulation of MC4R expression in cholinergic neurons in several extrahypothalamic brain regions implicate MC4R in regulation of energy balance and glucose homeostasis [28,62]. A recent study shows that constitutive activity of MC4R inhibits L-type voltage-gated calcium channels in cultured neurons [63]. Activation of MC4R shows neuroprotective and neuroregenerative effects in several models of neurodegenerative diseases [52], including neurogenesis and cognitive recovery in an animal model of Alzheimer's disease [64]. We reported that the MCR agonist ACTH1-39 protects cultured rat forebrain neurons from excitotoxic, apoptotic, oxidative and inflammation related insults [65], but the specific MCR subtypes involved are not known.

Melanocortins are increasingly being investigated for their effects on synaptic remodeling [3] (review). For example, in the hippocampal C1 region, activation of MC4R at the postsynaptic ending increases cAMP levels and activates PKA, thus modulating long-term potentiation and long-term depression. In dopaminergic neurons, cross talk between MC4R and dopamine receptors regulates increased expression of AMPA receptors to increase dopamine responsiveness, or promotes endocytosis of AMPA receptors to reduce long-term depression [3].

6.3. Effects on Glia

MCR expression and function has been previously characterized in astroglia, and to a lesser extent in microglia and Schwann cells. Our recent studies, summarized in later Sections 7 and 8 of this review, are the first to examine MCR expression and subtypes in OL and their precursors. Melanocortin effects on astroglia point to their roles in inflammation, obesity and regeneration [66] (review). Experiments to date in astroglia indicate that message for MC1R and MC4R but not MC3R is expressed, as analyzed by RT-PCR [7,67]. An early study on astroglia with a panel of MCR agonists showed that morphologic changes, including rounding of the cell body and process extension, were mediated by a cAMP mediated pathway, while proliferation was stimulated by an alternative pathway independent of cAMP [68]. More recently, MC4R activation in astroglia with the long acting α-MSH analogue NDP-MSH was shown to increase expression of brain-derived neuronotrophic factor [69] and stimulate the release of the anti-inflammatory TGF-β [70], in part via the ERK-cFos pathway. Microglia, the endogenous macrophages of the CNS, express MC4R [70] as well as MC1R, MC3R and MC5R [7,8]. Melanocortin peptides decrease microglial production of nitric oxide (NO) and the proinflammatory cytokines TNF-α and IL-6 [43,71], but increase release of the anti-inflammatory cytokine IL-10 [70]. In addition, NDP-MSH promotes an M2-like phenotype in microglia and inhibits microglial activation induced by Toll-Like Receptors 2 and 4 [72]. MCR expression and function have been less well studied in Schwann cells in the PNS. ACTH promotes peripheral nerve regeneration and axonal growth in vivo [73], while α-MSH inhibits inflammatory signaling in cultured Schwann cells [74]. The melanocortin analogue Org2766 as well as α-MSH stimulates Schwann cell proliferation, upregulates the NGF low-affinity receptor p75, induces release of an unidentified neurotrophic activity [75] and enhances nerve regeneration [76].

7. Effects on Endogenous Cells of the CNS with Potential Protective and Reparative Importance in MS and Other CNS Disorders

As noted in Section 6, MCR are expressed by neurons, astroglia, microglia and OL with some differential expression in different regions of the brain. Additionally there is in vitro and in vivo evidence, cited earlier, that these receptors are functional. Since MCR are known to be present within the CNS, some of the therapeutic effects in EAE and in the PNS experimental model experimental autoimmune neuritis (EAN) as well as in other animal models and in human diseases might be independent of the stimulation of endogenous corticosteroid production via MC2R signaling in the adrenal gland. While some of these non-corticosteroid disease modifying and anti-inflammatory effects might be due to stimulation of MC1R and other MCR expressed by peripheral immune cells, including lymphocytes and monocytes, direct effects on endogenous CNS cells may also occur.

An important animal model of MS is experimental autoimmune encephalomyelitis (EAE, originally called experimental allergic encephalomyelitis), which is induced by sensitization of experimental animals with CNS tissue, CNS myelin or specific constituents of myelin including myelin basic protein (MBP, originally called basic protein or encephalitogenic protein), proteolipid protein (PLP) and myelin oligodendrocyte glycoprotein (MOG). Alternatively, EAE can be induced in naive animals by passive transfer of lymphocytes, T cells or T cell lines or clones. Depending on the sensitizing antigen, the sensitization protocols, the species and strain of animals employed acute, hyperacute, chronic and relapsing courses of EAE can develop. Prevention and treatment of EAE can be achieved with many agents and treatments with EAE serving as a test system to screen for potential treatments for MS, both relapses and as disease modifying therapies [77,78]. ACTH was among the first experimental therapy employed in EAE [79] and as treatment for relapses (exacerbations) of MS; see Section 9, Treatment of Neurologic Diseases with Melanocortins.

Inhibition of EAE by ACTH may involve both corticosteroid and non-corticosteroid effects on the immune system. As noted there is indirect evidence of protective effects within the CNS including less demyelination along with evidence of repair; i.e., remyelination [80]. α-MSH, which cannot signal through MC2R expressed in the adrenals, inhibits the development of EAE [81,82] and EAN [83], a peripheral neuropathy, that serves as a model for some variants of Guillain-Barre Syndrome (GBS). An α-MSH analog, SValpha-MSH also inhibits EAE, acting to inhibit CD4+ T cells [84]. Since α-MSH cannot increase endogenous corticosteroids and yet inhibits development of EAN, this supports the idea that inhibition of EAN is due to the direct effects of α-MSH on immune cells and/or Schwann cells, the myelin forming cells of the PNS. Indeed it has been shown that a-MSH inhibits the translocation of NFκB in Schwann cells in vitro [74] demonstrating the potential for a direct effect of ACTH on Schwann cells in ameliorating EAN. Similarly, NDP-MSH, a long-lived analog of α-MSH, ameliorates EAE and restores BBB in mice; in vitro, protection of mouse and human neurons from excitotoxicity occurred via MC1R activation [85]. Melanocortins have been reported to be neuroprotective in animal models of excitotoxic injury [86], subarachnoid hemorrhage [87], traumatic CNS injury [88], an animal model of Alzheimer's disease [64] and in peripheral nerve injury and repair [76,89,90] as well as the previously mentioned EAN and EAE.

Melanocortins have been shown to provide protection in vitro for neurons from toxic effects of cisplatin [91], protect neuronal cell lines from serum-induced apoptosis [92], provide trophic effects to neurons [93] and enhance neurite outgrowth in vitro [94]. Melanocortins can inhibit production of proinflammatory molecules by microglia [71]. Prior to our investigations, there has been little work done on MCR expression and effects of melanocortins on OL function. MCR expression in vitro has been discussed earlier and importantly all MCR are expressed by OL (Figure 2) [95]. The expression of MC4R in differentiated OL is shown in Figure 2A.

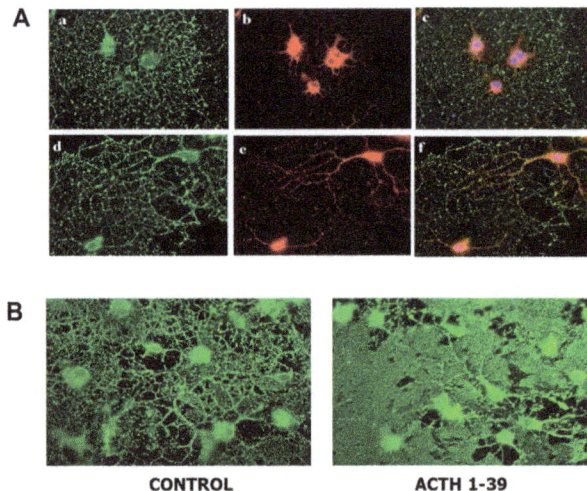

Figure 2. Oligodendroglia express melanocortin receptors and produce larger membrane sheets in response to ACTH. (**A**) Mixed glial cultures from rat brain were immunostained with antibody for MC4R (red) before (**a–c**) and after (**d–f**) permeabilization to visualize surface and total MC4R respectively; differentiated oligodendroglia were immunostained for surface galactolipid with antibody A007 (green) [96]. (**B**) Oligodendroglia were treated with 200 nM ACTH 1-39 for 3 days, then immunostained with O1 antibody to detect surface galactolipids; ACTH induced larger, more dense membrane sheets [97].

In order to dissect the mechanisms that may be involved in interactions between ACTH and other melanocortins and cells of the OL lineage, we undertook a series of experiments employing glial and neuronal cultures. ACTH inhibits death of OL and OPC induced by several mechanisms that are involved in damage to the CNS in MS as well as other disorders of the CNS, including glutamate (excitotoxicity) [96,97], apoptosis (induced by staurosporine, a widely employed molecule in apoptosis research) [96,97], reactive oxygen species (ROS) induced by hydrogen peroxide (H_2O_2) [96,97] and inflammation mediated by quinolinic acid (QA), a product downstream of kynurenic acid in the tryptophan indoleamine pathway [96,97] (Table 1). In the case of glutamate induced OL and OPC death, ACTH inhibited cell death mediated through all three of the ionotropic glutamate channels, NMDA, AMPA and kainate [96,97]. There was no protection of OL and OPC from toxicity induced by kynurenic acid, an earlier metabolite in the indoleamine tryptophan inflammatory pathway [96,97]. ACTH did not protect OL from either slow or rapid release of nitric oxide (NO) but provided modest protection of OPC from slow but not rapid release of NO [96,97].

Table 1. ACTH Protects Oligodendroglia, Oligodendroglial Progenitors and Neurons from Multiple Toxic Agents.

Toxic Agent	OL	OPC	Neurons
Glutamate	+	+	+
Staurosporine	+	+	+
Quinolinic acid	+	+	+
Kynurenic acid	none	none	none
H_2O_2 (reactive oxygen species	+	+	+
Nitric oxide (slow release)	none	slight	none
Nitric oxide (rapid release)	none	none	slight

ACTH at 200 nM protects cultured rat oligodendroglia (OL), oligodendroglial progenitors (OPC), and neurons from excitotoxic, apoptotic and inflammatory insults, as well as from reactive oxygen species [65,96,97]. No protection was found against kynurenic acid or nitric oxide, except for modest protection for OPC (slow release NO), and for neurons (rapid release NO). Cells were treated for 24 h with the toxic agents in the absence or presence of 200 nM ACTH, the concentration shown to cause maximal protection in these cultures. +, ACTH significantly protected cells from death induced by the toxic agents.

Neurons and axons are also targets of pathologic processes in MS and therefore we also examined whether ACTH could protect neurons from these same pathologic mechanisms. As noted, neurons in several regions of the CNS are known to express MCR. ACTH inhibited neuronal death induced by staurosporine, quinolinic acid and ROS induced by H_2O_2 as well as glutamate, including via NMDA, AMPA and kainate (Table 1). ACTH protected neurons from death induced by rapid release of NO but not slow release, the opposite of the findings in OPC. As with OL and OPC, ACTH failed to provide any protection from cell death induced by kynurenic acid [65].

In order to determine whether protection of OL was a result of direct effects of ACTH on OL, or whether astrocytes or microglia, which express MCR and are present in our mixed glial cultures, were potentially involved in protection of OL from the different toxic molecules, we undertook another series of experiments. Using highly purified OL cultures, we found that ACTH directly protected OL from staurosporine (apoptosis), H_2O_2 (ROS), glutamate including NMDA, AMPA and kainate, and quinolinic acid. As with the mixed glial cell cultures, OL in purified cultures were not protected from kynurenic acid or NO. Conditioned medium from astrocytes treated with ACTH was able to protect purified OL from glutamate, AMPA, quinolinic acid, and ROS but not from kainate, staurosporine, kynurenic acid or NO [98]. Thus, astrocytes may contribute to protection of OL from some but not all toxic molecules. Similar experiments with conditioned medium from microglia treated with ACTH failed to provide protection, suggesting that if microglia are also providing help in protecting OL from these molecules it is not through secreted molecules, but potentially could occur through cell-cell interactions.

In addition to MCR being important in inhibiting inflammation via stimulation of endogenous corticosteroid production, acting directly on cells of the immune system and providing protection for OL, OPC and neurons, we have been interested in the potential of MCR signaling to contribute to repair in the CNS in MS, as well as in other diseases of the CNS. Incubation of mixed glial cell cultures, containing mature OL, with ACTH resulted in striking extension of the OL membrane [97] (Figure 2). Incubation of OPC with ACTH resulted in both an increase in the rate of OPC proliferation and rate of maturation from OPC (expressing platelet derived growth factor alpha receptor; PDGFαR) to cells expressing both PDGFαR and O1, a marker of galactolipids (a phenotypic marker of mature OL) and to cells expressing only O1, i.e., mature OL. Since the OPC themselves, not being a clone but rather primary cultures, may be at different stages of maturation, this likely explains the effect of both enhanced OPC proliferation and maturation. An increase in the number of OPC and more rapid maturation of OPC into mature OL both have the potential to enhance repair by increasing remyelination.

8. Melanocortin Receptor Signaling in Oligodendroglial Protection

As described earlier, we reported that OL express MC4R in vitro and more recently, we have shown that OL also express MC1R, MC3R and MC5R but not MC2R [95]. Employing agonists and antagonists, we have found that MC1R, MC3R, MC4R and MC5R are functional in protection of OL from the toxic effects of staurosporine, glutamate, quinolinic acid and ROS. We have also shown that MC4R is functional in protecting OPC and stimulating their proliferation, but have not yet investigated the function of the other MCR subtypes in OPC. Additional studies employing other strategies including silencing RNA will be required to further characterize the relative roles of each of these receptors in signaling for protection as well as in OPC proliferation and maturation.

To further understand the role of MCR in protection of OL from cytotoxic mechanisms important in the pathogenesis of the MS lesion, we examined different signaling pathways activated by ACTH, which is known to bind and signal via all 5 MCR. To do this, we tested the ability of inhibitors of several intracellular signaling pathways to block the ACTH inhibition of toxicity of the different cytotoxic molecules. Purified OL cultures were incubated with inhibitors of PI3 kinase, MAP kinase (MAPK) and protein kinase C α,β isoforms (PKCα,β) followed by ACTH and the test molecules with the controls of the toxic molecules and toxic molecules and ACTH without prior incubation with the kinase pathway inhibitors [65,99]. PI3 kinase is used for ACTH protection from staurosporine (apoptosis), quinolinic acid (inflammation), glutamate including NMDA, AMPA and kainate (excitotoxicity). The MAP kinase pathway was also used for protection from staurosporine, and glutamate. Neither pathway was involved in ACTH induced protection from ROS. MCR are known to signal by activating adenylyl cyclase and upregulating intracellular cyclic adenosine monophosphate (cAMP). Inhibition of adenylyl cyclase prevented ACTH from protecting OL from the toxic effects of glutamate, quinolinic acid, ROS and staurosporine, demonstrating that in our system ACTH signaling involves activation of adenylyl cyclase [99]. PKCα,β inhibition did not block or enhance ACTH protection from any of the toxic molecules but inhibition of PKCα,β *per se* protected OL from the same molecules as ACTH, suggesting that cell death from those molecules involves the PKCα,β pathway or alternatively that inhibiting PKCα,β activates adenylyl cyclase.

9. Treatment of Human Neurologic Diseases with Melanocortins

ACTH has been used as treatment for a wide variety of non-neurological diseases including nephrotic syndrome, sarcoidosis, and rheumatologic disorders [4], but has been less explored for treatment of neurologic diseases. ACTH in a depo form is called ACTHar gel. It is prepared from pituitary extract and likely contains other peptides and melanocortins, including α-MSH as a breakdown product of ACTH. ACTH is used as treatment for West syndrome, which is characterized by infantile spasms and an EEG pattern referred to as hypsyrrythmia. In several studies, ACTH has been found to be more effective than corticosteroids [100–102], suggesting that ACTH may act, in part, independent of the ability to increase levels of endogenous corticosteroids. Studies have suggested that exogenous ACTH does not readily cross the blood CSF barrier but this may be different than the blood brain barrier (BBB), and CSF levels are what have been examined in patients; see Section 10, Future Directions. Endogenous ACTH appears to be lower in the CSF of patients with West syndrome but treatment with ACTH does not seem to increase concentrations of total CSF ACTH [103–105]. ACTH also stimulates production of the mineralocorticoid deoxycorticosterone by the adrenal cortex. This molecule can be metabolized to allotetrahyrodeoxycorticostereone, a neurosteroid that is known to cross the BBB [106]. In the case of activating brain stem neurons, the BBB is absent in parts of the brain stem. Direct effects on abnormally firing cortical neurons may be possible, since ACTH and other melanocortins are small polypeptides and the BBB is not fully developed in infants. There is also a report on higher levels of CSF corticosteroids in patients with opsoclonus myoclonus treated with ACTH than treated with corticosteroids [107].

ACTH is used for the treatment of relapses of MS and was the first agent found to be effective in shortening the duration of relapses [108–113]. It is now administered intramuscularly as ACTHar gel. ACTH has been mainly replaced by very high doses of corticosteroids administered intravenously or by mouth, resulting in much higher but shorter lived blood levels of corticosteroids when compared to blood levels resulting from ACTH [114]. In one head to head study, ACTH has been shown to be equally effective in reducing duration of relapses compared to corticosteroids [113]. In a small study, dexamethasone was superior to methylprednisolone and ACTH in shortening duration of relapses, but there were only 30 patients in that study [115]. More recently, a small randomized open label rater blinded study demonstrated that ACTH was more effective than intravenous methylprednisolone for relapses and had greater effects on plasma cytokines, when added to interferon beta [116]. ACTH as treatment for relapses is generally reserved for patients who are allergic to corticosteroids, develop psychosis with corticosteroid therapy or who fail to respond to treatment with corticosteroids. Whether the beneficial effect on relapses is due to corticosteroids, direct effects of ACTH on immune cells and/or effects on endogenous cells of the CNS is not clear and may well involve all of these mechanisms. There are no studies on ACTH entry into the CNS in any animal models but again ACTH and other melanocortins are relatively small molecules and in relapses it is clear that large proteins, including serum albumin and immunoglobulins (Ig) enter the CNS. ACTH and ACTH followed by prednisone were more effective in reducing CSF IgG synthesis rate than oral prednisone alone, dexamethasone or intrathecal hydrocortisone. However, oligoclonal bands persisted and there was no clinical effect in a group of patients who were in progressive stage of MS [117]. What is needed is a large study comparing the longer term effects of ACTH with corticosteroids for relapses using both clinical outcomes as well as MRI, VEP and OCT to determine if the use of ACTH, which has both steroidogenic and non-steroidogenic effects, is superior to corticosteroids for treatment of relapses. A study of secondary progressive MS (SPMS), for which there is currently no approved therapy other than the mitoxantrone with the problems of sterility, congestive heart failure and leukemia, would also be of interest, again using imaging as one of the outcomes. There are currently a large number of studies underway testing the effect of ACTH activation of MCR in different diseases including several neurologic disorders [4].

10. Future Studies

There is a need to further characterize the relative roles of MC1R, MC3R, MC4R and MC5R in vitro and in vivo so that one might develop new small molecules that bind with high avidity to the specific receptors most important for protection and repair. Silencing genes specific for each receptor and testing both the in vitro and in vivo effects would be a useful approach to this problem.

While there are many studies on the effects of ACTH on EAE, more studies are needed examining the effects on passive EAE and chronic EAE; the former to look at effects in MCR signaling limited to the effector stage of EAE and the latter to see if there are protective and reparative effects during a more chronic stage of EAE at a time the inflammatory phase of the disease with influx of cells from the peripheral immune system into the CNS is less marked. Chronic EAE is a better model for progressive phases of MS than acute EAE. In addition, studying the effects of natural melanocortins like ACTH and α-MSH on non-immune mediated models of demyelination including acute and chronic cuprizone poisoning and CNS lysolecithin injections would allow one to separate demyelination and remyelination from the role of exogenous inflammatory cells as in these models where the inflammation is basically activation of the endogenous microglia.

Studies using the available ACTH as ACTHar gel looking the effects at one and two years post-treatment of a relapse in comparison to corticosteroids employing advanced MRI metrics, MRS and/or OCT in a phase 2 study would be of interest and might help separate out the effects of melanocortins directly on the CNS and immune system from those of corticosteroids. A study of progressive MS, SPMS and/or PPMS should also be considered using the same type of metrics given the limited availability of highly effective treatments for PPMS and the lack of currently approved treatments for SPMS.

The issue of whether ACTH and perhaps even the smaller melanocortins can get through the BBB when there is not a major breakdown of the barrier and thus interact directly with MCR on glia and neurons deserves additional attention. The breakdown of the BBB in MS is likely underestimated since we know that using triple dose gadolinium compared to the single dose used for clinical MRI scans reveals many more enhancing lesions [118–120]. As pointed out, proteins considerably larger than ACTH and α-MSH do get through the BBB to a limited extent during periods of clinical stability and to a greater extent during relapses accompanied by gadolinium enhancing lesions on MRI scans. It is now well established that in patients with secondary progressive MS (SPMS) and primary progressive MS (PPMS), gadolinium enhancing lesions occur in as many as 13–25% of patients [120–125]. It should also be remembered that most studies of gadolinium enhancing lesions in patients with MS employ standard doses of gadolinium and often 1.5 Tesla magnets. It has been shown using triple dose gadolinium infusions and 3 Tesla magnets that routine single dose gadolinium and 1.0 or 1.5 Tesla magnets greatly underestimate the number of lesions with changes in the BBB [126,127]. There are differences in the BBB and the blood-CSF and blood-meningeal barriers and studies of ACTH within the CNS measure CSF levels, not brain/spinal cord or meningeal levels [128–135]. Thus, even with the rapid breakdown of the peptides, the amount of ACTH and α-MSH that is in the brain and spinal cord may be greater than might be suspected. Additional studies to examine this question are clearly needed.

Exosomes are known to cross the blood brain barrier, and the use of exosomes as a form of delivery has been suggested as a strategy for delivery of disease modifying treatments to the CNS [136–138] as has delivery employing nanotechnology [139–142]. In addition, the problem of breakdown of smaller melanocortins like α-MSH has been approached by modifying the peptide as with NDP-MSH [143], allowing the MSH to better stimulate MCR in vivo. Finally, development of non-peptide ligands for the different MCR that would readily cross the BBB would also have the potential to mediate protection and repair within the CNS in MS and other neurodegenerative diseases.

11. Conclusions

It is clear that signaling by melanocortins through their receptors in the CNS has potential for neuroprotection and repair in diseases like MS. This concept is reinforced by our published results showing that the MCR agonist ACTH protects OL, OPC and neurons from excitotoxic, apoptotic, oxidative and inflammation-related effects likely to play a role in CNS damage in MS and other neurodegenerative diseases. While effects on the immune system by direct effects on the circulating cells (lymphocytes and monocytes) and by signaling through CNS cells in regions lacking a mature BBB are clear, additional studies are needed to develop highly effective therapies that directly affect endogenous cells of the CNS, particularly OL, OPC and neurons.

Acknowledgments: Some of our previously published studies were supported in part by an Investigator Initiated Research Award (R.P.L.) from Mallinckrodt Pharmaceuticals, Autoimmune and Rare Diseases (formerly Questcor Pharmaceuticals), and the Parker Webber Chair in Neurology (DMC Foundation/Wayne State University) (R.P.L.).

Author Contributions: R.P.L. and J.A.B. contributed equally to this review.

Conflicts of Interest: Robert P. Lisak has received funding in the past for investigator-initiated research from Mallinckrodt Pharmaceuticals, Autoimmune and Rare Disorders (formerly Questcor Pharmaceuticals), Avanir and Teva Pharmaceuticals and has been a site investigator or site PI for clinical trials funded by Teva Pharmaceuticals, Novartis, Sanofi, Genentech, Accorda, Serono and Medimmune. In the last 2 years, he has served as a consultant for Syntimmune and has given lectures sponsored by Teva (non-drug related) and Chairs the Ajudication Committee for a clinical trial of biotin in secondary and primary progressive multiple sclerosis sponsored by PAREXEL and medDay Pharmaceuticals. Joyce A. Benjamins has received funding in the past for investigator-initiated research from Mallinckrodt Pharmaceuticals, Autoimmune and Rare Disorders (formerly Questcor Pharmaceuticals), Avanir and Teva Pharmaceuticals.

References

1. Adan, R.A.; Gispen, W.H. Brain melanocortin receptors: From cloning to function. *Peptides* **1997**, *18*, 1279–1287. [CrossRef]
2. Cone, R.D. Studies on the physiological functions of the melanocortin system. *Endocr. Rev.* **2006**, *27*, 736–749. [CrossRef] [PubMed]
3. Caruso, V.; Lagerstrom, M.C.; Olszewski, P.K.; Fredriksson, R.; Schioth, H.B. Synaptic changes induced by melanocortin signalling. *Nat. Rev. Neurosci.* **2014**, *15*, 98–110. [CrossRef] [PubMed]
4. Montero-Melendez, T. ACTH: The forgotten therapy. *Semin. Immunol.* **2015**, *27*, 216–226. [CrossRef] [PubMed]
5. Shen, W.J.; Yao, T.; Kong, X.; Williams, K.W.; Liu, T. Melanocortin neurons: Multiple routes to regulation of metabolism. *Biochim. Biophys. Acta* **2017**. [CrossRef] [PubMed]
6. Arnason, B.G.; Berkovich, R.; Catania, A.; Lisak, R.P.; Zaidi, M. Mechanisms of action of adrenocorticotropic hormone and other melanocortins relevant to the clinical management of patients with multiple sclerosis. *Mult. Scler.* **2013**, *19*, 130–136. [CrossRef] [PubMed]
7. Brzoska, T.; Luger, T.A.; Maaser, C.; Abels, C.; Bohm, M. Alpha-melanocyte-stimulating hormone and related tripeptides: Biochemistry, antiinflammatory and protective effects in vitro and in vivo, and future perspectives for the treatment of immune-mediated inflammatory diseases. *Endocr. Rev.* **2008**, *29*, 581–602. [CrossRef] [PubMed]
8. Catania, A.; Gatti, S.; Colombo, G.; Lipton, J.M. Targeting melanocortin receptors as a novel strategy to control inflammation. *Pharmacol. Rev.* **2004**, *56*, 1–29. [CrossRef] [PubMed]
9. Catania, A.; Lonati, C.; Sordi, A.; Carlin, A.; Leonardi, P.; Gatti, S. The melanocortin system in control of inflammation. *Sci. World J.* **2010**, *10*, 1840–1853. [CrossRef] [PubMed]
10. Muceniece, R.; Dambrova, M. Melanocortins in brain inflammation: The role of melanocortin receptor subtypes. *Adv. Exp. Med. Biol.* **2010**, *681*, 61–70. [PubMed]
11. Bertolini, A.; Tacchi, R.; Vergoni, A.V. Brain effects of melanocortins. *Pharmacol. Res.* **2009**, *59*, 13–47. [CrossRef] [PubMed]
12. Catania, A. Neuroprotective actions of melanocortins: A therapeutic opportunity. *Trends Neurosci.* **2008**, *31*, 353–360. [CrossRef] [PubMed]
13. Dores, R.M.; Londraville, R.L.; Prokop, J.; Davis, P.; Dewey, N.; Lesinski, N. Molecular evolution of GPCRs: Melanocortin/melanocortin receptors. *J. Mol. Endocrinol.* **2014**, *52*, T29–T42. [CrossRef] [PubMed]
14. Mountjoy, K.G. Distribution and Function of Melanocortin Receptors within the Brain. *Adv. Exp. Med. Biol.* **2010**, *681*, 29–48. [PubMed]
15. Schioth, H.B.; Haitina, T.; Ling, M.K.; Ringholm, A.; Fredriksson, R.; Cerda-Reverter, J.M.; Klovins, J. Evolutionary conservation of the structural, pharmacological, and genomic characteristics of the melanocortin receptor subtypes. *Peptides* **2005**, *26*, 1886–1900. [CrossRef] [PubMed]
16. Yang, Y. Structure, function and regulation of the melanocortin receptors. *Eur. J. Pharmacol.* **2011**, *660*, 125–130. [CrossRef] [PubMed]
17. Yang, Y.; Harmon, C.M. Molecular signatures of human melanocortin receptors for ligand binding and signaling. *Biochim. Biophys. Acta* **2017**. [CrossRef] [PubMed]

18. Switonski, M.; Mankowska, M.; Salamon, S. Family of melanocortin receptor (MCR) genes in mammals-mutations, polymorphisms and phenotypic effects. *J. Appl. Genet.* **2013**, *54*, 461–472. [CrossRef] [PubMed]

19. Abdel-Malek, Z.A.; Swope, V.B.; Starner, R.J.; Koikov, L.; Cassidy, P.; Leachman, S. Melanocortins and the melanocortin 1 receptor, moving translationally towards melanoma prevention. *Arch. Biochem. Biophys.* **2014**, *563*, 4–12. [CrossRef] [PubMed]

20. Demidowich, A.P.; Jun, J.Y.; Yanovski, J.A. Polymorphisms and mutations in the melanocortin-3 receptor and their relation to human obesity. *Biochim. Biophys. Acta* **2017**. [CrossRef] [PubMed]

21. Herraiz, C.; Garcia-Borron, J.C.; Jimenez-Cervantes, C.; Olivares, C. MC1R signaling.Intracellular partners and pathophysiological implications. *Biochim. Biophys. Acta* **2017**. [CrossRef]

22. Friedman, A.P. Do hyporesponsive genetic variants of the melanocortin 1 receptor contribute to the etiology of multiple sclerosis? *Med. Hypotheses* **2004**, *62*, 49–52. [CrossRef]

23. Partridge, J.M.; Weatherby, S.J.; Woolmore, J.A.; Highland, D.J.; Fryer, A.A.; Mann, C.L.; Boggild, M.D.; Ollier, W.E.; Strange, R.C.; Hawkins, C.P. Susceptibility and outcome in MS: Associations with polymorphisms in pigmentation-related genes. *Neurology* **2004**, *62*, 2323–2325. [CrossRef] [PubMed]

24. Strange, R.C.; Ramachandran, S.; Zeegers, M.P.; Emes, R.D.; Abraham, R.; Raveendran, V.; Boggild, M.; Gilford, J.; Hawkins, C.P. The Multiple Sclerosis Severity Score: Associations with MC1R single nucleotide polymorphisms and host response to ultraviolet radiation. *Mult. Scler.* **2010**, *16*, 1109–1116. [CrossRef] [PubMed]

25. Ding, Y.X.; Zou, L.P.; He, B.; Yue, W.H.; Liu, Z.L.; Zhang, D. ACTH receptor (MC2R) promoter variants associated with infantile spasms modulate MC2R expression and responsiveness to ACTH. *Pharmacogenet Genom.* **2010**, *20*, 71–76. [CrossRef] [PubMed]

26. Liu, Z.L.; He, B.; Fang, F.; Tang, C.Y.; Zou, L.P. Genetic polymorphisms of MC2R gene associated with responsiveness to adrenocorticotropic hormone therapy in infantile spasms. *Chin. Med. J.* **2008**, *121*, 1627–1632. [PubMed]

27. Farooqi, I.S.; Yeo, G.S.; Keogh, J.M.; Aminian, S.; Jebb, S.A.; Butler, G.; Cheetham, T.; O'Rahilly, S. Dominant and recessive inheritance of morbid obesity associated with melanocortin 4 receptor deficiency. *J. Clin. Investig.* **2000**, *106*, 271–279. [CrossRef] [PubMed]

28. Hinney, A.; Volckmar, A.L.; Knoll, N. Melanocortin-4 receptor in energy homeostasis and obesity pathogenesis. *Prog. Mol. Biol. Transl. Sci.* **2013**, *114*, 147–191. [PubMed]

29. Miller, C.L.; Murakami, P.; Ruczinski, I.; Ross, R.G.; Sinkus, M.; Sullivan, B.; Leonard, S. Two complex genotypes relevant to the kynurenine pathway and melanotropin function show association with schizophrenia and bipolar disorder. *Schizophr. Res.* **2009**, *113*, 259–267. [CrossRef] [PubMed]

30. Valli-Jaakola, K.; Suviolahti, E.; Schalin-Jantti, C.; Ripatti, S.; Silander, K.; Oksanen, L.; Salomaa, V.; Peltonen, L.; Kontula, K. Further evidence for the role of ENPP1 in obesity: Association with morbid obesity in Finns. *Obesity* **2008**, *16*, 2113–2119. [CrossRef] [PubMed]

31. Eves, P.C.; Haycock, J.W. Melanocortin signalling mechanisms. *Adv. Exp. Med. Biol.* **2010**, *681*, 19–28. [PubMed]

32. Rodrigues, A.R.; Almeida, H.; Gouveia, A.M. Intracellular signaling mechanisms of the melanocortin receptors: Current state of the art. *Cell. Mol. Life Sci.* **2015**, *72*, 1331–1345. [CrossRef] [PubMed]

33. Huang, H.; Wang, W.; Tao, Y.X. Pharmacological chaperones for the misfolded melanocortin-4 receptor associated with human obesity. *Biochim. Biophys. Acta* **2017**. [CrossRef] [PubMed]

34. Rouault, A.A.J.; Srinivasan, D.K.; Yin, T.C.; Lee, A.A.; Sebag, J.A. Melanocortin Receptor Accessory Proteins (MRAPs): Functions in the melanocortin system and beyond. *Biochim. Biophys. Acta* **2017**. [CrossRef] [PubMed]

35. Ericson, M.D.; Lensing, C.J.; Fleming, K.A.; Schlasner, K.N.; Doering, S.R.; Haskell-Luevano, C. Bench-top to clinical therapies: A review of melanocortin ligands from 1954 to 2016. *Biochim. Biophys. Acta* **2017**. [CrossRef] [PubMed]

36. Pantel, J.; Williams, S.Y.; Mi, D.; Sebag, J.; Corbin, J.D.; Weaver, C.D.; Cone, R.D. Development of a high throughput screen for allosteric modulators of melanocortin-4 receptor signaling using a real time cAMP assay. *Eur. J. Pharmacol.* **2011**, *660*, 139–147. [CrossRef] [PubMed]

37. O'Callaghan, K.; Kuliopulos, A.; Covic, L. Turning receptors on and off with intracellular pepducins: New insights into G-protein-coupled receptor drug development. *J. Biol. Chem.* **2012**, *287*, 12787–12796. [CrossRef] [PubMed]

38. Breit, A.; Buch, T.R.; Boekhoff, I.; Solinski, H.J.; Damm, E.; Gudermann, T. Alternative G protein coupling and biased agonism: New insights into melanocortin-4 receptor signalling. *Mol. Cell. Endocrinol.* **2011**, *331*, 232–240. [CrossRef] [PubMed]

39. Tao, Y.X. Constitutive activity in melanocortin-4 receptor: Biased signaling of inverse agonists. *Adv. Pharmacol.* **2014**, *70*, 135–154. [PubMed]

40. Clark, A.J.; Forfar, R.; Hussain, M.; Jerman, J.; McIver, E.; Taylor, D.; Chan, L. ACTH Antagonists. *Front. Endocrinol.* **2016**, *7*, 101. [CrossRef] [PubMed]

41. Ghaddhab, C.; Vuissoz, J.M.; Deladoey, J. From Bioinactive ACTH to ACTH Antagonist: The Clinical Perspective. *Front. Endocrinol.* **2017**, *8*, 17. [CrossRef] [PubMed]

42. Catania, A. The melanocortin system in leukocyte biology. *J. Leukoc. Biol.* **2007**, *81*, 383–392. [CrossRef] [PubMed]

43. Galimberti, D.; Baron, P.; Meda, L.; Prat, E.; Scarpini, E.; Delgado, R.; Catania, A.; Lipton, J.M.; Scarlato, G. Alpha-MSH peptides inhibit production of nitric oxide and tumor necrosis factor-alpha by microglial cells activated with beta-amyloid and interferon gamma. *Biochem. Biophys. Res. Commun.* **1999**, *263*, 251–256. [CrossRef] [PubMed]

44. Lipton, J.M.; Catania, A. Mechanisms of antiinflammatory action of the neuroimmunomodulatory peptide alpha-MSH. *Ann. N. Y. Acad. Sci.* **1998**, *840*, 373–380. [CrossRef] [PubMed]

45. Taylor, A.; Namba, K. In vitro induction of CD25+ CD4+ regulatory T cells by the neuropeptide alpha-melanocyte stimulating hormone (alpha-MSH). *Immunol. Cell Biol.* **2001**, *79*, 358–367. [CrossRef] [PubMed]

46. Taylor, A.W.; Yee, D.G.; Nishida, T.; Namba, K. Neuropeptide regulation of immunity. The immunosuppressive activity of alpha-melanocyte-stimulating hormone (alpha-MSH). *Ann. N. Y. Acad. Sci.* **2000**, *917*, 239–247. [CrossRef] [PubMed]

47. Reder, A.; Birnbaum, G. B-cell differentiation in multiple sclerosis and the effect of intravenous ACTH. *Neurology* **1983**, *33*, 442–446. [CrossRef] [PubMed]

48. Taylor, A.W.; Kitaichi, N.; Biros, D. Melanocortin 5 receptor and ocular immunity. *Cell. Mol. Biol.* **2006**, *52*, 53–59. [PubMed]

49. Biros, D.J.; Namba, K.; Taylor, A.W. Alpha-MSH regulates protein ubiquitination in T cells. *Cell. Mol. Biol.* **2006**, *52*, 33–38. [PubMed]

50. Getting, S.J.; Christian, H.C.; Flower, R.J.; Perretti, M. Activation of melanocortin type 3 receptor as a molecular mechanism for adrenocorticotropic hormone efficacy in gouty arthritis. *Arthritis Rheum.* **2002**, *46*, 2765–2775. [CrossRef] [PubMed]

51. Gautron, L.; Lee, C.; Funahashi, H.; Friedman, J.; Lee, S.; Elmquist, J. Melanocortin-4 receptor expression in a vago-vagal circuitry involved in postprandial functions. *J. Comp. Neurol.* **2010**, *518*, 6–24. [CrossRef] [PubMed]

52. Giuliani, D.; Ottani, A.; Neri, L.; Zaffe, D.; Grieco, P.; Jochem, J.; Cavallini, G.M.; Catania, A.; Guarini, S. Multiple beneficial effects of melanocortin MC4 receptor agonists in experimental neurodegenerative disorders: Therapeutic perspectives. *Prog. Neurobiol.* **2017**, *148*, 40–56. [CrossRef] [PubMed]

53. Gee, C.E.; Chen, C.L.; Roberts, J.L.; Thompson, R.; Watson, S.J. Identification of proopiomelanocortin neurones in rat hypothalamus by in situ cDNA-mRNA hybridization. *Nature* **1983**, *306*, 374–376. [CrossRef]

54. Kishi, T.; Aschkenasi, C.J.; Lee, C.E.; Mountjoy, K.G.; Saper, C.B.; Elmquist, J.K. Expression of melanocortin 4 receptor mRNA in the central nervous system of the rat. *J. Comp. Neurol.* **2003**, *457*, 213–235. [CrossRef] [PubMed]

55. Mountjoy, K.G.; Mortrud, M.T.; Low, M.J.; Simerly, R.B.; Cone, R.D. Localization of the melanocortin-4 receptor (MC4-R) in neuroendocrine and autonomic control circuits in the brain. *Mol. Endocrinol.* **1994**, *8*, 1298–1308. [PubMed]

56. Spaccapelo, L.; Bitto, A.; Galantucci, M.; Ottani, A.; Irrera, N.; Minutoli, L.; Altavilla, D.; Novellino, E.; Grieco, P.; Zaffe, D.; et al. Melanocortin MC(4) receptor agonists counteract late inflammatory and apoptotic responses and improve neuronal functionality after cerebral ischemia. *Eur. J. Pharmacol.* **2011**, *670*, 479–486. [CrossRef] [PubMed]

57. Wan, S.; Browning, K.N.; Coleman, F.H.; Sutton, G.; Zheng, H.; Butler, A.; Berthoud, H.R.; Travagli, R.A. Presynaptic melanocortin-4 receptors on vagal afferent fibers modulate the excitability of rat nucleus tractus solitarius neurons. *J. Neurosci.* **2008**, *28*, 4957–4966. [CrossRef] [PubMed]

58. Williams, D.L.; Kaplan, J.M.; Grill, H.J. The role of the dorsal vagal complex and the vagus nerve in feeding effects of melanocortin-3/4 receptor stimulation. *Endocrinology* **2000**, *141*, 1332–1337. [CrossRef] [PubMed]

59. Wang, H.; Yu, M.; Ochani, M.; Amella, C.A.; Tanovic, M.; Susarla, S.; Li, J.H.; Yang, H.; Ulloa, L.; Al-Abed, Y.; et al. Nicotinic acetylcholine receptor alpha7 subunit is an essential regulator of inflammation. *Nature* **2003**, *421*, 384–388. [CrossRef] [PubMed]

60. Catania, A.; Lipton, J.M. Peptide modulation of fever and inflammation within the brain. *Ann. N. Y. Acad. Sci.* **1998**, *856*, 62–68. [CrossRef] [PubMed]

61. Tanida, M.; Shintani, N.; Hashimoto, H. The melanocortin system is involved in regulating autonomic nerve activity through central pituitary adenylate cyclase-activating polypeptide. *Neurosci. Res.* **2011**, *70*, 55–61. [CrossRef] [PubMed]

62. Rossi, J.; Balthasar, N.; Olson, D.; Scott, M.; Berglund, E.; Lee, C.E.; Choi, M.J.; Lauzon, D.; Lowell, B.B.; Elmquist, J.K. Melanocortin-4 receptors expressed by cholinergic neurons regulate energy balance and glucose homeostasis. *Cell Metab.* **2011**, *13*, 195–204. [CrossRef] [PubMed]

63. Agosti, F.; Cordisco Gonzalez, S.; Martinez Damonte, V.; Tolosa, M.J.; Di Siervi, N.; Schioth, H.B.; Davio, C.; Perello, M.; Raingo, J. Melanocortin 4 receptor constitutive activity inhibits L-type voltage-gated calcium channels in neurons. *Neuroscience* **2017**, *346*, 102–112. [CrossRef] [PubMed]

64. Giuliani, D.; Neri, L.; Canalini, F.; Calevro, A.; Ottani, A.; Vandini, E.; Sena, P.; Zaffe, D.; Guarini, S. NDP-alpha-MSH induces intense neurogenesis and cognitive recovery in Alzheimer transgenic mice through activation of melanocortin MC4 receptors. *Mol. Cell. Neurosci.* **2015**, *67*, 13–21. [CrossRef] [PubMed]

65. Lisak, R.P.; Nedelkoska, L.; Bealmear, B.; Benjamins, J.A. Melanocortin receptor agonist ACTH 1–39 protects rat forebrain neurons from apoptotic, excitotoxic and inflammation-related damage. *Exp. Neurol.* **2015**, *273*, 161–167. [CrossRef] [PubMed]

66. Caruso, C.; Carniglia, L.; Durand, D.; Scimonelli, T.N.; Lasaga, M. Astrocytes: New targets of melanocortin 4 receptor actions. *J. Mol. Endocrinol.* **2013**, *51*, R33–R50. [CrossRef] [PubMed]

67. Caruso, C.; Durand, D.; Schioth, H.B.; Rey, R.; Seilicovich, A.; Lasaga, M. Activation of melanocortin 4 receptors reduces the inflammatory response and prevents apoptosis induced by lipopolysaccharide and interferon-gamma in astrocytes. *Endocrinology* **2007**, *148*, 4918–4926. [CrossRef] [PubMed]

68. Zohar, M.; Salomon, Y. Melanocortins stimulate proliferation and induce morphological changes in cultured rat astrocytes by distinct transducing mechanisms. *Brain Res.* **1992**, *576*, 49–58. [CrossRef]

69. Ramirez, D.; Saba, J.; Carniglia, L.; Durand, D.; Lasaga, M.; Caruso, C. Melanocortin 4 receptor activates ERK-cFos pathway to increase brain-derived neurotrophic factor expression in rat astrocytes and hypothalamus. *Mol. Cell. Endocrinol.* **2015**, *411*, 28–37. [CrossRef] [PubMed]

70. Carniglia, L.; Durand, D.; Caruso, C.; Lasaga, M. Effect of NDP-alpha-MSH on PPAR-gamma and -beta expression and anti-inflammatory cytokine release in rat astrocytes and microglia. *PLoS ONE* **2013**, *8*, e57313. [CrossRef] [PubMed]

71. Delgado, R.; Carlin, A.; Airaghi, L.; Demitri, M.T.; Meda, L.; Galimberti, D.; Baron, P.; Lipton, J.M.; Catania, A. Melanocortin peptides inhibit production of proinflammatory cytokines and nitric oxide by activated microglia. *J. Leukoc. Biol.* **1998**, *63*, 740–745. [PubMed]

72. Carniglia, L.; Ramirez, D.; Durand, D.; Saba, J.; Caruso, C.; Lasaga, M. [Nle4, D-Phe7]-alpha-MSH Inhibits Toll-Like Receptor (TLR)2- and TLR4-Induced Microglial Activation and Promotes a M2-Like Phenotype. *PLoS ONE* **2016**, *11*, e0158564. [CrossRef] [PubMed]

73. Mohammadi, R.; Yadegarazadi, M.J.; Amini, K. Peripheral nerve regeneration following transection injury to rat sciatic nerve by local application of adrenocorticotropic hormone. *J. Craniomaxillofac. Surg.* **2014**, *42*, 784–789. [CrossRef] [PubMed]

74. Teare, K.A.; Pearson, R.G.; Shakesheff, K.M.; Haycock, J.W. Alpha-MSH inhibits inflammatory signalling in Schwann cells. *Neuroreport* **2004**, *15*, 493–498. [CrossRef] [PubMed]

75. Dyer, J.K.; Philipsen, H.L.; Tonnaer, J.A.; Hermkens, P.H.; Haynes, L.W. Melanocortin analogue Org2766 binds to rat Schwann cells, upregulates NGF low-affinity receptor p75, and releases neurotrophic activity. *Peptides* **1995**, *16*, 515–522. [CrossRef]

76. Van der Zee, C.E.; Brakkee, J.H.; Gispen, W.H. alpha-MSH and Org.2766 in peripheral nerve regeneration: Different routes of delivery. *Eur. J. Pharmacol.* **1988**, *147*, 351–357. [CrossRef]
77. Lisak, R.; Kies, M. Experimental allergic encephalomyelitis as a tool for evaluating immunsuppressant activity of drugs. In *Immunopharmacology*; Rosenthale, M., Mansmann, H., Jr., Eds.; Spectrum Publications: New York, NY, USA, 1976; pp. 173–185.
78. Steinman, L.; Zamvil, S.S. Virtues and pitfalls of EAE for the development of therapies for multiple sclerosis. *Trends Immunol.* **2005**, *26*, 565–571. [CrossRef] [PubMed]
79. Moyer, A.W.; Jervis, G.A.; Black, J.; Koprowski, H.; Cox, H.R. Action of adrenocorticotropic hormone (ACTH) in experimental allergic encephalomyelitis of the guinea pig. *Proc. Soc. Exp. Biol. Med.* **1950**, *75*, 387–390. [CrossRef] [PubMed]
80. Cusick, M.F.; Libbey, J.E.; Oh, L.; Jordan, S.; Fujinami, R.S. Acthar gel treatment suppresses acute exacerbations in a murine model of relapsing-remitting multiple sclerosis. *Autoimmunity* **2015**, *48*, 222–230. [CrossRef] [PubMed]
81. Taylor, A.W.; Kitaichi, N. The diminishment of experimental autoimmune encephalomyelitis (EAE) by neuropeptide alpha-melanocyte stimulating hormone (alpha-MSH) therapy. *Brain Behav. Immun.* **2008**, *22*, 639–646. [CrossRef] [PubMed]
82. Yin, P.; Luby, T.M.; Chen, H.; Etemad-Moghadam, B.; Lee, D.; Aziz, N.; Ramstedt, U.; Hedley, M.L. Generation of expression constructs that secrete bioactive alphaMSH and their use in the treatment of experimental autoimmune encephalomyelitis. *Gene Ther.* **2003**, *10*, 348–355. [CrossRef] [PubMed]
83. Duckers, H.J.; Verhaagen, J.; de Bruijn, E.; Gispen, W.H. Effective use of a neurotrophic ACTH4-9 analogue in the treatment of a peripheral demyelinating syndrome (experimental allergic neuritis). An intervention study. *Brain* **1994**, *117*, 365–374. [CrossRef] [PubMed]
84. Fang, J.; Han, D.; Hong, J.; Zhang, H.; Ying, Y.; Tian, Y.; Zhang, L.; Lin, J. SValpha-MSH, a novel alpha-melanocyte stimulating hormone analog, ameliorates autoimmune encephalomyelitis through inhibiting autoreactive CD4(+) T cells activation. *J. Neuroimmunol.* **2014**, *269*, 9–19. [CrossRef] [PubMed]
85. Mykicki, N.; Herrmann, A.M.; Schwab, N.; Deenen, R.; Sparwasser, T.; Limmer, A.; Wachsmuth, L.; Klotz, L.; Köhrer, K.; Faber, C.; et al. Melanocortin-1 receptor activation is neuroprotective in mouse models of neuroinflammatory disease. *Sci. Transl. Med.* **2016**, *8*, 362ra146. [CrossRef] [PubMed]
86. Forslin Aronsson, A.; Spulber, S.; Oprica, M.; Winblad, B.; Post, C.; Schultzberg, M. Alpha-MSH rescues neurons from excitotoxic cell death. *J. Mol. Neurosci.* **2007**, *33*, 239–251. [CrossRef] [PubMed]
87. Gatti, S.; Lonati, C.; Acerbi, F.; Sordi, A.; Leonardi, P.; Carlin, A.; Gaini, S.M.; Catania, A. Protective action of NDP-MSH in experimental subarachnoid hemorrhage. *Exp. Neurol.* **2012**, *234*, 230–238. [CrossRef] [PubMed]
88. Schaible, E.V.; Steinsträsser, A.; Jahn-Eimermacher, A.; Luh, C.; Sebastiani, A.; Kornes, F.; Pieter, D.; Schafer, M.K.; Engelhard, K.; Thal, S.C. Single administration of tripeptide alpha-MSH(11–13) attenuates brain damage by reduced inflammation and apoptosis after experimental traumatic brain injury in mice. *PLoS ONE* **2013**, *8*, e71056. [CrossRef] [PubMed]
89. Bijlsma, W.A.; Schotman, P.; Jennekens, F.G.; Gispen, W.H.; De Wied, D. The enhanced recovery of sensorimotor function in rats is related to the melanotropic moiety of ACTH/MSH neuropeptides. *Eur. J. Pharmacol.* **1983**, *92*, 231–236. [CrossRef]
90. Gispen, W.H.; Verhaagen, J.; Bar, D. ACTH/MSH-derived peptides and peripheral nerve plasticity: Neuropathies, neuroprotection and repair. *Prog. Brain Res.* **1994**, *100*, 223–229. [PubMed]
91. Bar, P.R.; Mandys, V.; Turecek, R.; Gispen, W.H. Alpha-melanocyte-stimulating hormone has protective properties against the toxic effect of cisplatin on cultured dorsal root ganglia. *Ann. N. Y. Acad. Sci.* **1993**, *680*, 649–651. [CrossRef] [PubMed]
92. Chai, B.; Li, J.Y.; Zhang, W.; Newman, E.; Ammori, J.; Mulholland, M.W. Melanocortin-4 receptor-mediated inhibition of apoptosis in immortalized hypothalamic neurons via mitogen-activated protein kinase. *Peptides* **2006**, *27*, 2846–2857. [CrossRef] [PubMed]
93. Bar, P.R.; Hol, E.M.; Gispen, W.H. Trophic effects of melanocortins on neuronal cells in culture. *Ann. N. Y. Acad. Sci.* **1993**, *692*, 284–286. [CrossRef] [PubMed]
94. Joosten, E.A.; Verhaagh, S.; Martin, D.; Robe, P.; Franzen, R.; Hooiveld, M.; Doornbos, R.; Bar, P.R.; Moonen, G. Alpha-MSH stimulates neurite outgrowth of neonatal rat corticospinal neurons in vitro. *Brain Res* **1996**, *736*, 91–98. [CrossRef]

95.   Benjmains, J.; Nedelkoska, L.; Lisak, R. Melanocortin receptor subtypes are expressed on cells in the oligodendroglial lineage and signal ACTH protection. *J. Neurosci. Res.* **2017**, in press.

96.   Benjamins, J.A.; Nedelkoska, L.; Bealmear, B.; Lisak, R.P. ACTH protects mature oligodendroglia from excitotoxic and inflammation-related damage in vitro. *Glia* **2013**, *61*, 1206–1217. [CrossRef] [PubMed]

97.   Benjamins, J.A.; Nedelkoska, L.; Lisak, R.P. Adrenocorticotropin hormone 1–39 promotes proliferation and differentiation of oligodendroglial progenitor cells and protects from excitotoxic and inflammation-related damage. *J. Neurosci. Res.* **2014**, *92*, 1243–1251. [CrossRef] [PubMed]

98.   Lisak, R.P.; Nedelkoska, L.; Benjamins, J.A. The melanocortin ACTH 1–39 promotes protection of oligodendrocytes by astroglia. *J. Neurol. Sci.* **2016**, *362*, 21–26. [CrossRef] [PubMed]

99.   Lisak, R.; Nedellkoska, L.; Benjamins, J. Melanocortin receptor ACTH 1–39 may protect oigodendroglia by inhibiting prtein kinase C. In Proceedings of the Annual Meeting of the American Academy of Neurology, Vancouver, BC, Canada, 15–21 April 2016.

100.   Baram, T.Z.; Mitchell, W.G.; Tournay, A.; Snead, O.C.; Hanson, R.A.; Horton, E.J. High-dose corticotropin (ACTH) versus prednisone for infantile spasms: A prospective, randomized, blinded study. *Pediatrics* **1996**, *97*, 375–379. [PubMed]

101.   Stafstrom, C.E.; Arnason, B.G.; Baram, T.Z.; Catania, A.; Cortez, M.A.; Glauser, T.A.; Pranzatelli, M.R.; Riikonen, R.; Rogawski, M.A.; Shinnar, S.; et al. Treatment of infantile spasms: Emerging insights from clinical and basic science perspectives. *J. Child Neurol.* **2011**, *26*, 1411–1421. [CrossRef] [PubMed]

102.   Shumiloff, N.A.; Lam, W.M.; Manasco, K.B. Adrenocorticotropic hormone for the treatment of West Syndrome in children. *Ann. Pharmacother.* **2013**, *47*, 744–754. [CrossRef] [PubMed]

103.   Nalin, A.; Facchinetti, F.; Galli, V.; Petraglia, F.; Storchi, R.; Genazzani, A.R. Reduced ACTH content in cerebrospinal fluid of children affected by cryptogenic infantile spasms with hypsarrhythmia. *Epilepsia* **1985**, *26*, 446–449. [CrossRef] [PubMed]

104.   Baram, T.Z.; Mitchell, W.G.; Snead, O.C., 3rd; Horton, E.J.; Saito, M. Brain-adrenal axis hormones are altered in the CSF of infants with massive infantile spasms. *Neurology* **1992**, *42*, 1171–1175. [CrossRef] [PubMed]

105.   Nagamitsu, S.; Matsuishi, T.; Yamashita, Y.; Shimizu, T.; Iwanaga, R.; Murakami, Y.; Miyazaki, M.; Hashimoto, T.; Kato, H. Decreased cerebrospinal fluid levels of beta-endorphin and ACTH in children with infantile spasms. *J. Neural Transm.* **2001**, *108*, 363–371. [CrossRef] [PubMed]

106.   Reddy, D.S.; Rogawski, M.A. Stress-induced deoxycorticosterone-derived neurosteroids modulate GABA(A) receptor function and seizure susceptibility. *J. Neurosci.* **2002**, *22*, 3795–3805. [PubMed]

107.   Pranzatelli, M.R.; Chun, K.Y.; Moxness, M.; Tate, E.D.; Allison, T.J. Cerebrospinal fluid ACTH and cortisol in opsoclonus-myoclonus: Effect of therapy. *Pediatr. Neurol.* **2005**, *33*, 121–126. [CrossRef] [PubMed]

108.   Glaser, G.H.; Merritt, H.H. Effects of ACTH and cortisone in multiple sclerosis. *Trans. Am. Neurol. Assoc.* **1951**, *56*, 130–133. [PubMed]

109.   Miller, H.; Newell, D.J.; Ridley, A. Multiple sclerosis. Treatment of acute exacerbations with corticotrophin (A.C.T.H.). *Lancet* **1961**, *2*, 1120–1122. [CrossRef]

110.   Filippini, G.; Brusaferri, F.; Sibley, W.A.; Citterio, A.; Ciucci, G.; Midgard, R.; Candelise, L. Corticosteroids or ACTH for acute exacerbations in multiple sclerosis. *Cochrane Database Syst. Rev.* **2000**. [CrossRef]

111.   Rose, A.S.; Kuzma, J.W.; Kurtzke, J.F.; Namerow, N.S.; Sibley, W.A.; Tourtellotte, W.W. Cooperative study in the evaluation of therapy in multiple sclerosis. ACTH vs. placebo—Final report. *Neurology* **1970**, *20*, 1–59. [PubMed]

112.   Rose, A.S.; Kuzma, J.W.; Kurtzke, J.F.; Sibley, W.A.; Tourtellotte, W.W. Cooperative study in the evaluation of therapy in multiple sclerosis; ACTH vs. placebo in acute exacerbations. Preliminary report. *Neurology* **1968**, *18*, 1–10.

113.   Thompson, A.J.; Kennard, C.; Swash, M.; Summers, B.; Yuill, G.M.; Shepherd, D.I.; Roche, S.; Perkin, G.D.; Loizou, L.A.; Ferner, R.; et al. Relative efficacy of intravenous methylprednisolone and ACTH in the treatment of acute relapse in MS. *Neurology* **1989**, *39*, 969–971. [CrossRef] [PubMed]

114.   Lal, R.; Bell, S.; Challenger, R.; Hammock, V.; Nyberg, M.; Decker, D.; Becker, P.M.; Young, D. Pharmacodynamics and tolerability of repository corticotropin injection in healthy human subjects: A comparison with intravenous methylprednisolone. *J. Clin. Pharmacol.* **2016**, *56*, 195–202. [CrossRef] [PubMed]

115. Milanese, C.; La Mantia, L.; Salmaggi, A.; Campi, A.; Eoli, M.; Scaioli, V.; Nespolo, A.; Corridori, F. Double-blind randomized trial of ACTH versus dexamethasone versus methylprednisolone in multiple sclerosis bouts. Clinical, cerebrospinal fluid and neurophysiological results. *Eur. Neurol.* **1989**, *29*, 10–14. [CrossRef] [PubMed]

116. Berkovich, R.; Bakshi, R.; Amezcua, L.; Axtell, R.C.; Cen, S.Y.; Tauhid, S.; Neema, M.; Steinman, L. Adrenocorticotropic hormone versus methylprednisolone added to interferon beta in patients with multiple sclerosis experiencing breakthrough disease: A randomized, rater-blinded trial. *Ther. Adv. Neurol. Disord.* **2017**, *10*, 3–17. [CrossRef] [PubMed]

117. Tourtellotte, W.W.; Baumhefner, R.W.; Potvin, A.R.; Ma, B.I.; Potvin, J.H.; Mendez, M.; Syndulko, K. Multiple sclerosis de novo CNS IgG synthesis: Effect of ACTH and corticosteroids. *Neurology* **1980**, *30*, 1155–1162. [CrossRef] [PubMed]

118. Tortorella, C.; Codella, M.; Rocca, M.A.; Gasperini, C.; Capra, R.; Bastianello, S.; Filippi, M. Disease activity in multiple sclerosis studied by weekly triple-dose magnetic resonance imaging. *J. Neurol.* **1999**, *246*, 689–692. [CrossRef] [PubMed]

119. Cadavid, D.; Wolansky, L.J.; Skurnick, J.; Lincoln, J.; Cheriyan, J.; Szczepanowski, K.; Kamin, S.S.; Pachner, A.R.; Halper, J.; Cook, S.D. Efficacy of treatment of MS with IFNbeta-1b or glatiramer acetate by monthly brain MRI in the BECOME study. *Neurology* **2009**, *72*, 1976–1983. [CrossRef] [PubMed]

120. Cook, S.D.; Dhib-Jalbut, S.; Dowling, P.; Durelli, L.; Ford, C.; Giovannoni, G.; Halper, J.; Harris, C.; Herbert, J.; Li, D.; et al. Use of Magnetic Resonance Imaging as Well as Clinical Disease Activity in the Clinical Classification of Multiple Sclerosis and Assessment of Its Course: A Report from an International CMSC Consensus Conference, March 5–7, 2010. *Int. J. MS Care* **2012**, *14*, 105–114. [CrossRef] [PubMed]

121. Filippi, M.; Campi, A.; Martinelli, V.; Colombo, B.; Yousry, T.; Canal, N.; Scotti, G.; Comi, G. Comparison of triple dose versus standard dose gadolinium-DTPA for detection of MRI enhancing lesions in patients with primary progressive multiple sclerosis. *J. Neurol. Neurosurg. Psychiatry* **1995**, *59*, 540–544. [CrossRef] [PubMed]

122. Wolinsky, J.S.; Narayana, P.A.; O'Connor, P.; Coyle, P.K.; Ford, C.; Johnson, K.; Miller, A.; Pardo, L.; Kadosh, S.; Ladkani, D. Glatiramer acetate in primary progressive multiple sclerosis: Results of a multinational, multicenter, double-blind, placebo-controlled trial. *Ann. Neurol.* **2007**, *61*, 14–24. [CrossRef] [PubMed]

123. Hawker, K.; O'Connor, P.; Freedman, M.S.; Calabresi, P.A.; Antel, J.; Simon, J.; Hauser, S.; Waubant, E.; Vollmer, T.; Panitch, H.; et al. Rituximab in patients with primary progressive multiple sclerosis: Results of a randomized double-blind placebo-controlled multicenter trial. *Ann. Neurol.* **2009**, *66*, 460–471. [CrossRef] [PubMed]

124. Lublin, F.; Miller, D.H.; Freedman, M.S.; Cree, B.A.; Wolinsky, J.S.; Weiner, H.; Lubetzki, C.; Hartung, H.P.; Montalban, X.; Uitdehaag, B.M.; et al. Oral fingolimod in primary progressive multiple sclerosis (INFORMS): A phase 3, randomised, double-blind, placebo-controlled trial. *Lancet* **2016**, *387*, 1075–1084. [CrossRef]

125. Montalban, X.; Hauser, S.L.; Kappos, L.; Arnold, D.L.; Bar-Or, A.; Comi, G.; de Seze, J.; Giovannoni, G.; Hartung, H.P.; Hemmer, B.; et al. Ocrelizumab versus Placebo in Primary Progressive Multiple Sclerosis. *N. Engl. J. Med.* **2017**, *376*, 209–220. [CrossRef] [PubMed]

126. Stankiewicz, J.M.; Glanz, B.I.; Healy, B.C.; Arora, A.; Neema, M.; Benedict, R.H.; Guss, Z.D.; Tauhid, S.; Buckle, G.J.; Houtchens, M.K.; et al. Brain MRI lesion load at 1.5T and 3T versus clinical status in multiple sclerosis. *J. Neuroimaging* **2011**, *21*, e50–e56. [CrossRef] [PubMed]

127. Rocca, M.; Gerevini, S.; Filippi, M.; Falini, A. HIgh-field strength MRI (3.0T or more) in white matter diseases. In *High Field Brain MRI. Use in Clincal Practice*; Scarabino, T., Pollice, S., Popolizio, T., Eds.; Springer International PUblishing: Basel, Switzerland, 2017.

128. Vorbrodt, A.W.; Lassmann, H.; Wisniewski, H.M.; Lossinsky, A.S. Ultracytochemical studies of the blood-meningeal barrier (BMB) in rat spinal cord. *Acta Neuropathol.* **1981**, *55*, 113–123. [CrossRef] [PubMed]

129. Angelov, D.N. Ultrastructural investigation of the meningeal compartment of the blood-cerebrospinal fluid-barrier in rats and cats. A horseradish peroxidase study. *Z. Mikrosk. Anat. Forsch.* **1990**, *104*, 1–16. [PubMed]

130. Zheng, W.; Zhao, Q.; Graziano, J.H. Primary culture of choroidal epithelial cells: Characterization of an in vitro model of blood-CSF barrier. *In Vitro Cell. Dev. Biol. Anim.* **1998**, *34*, 40–45. [CrossRef] [PubMed]

131. Drewes, L.R. What is the blood-brain barrier? A molecular perspective. Cerebral vascular biology. *Adv. Exp. Med. Biol.* **1999**, *474*, 111–122. [PubMed]

132. Prat, A.; Biernacki, K.; Wosik, K.; Antel, J.P. Glial cell influence on the human blood-brain barrier. *Glia* **2001**, *36*, 145–155. [CrossRef] [PubMed]

133. Dore-Duffy, P. Pericytes: Pluripotent cells of the blood brain barrier. *Curr. Pharm. Des.* **2008**, *14*, 1581–1593. [CrossRef] [PubMed]

134. Benarroch, E.E. Blood-brain barrier: Recent developments and clinical correlations. *Neurology* **2012**, *78*, 1268–1276. [CrossRef] [PubMed]

135. Monnot, A.D.; Zheng, W. Culture of choroid plexus epithelial cells and in vitro model of blood-CSF barrier. *Methods Mol. Biol.* **2013**, *945*, 13–29. [PubMed]

136. Chaput, N.; Thery, C. Exosomes: Immune properties and potential clinical implementations. *Semin. Immunopathol.* **2011**, *33*, 419–440. [CrossRef] [PubMed]

137. Pusic, A.; Lusic, K.; Kraig, R. What are exosomes and how can they be used in multiple sclerosis therapy? *Expert Rev. Neurother.* **2014**, *14*, 353–355. [CrossRef] [PubMed]

138. Selmaj, I.; Mycko, M.P.; Raine, C.S.; Selmaj, K.W. The role of exosomes in CNS inflammation and their involvement in multiple sclerosis. *J. Neuroimmunol.* **2017**, *306*, 1–10. [CrossRef] [PubMed]

139. Kreuter, J. Nanoparticulate systems for brain delivery of drugs. *Adv. Drug Deliv. Rev.* **2001**, *47*, 65–81. [CrossRef]

140. Kreuter, J.; Shamenkov, D.; Petrov, V.; Ramge, P.; Cychutek, K.; Koch-Brandt, C.; Alyautdin, R. Apolipoprotein-mediated transport of nanoparticle-bound drugs across the blood-brain barrier. *J. Drug Target.* **2002**, *10*, 317–325. [CrossRef] [PubMed]

141. Gelperina, S.E.; Khalansky, A.S.; Skidan, I.N.; Smirnova, Z.S.; Bobruskin, A.I.; Severin, S.E.; Turowski, B.; Zanella, F.E.; Kreuter, J. Toxicological studies of doxorubicin bound to polysorbate 80-coated poly(butyl cyanoacrylate) nanoparticles in healthy rats and rats with intracranial glioblastoma. *Toxicol. Lett.* **2002**, *126*, 131–141. [CrossRef]

142. LaVan, D.A.; McGuire, T.; Langer, R. Small-scale systems for in vivo drug delivery. *Nat. Biotechnol.* **2003**, *21*, 1184–1191. [CrossRef] [PubMed]

143. Sawyer, T.K.; Castrucci, A.M.; Staples, D.J.; Affholter, J.A.; De Vaux, A.; Hruby, V.J.; Hadley, M.E. Structure-activity relationships of [Nle4, D-Phe7]alpha-MSH. Discovery of a tripeptidyl agonist exhibiting sustained bioactivity. *Ann. N. Y. Acad. Sci.* **1993**, *680*, 597–599. [CrossRef]

**brain
sciences**

MDPI

Review

Pattern Recognition of the Multiple Sclerosis Syndrome

Rana K. Zabad [1,*] [ORCID]**, Renee Stewart [2] and Kathleen M. Healey [1]**

[1] Department of Neurological Sciences, University of Nebraska Medical Center College of Medicine, Omaha, NE 68198-8440, USA; khealey@unmc.edu

[2] University of Nebraska Medical Center College of Nursing, Omaha, NE 68198-5330, USA; renee.stewart@unmc.edu

* Correspondence: rzabad@unmc.edu; Tel.: +1-402-559-6591

Received: 29 August 2017; Accepted: 17 October 2017; Published: 24 October 2017

Abstract: During recent decades, the autoimmune disease neuromyelitis optica spectrum disorder (NMOSD), once broadly classified under the umbrella of multiple sclerosis (MS), has been extended to include autoimmune inflammatory conditions of the central nervous system (CNS), which are now diagnosable with serum serological tests. These antibody-mediated inflammatory diseases of the CNS share a clinical presentation to MS. A number of practical learning points emerge in this review, which is geared toward the pattern recognition of optic neuritis, transverse myelitis, brainstem/cerebellar and hemispheric tumefactive demyelinating lesion (TDL)-associated MS, aquaporin-4-antibody and myelin oligodendrocyte glycoprotein (MOG)-antibody NMOSD, overlap syndrome, and some yet-to-be-defined/classified demyelinating disease, all unspecifically labeled under *MS syndrome*. The goal of this review is to increase clinicians' awareness of the clinical nuances of the autoimmune conditions for MS and NMSOD, and to highlight highly suggestive patterns of clinical, paraclinical or imaging presentations in order to improve differentiation. With overlay in clinical manifestations between MS and NMOSD, magnetic resonance imaging (MRI) of the brain, orbits and spinal cord, serology, and most importantly, high index of suspicion based on pattern recognition, will help lead to the final diagnosis.

Keywords: MS; NMOSD; clinically isolated syndrome (CIS); optic neuritis; transverse myelitis; brainstem syndrome; tumefactive demyelinating lesions; AQP4 antibodies; MOG antibodies

1. Introduction

A multiple sclerosis (MS) diagnosis is at the forefront when a woman or man aged 20–50 years presents neurological symptoms and/or white matter lesions on magnetic resonance imaging (MRI) of the brain. Although MS remains the most common etiology for inflammatory demyelinating diseases of the central nervous system (CNS), the autoimmune disease neuromyelitis optica spectrum disorder (NMOSD) is a major differential diagnosis. The discovery of autoantibodies, such as aquaporin 4-IgG (AQP4-IgG) followed by myelin oligodendrocyte glycoprotein-IgG (MOG-IgG), and likely more to come [1,2], has further broadened the differential diagnosis of inflammatory demyelinating diseases. This is with the understanding that AQP4-antibody-associated NMOSD is frequently added to primarily inflammatory demyelinating diseases, although it is an astrocytopathy followed by oligodendrocytopathy and demyelination [3,4]. With more literature being published on MS and NMOSD, pattern recognition emerges. Pattern recognition not only affects the clinical manifestations of MS and NMOSD, such as recognizing the spectrum of optic neuritis, transverse myelitis, and brainstem syndrome, but also affects MRI findings in the brain, brainstem, spinal cord and the orbits. This review focuses on pattern recognition of these clinical presentations therefore our descriptive designation as the MS syndrome.

2. Brief Historical Overview of Multiple Sclerosis (MS) Diagnosis

The first diagnostic criteria for MS were introduced by Allison and Millar in 1954, followed by McAlpine in 1965. That same year, the Schumacher Committee formally published the first MS diagnostic criteria, heralding a half-century of intense research in the field of MS diagnosis, prognosis, pathophysiology, immunopathology, and treatment [5,6]. Due to the absence of a gold standard for unequivocally diagnosing MS, such as blood or cerebrospinal fluid (CSF) tests, the patterns of dissemination in time (DIT) (i.e., progression in time for primary progressive disease) and dissemination in space (DIS) have been considered diagnostic of the disease. These patterns at first relied on clinical data, limited paraclinical criteria [5,7], and subsequently on MRI [8–11]. Since the publication of the first McDonald criteria in 2001 [8], these diagnostic criteria have undergone numerous modifications but the criteria of DIS and DIT by clinical and/or MRI remain paramount to the diagnosis (Supplementary Material Table S1). Today, MRI of the brain and spinal cord is used to diagnose and prognosticate MS pre- and post-treatment. The emergence of disease-modifying therapies, with proven effectiveness in clinically isolated syndrome and MS, has called for further refinement of MRI criteria with exceptional sensitivity, specificity, and accuracy, thus allowing for an earlier diagnosis of the disease. Nevertheless, confusion of other inflammatory demyelinating diseases with MS remains problematic, particularly with practitioners who do not commonly see demyelinating diseases.

3. Overview of Neuromyelitis Optica Spectrum Disorder

The presence of a longitudinally extensive transverse myelitis (LETM) typically alerts the neurologist to the diagnosis of NMOSD, which is confirmed by testing positive for the neuromyelitis optica or the APQ4 antibody [12]. However, short segment spinal cord lesions (SSSCLs), that might not be unusual in early [13] and seronegative NMOSD can be easily confused with MS. Clinical presentation with bilateral simultaneous or sequential optic neuritis, with or without transverse myelitis, is highly suggestive of NMOSD. However, longitudinally extensive optic neuritis (LEON) might be overlooked because of the lack of routine use of MRI for the orbits in the diagnosis of optic neuritis. The differential diagnosis of a large edematous corpus callosal lesion is broad, and includes lymphomas, tumors, trauma, infections, metabolic (Marchiafava-Bignami) and vascular abnormalities, to cite a few [14], but the pattern is increasingly recognized in NMOSD (Table S2 and Figure S3a,b) [15,16]. Area postrema syndrome (Figure S4a,b), another core clinical presentation of NMOSD can be easily mistaken for a gastrointestinal illness in the hands of non-neurologists. Because of the pleomorphic presentation of demyelination and its variable outcome, there is a lack of unanimity between MS/NMOSD experts. A study by Jurynczyk et al. evaluated the agreement between different MS and NMOSD experts on the diagnosis of the seronegative AQP4-antibody NMOSD, MS and overlapping syndrome. Not surprisingly, the mean proportion of agreement for the diagnosis was low ($\rho_0 = 0.51$) and ranged from 0.25 to 0.73 for individual patients. Clinical presentations associated with very low agreement ($\rho_0 < 0.5$) included optic neuritis with limited recovery and short transverse myelitis, mild optic neuritis with short transverse myelitis and normal brain MRI, optic neuritis and borderline LETM, optic neuritis and transverse myelitis with brain lesions not fully typical of MS or NMO, and monophasic acute disseminated encephalomyelitis (ADEM)-like with optic neuritis and LETM [17].

Brain and spinal cord MRI have a significant role in differentiating MS from NMOSD, but there remains a group with demyelinating disease where the separation remains challenging. For instance, Barkhof's criteria for DIS have been fulfilled by 5–42% of patients with NMOSD [18–21]. This MRI overlap between MS and NMOSD extends to both the AQP4 antibody and MOG antibody-associated NMOSD [22]. Clinical, imaging and differentiating patterns that suggest and support NMOSD are examined below. A historical overview, pathophysiology, and pathogenicity of two important biomarkers that can differentiate the two conditions are included. Additionally, brain MRI findings characteristic of NMOSD are summarized in Supplementary Material Table S2 [18–20,23–29], Figures S1–S5 and Figures 1–4.

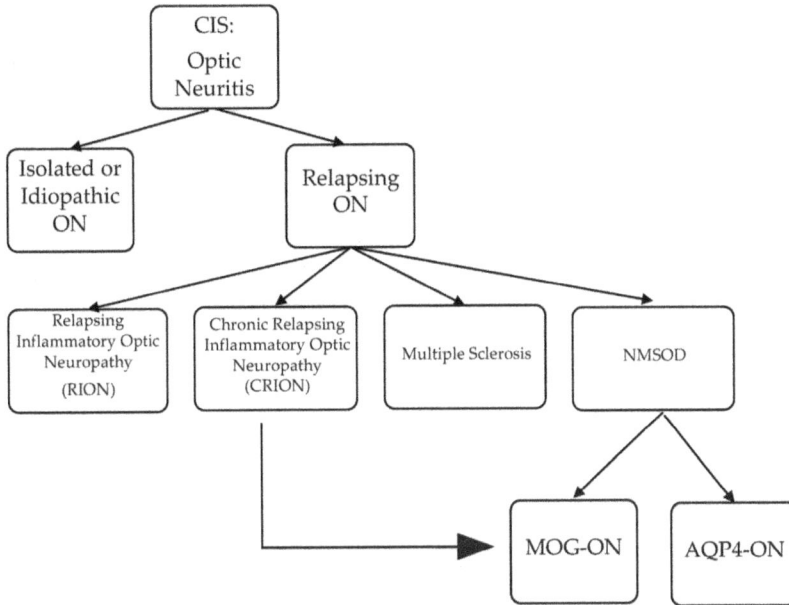

Figure 1. Evolution of clinically isolated syndrome (CIS) optic neuritis: The majority of patients with ON will evolve into RION, CRION, MS or NMOSD. A percentage of patients will stay as isolated ON. Furthermore, there is more data that at least a subset of patients with CRION are indeed MOG-antibody associated NMOSD. AQP4-ON: Aquaporin 4-antibody-associated optic neuritis; MOG-ON: Myelin oligodendrocyte glycoprotein-antibody-associated optic neuritis.

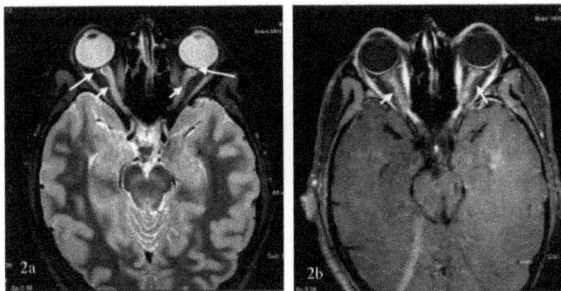

Figure 2. Axial short tau inversion recovery (STIR and T1 with contrast orbital MRI of an 18-year-old Caucasian male, with bilateral longitudinally extensive (small arrows) optic neuritis with anterior predominance, perineural sheath swelling (long arrows) and tilting and twisting of both optic nerves better seen on axial STIR (**2a,2b**). Serum MOG-antibody was positive.

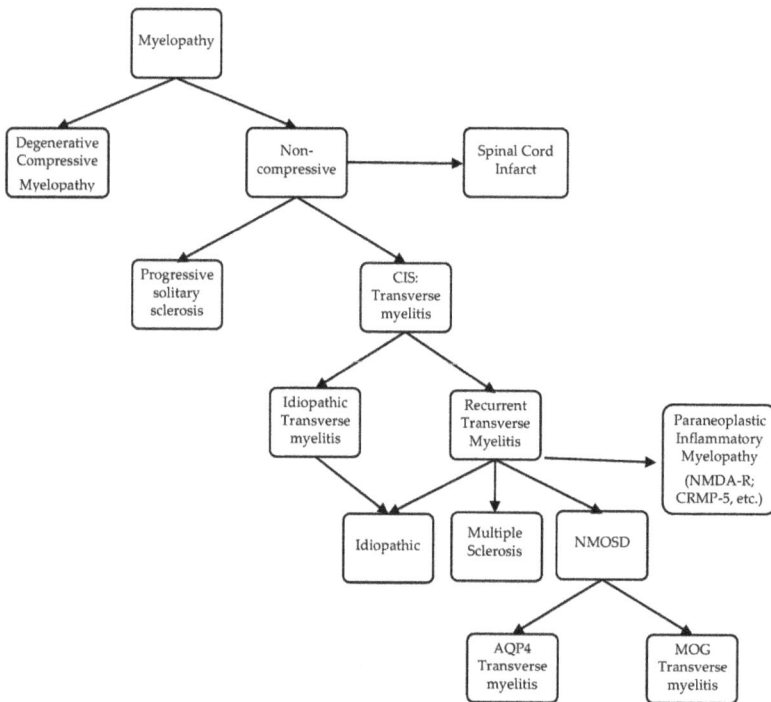

Figure 3. Evolution of clinically isolated syndrome (CIS) transverse myelitis.

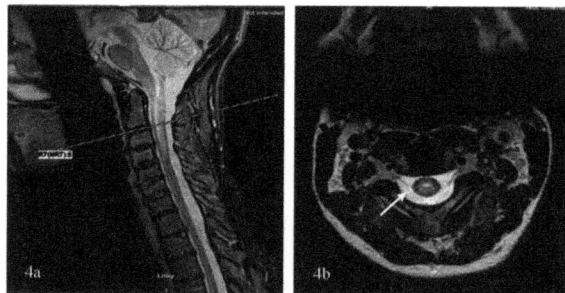

Figure 4. 55-year-old African-American female, with AQP4-NMOSD; sagittal STIR cervical MRI demonstrates simultaneous linear lesion and LETM from the medulla to C4 (**4a**) seen at the center of the cord on axial T2 (**4b**).

3.1. AQP4-Antibody Positive NMOSD: Pathophysiology and Pathogenicity of Aquaporin 4 Neuromyelitis Optica Spectrum Disorder (AQP4-NMOSD)

In 1999, Wingerchuk et al. described the clinical, MRI, and CSF features of 71 patients with NMO, with emphasis on the severity of the disease [30]. A B-cell-mediated pathology was suspected due to the association of NMO with autoantibodies and B-cell-mediated diseases. In 2004, Lennon et al. described a new antibody, NMO-IgG, localizing to the blood brain barrier (BBB) that was a specific serological biomarker of NMO and high-risk syndromes suggestive of NMO, such as LETM and recurring severe optic neuritis. A subsequent study by the same group demonstrated that the water channel aquaporin 4 (AQP4) was the substrate for the NMO-IgG [31]. To clarify, different types of aquaporins have been involved in water homeostasis in the brain and were associated with vasogenic and cytotoxic edema. APQs are comprised of highly conserved monomers or units that form homotetramers. Each unit, or monomer, has eight membrane-embedded domains, six transmembrane helices, and two short helical segments with a C- and N-terminus on the cytoplasmic side. These membrane-embedded domains surround a narrow aqueous pore [32]. On the extracellular side, there are three loops (i.e., A, C and E), and on the intracellular side, there are two loops (i.e., B and D). There is also a highly conserved asparagine-proline-alanine motif responsible for the selective orientation of the water transportation and an aromatic/arginine (AR/R) selectivity filter that prevents the entry of other molecules with water across the water channel [33]. An integral protein of the astrocytic plasma membrane [34], human AQP4 is expressed by astrocytes, and other AQP4-containing cells, by alternative splicing in the following two major isoforms: a long isoform called "M1", and a short isoform called "M23". In general, a highly homologous structure is characteristic of all members of the AQP family [35]. In the case of AQP4, however, M1-AQP4 and M23-AQP4 form heterotetramers that further aggregate in the cell plasma membrane in supramolecular crystalline assemblies called an orthogonal array of particles (OAPs). The size, shape, and composition of OAPs depend on the relative amounts of M1- vs. M23-AQP4, with larger particles formed at an increased M23:M1 ratio [36].

AQP4 is highly concentrated in the foot processes that make contact with micropapillary endothelia that form the BBB and in ependymal cells at brain cerebrospinal interfaces [34]. AQP4 is upregulated during astrocytosis and certain scar-forming pathologies but is considerably reduced in NMO. Other AQP4-expressing tissues include epithelial cells in kidneys, airways, gastrointestinal organs, and, at low levels, in musculoskeletal cells; thus, the most recent reported cases of acute myopathies involve AQP4 associated with NMOSD [37–39]. The AQP4 antibody binds to the extracellular surface of the AQP4 receptor in the three-dimensional form of the epitopes rather than their linear form, a pattern that is typical in human autoimmune disorders [34]. Despite its polyclonal production, the APQ4 antibody shows preferential binding and greater affinity to OAPs and thus to M23-AQP4. The binding of AQP4-IgG1 to AQP4 leads to complement-dependent cytotoxicity [36] and antibody-dependent cellular cytotoxicity [40]. Efficient complement-dependent cytotoxicity requires AQP4 assembly in OAPs, and therefore, this mechanism is minimal for M1-AQP4-expressing cells. Importantly, high concentrations of AQP4-IgG were reported not to inhibit AQP4 water permeability and not to lead to cellular internalization of AQP4 or AQP4-antibody binding [41]. In vivo consequences of AQP4 binding to the AQP4 antibody results in complement dependent cytotoxicity axonal injury followed by recruitment of granulocytes first and macrophages second, further disrupting the BBB [34]. Astrocyte loss and inflammation, with degranulation of neutrophils and eosinophils, and cytokine release culminate into secondary damage to the oligodendrocytes, with demyelination and neuronal/axonal loss. Thus, AQP4-NMOSD is not primarily a demyelinating disorder, but is nevertheless lumped under the MS syndrome due to clinical phenotypic similarities. In 2006, the NMO diagnostic criteria were updated to incorporate patients with NMO who had extra optico-spinal disease and NMO-IgG as a biomarker. Almost a decade later, newer NMOSD guidelines were published [42]. During that decade, numerous studies were published regarding (1) other clinical and inaugural manifestations of the disease, (2) best diagnostic techniques for the antibody, currently the approved technique is the cell-based essay, (3) pathogenicity of NMO IgG, and (4) brain, spinal cord, and orbit MRI findings of the disease, to cite a few. Attempts took place to find other suspected antibodies, resulting in the anti-myelin oligodendrocyte glycoprotein antibodies (MOG antibodies). It transpires that MOG antibody is not only associated with NMOSD (MOG-NMOSD), but also other inflammatory demyelinating disorders, such as pediatric acute disseminated encephalomyelitis (ADEM), pediatric multiphasic disseminated encephalomyelitis (MDEM), ADEM/MDEM-optic neuritis complex, benign unilateral cerebral cortical encephalitis with epilepsy and overlap syndrome or NMOSD-encephalitis complex that will be described later. The quest for further antibodies (such as AQP1, NMDA-R antibodies, etc.) associated with NMOSD and other inflammatory demyelinating disorders remains a work in progress and the future holds more antibodies to come [1,2].

3.2. Pathophysiology and Pathogenicity of Myelin Oligodendrocyte Glycoprotein (MOG) Antibodies

MOG is a mammalian glycoprotein exclusively expressed in the CNS. This glycoprotein is limited to the external surface of myelin and the plasma membranes of oligodendrocytes, with its highest antigen density in the outermost lamellae of myelin sheaths, thus making MOG accessible to autoantibodies [43]. MOG belongs to the Ig superfamily, with a single extracellular immunoglobulin variable (IgV) domain, a transmembrane domain, a cytoplasmic loop, a membrane-associated region, and a cytoplasmic tail. In humans, 15 different alternatively spliced MOG isoforms have been detected. These isoforms have been localized to the cell surface, in the endoplasmic reticulum, in the endocytic system, or found in secreted form. The secreted form could have important effects in triggering autoimmunity if released into the CSF and then drained into the bloodstream [44]. MOG antibodies isolated from animal models of MS target a denatured MOG protein. Similar autoantibodies (both IgG and IgM) for denatured proteins were present in MS patients with low titers and did not correlate with disease activity [45–47]. Importantly, the presence of cell-based assays has allowed for the isolation and quantification of MOG antibodies against the native or conformational epitope of MOG (nMOG), located on the extracellular domain of the protein [48]. Anti-MOG antibodies are likely relevant to the pathophysiology of MS, considering that they are present in early states of the disease and are not an epiphenomenon. Anti-MOG antibodies also persist during the disease course and are likely relevant to long-term pathophysiology. Brilot et al. investigated the occurrence and biological activity of IgG and IgM autoantibodies against nMOG in the serum and CSF of 47 children (mean age 7.63 years) during their first acute demyelinating syndrome (19 with ADEM and 28 with clinically isolated syndrome). The serum and CSF of these children were taken at the same time and prior to any treatment. Control groups included healthy children, children with other neurological diseases, children with type I diabetes mellitus, and adult MS patients. Native MOG antibodies were present in 47% of children with a demyelinating event (ADEM or clinically isolated syndrome), 6.9% of children with other neurological diseases, and absent in healthy controls as well as adults with MS and children with type I diabetes mellitus. The presence of MOG antibodies in pediatric demyelination and other neurological diseases and their absence in type I diabetes, highlights that these antibodies are markers of demyelination and not immune dysregulation. Native MOG antibodies were produced peripherally and in the CNS. The serum and CSF was simultaneously analyzed for a cohort of eight children (five with clinically isolated syndrome and three with ADEM). All patients, except three with clinically isolated syndrome, showed reduced antibody titers in the CSF. Native MOG antibodies were cytotoxic in demyelinating patients using an in vitro antibody-dependent cellular cytotoxicity assay. Furthermore, native MOG antibody titers inversely correlated with age ($r = -0.46$), suggesting a temporal evolution of the MOG antibody [49]. This possible temporal evolution was subsequently studied in 78 pediatric cases of CNS disease (27 with ADEM, 18 with clinically isolated syndrome, 18 with relapsing-remitting MS, and 15 with other general neurological diseases), 188 adult cases (71 with MS, 43 with other general neurological diseases, 20 with clinically isolated syndrome, and 7 with ADEM), and 43 healthy controls. Increased MOG antibodies serum titers were observed in pediatric ADEM, and these increased titers were associated with a younger age of onset. Recovery from ADEM was associated with a decrease in MOG antibody titers at last follow-up, with seroreversion in one patient. Incomplete recovery from ADEM was associated with a persistently increased antibody titer and reduced fluctuations in titer levels. Seroreversion was observed in patients with clinically isolated syndrome. Longitudinal analysis of nine patients with MS revealed stably low titers; three nevertheless became seropositive with time, with persistently low titers, which indicated ongoing CNS inflammation. Antibodies against nMOG were present in the CSF when serum titers were high, suggesting a peripheral production of MOG antibodies [50]. The persistence of MOG antibodies during remission suggests that, in isolation, these antibodies may be insufficient for disease activity [51]. Lately, MOG-antibody associated ADEM was reported in 2 adults. This further highlights the fact that the clinical spectrum of MOG-antibody associated diseases remains to be defined [52].

4. Optic Neuritis: From Clinically Isolated Syndrome to MS, NMOSD and Others

Although MS remains the most common etiology of ON, other etiologies are possible and summarized in Figure 1.

4.1. Single Inflammatory Optic Neuritis (SION), Relapsing Inflammatory Optic Neuritis (RION), and Chronic Relapsing Inflammatory Optic Neuropathy (CRION): "Formes frustes" of MS or NMOSD?

The term idiopathic [53] or isolated optic neuritis is somewhat loosely used in the literature to refer to optic neuritis without evidence of MS, NMOSD, or other diseases. The diagnosis of idiopathic or isolated optic neuritis cannot be made with certainty when a patient first presents with ON. Idiopathic/isolated optic neuritis is frequently limited to one episode [54] and referred to in the literature as single inflammatory/isolated optic neuritis (SION) or monophasic isolated optic neuritis [55]. However, isolated optic neuritis might recur outside the context of MS and NMOSD; while consensus regarding their presence is lacking, the following two forms of relapsing optic neuritis have been reported in the literature: relapsing inflammatory or isolated optic neuritis (RION) [55] and chronic relapsing inflammatory optic neuropathy (CRION). The existence of RION is debatable, considering that the conversion to MS, NMOSD, or other diseases may only be a matter of time. However, more studies on RION are being published. For example, analyzing the clinical and demographic criteria for 62 patients with RION, Benoilid et al. found that 40 patients (64.5%) did not convert to MS, NMO, or other autoimmune diseases over eight years of follow-up [56]. Furthermore, the natural history of RION was studied in 72 patients with two or more episodes of optic neuritis. Specifically, in a study by Pirko et al., the one-, five- and 10-year conversion rate of optic neuritis to MS was 2.8%, 14.4%, and 29.8%, respectively, and the conversion rate of optic neuritis to NMO was 5.6%, 12.5%, and 12.5%, respectively [57]. Predictors of RION converting to NMO included decreased visual acuity, shorter time to second relapse, more frequent relapses, and a significant female predilection. This study published in 2004 did not differentiate between AQP4- or MOG-antibody associated NMOSD.

CRION is differentiated from RION by the presence of progressive visual loss in between relapses and corticosteroid dependence, albeit there is no consensus regarding this latter criterion [56]. In 2003, the first case series of eight patients with CRION was published [58]. Subsequent reports in the literature were compiled in a systematic review of the clinical, laboratory, and imaging characteristics of 122 patients with CRION [59]. The age range for CRION is wide, spanning teenage to elderly years. Unilateral or bilateral, and simultaneous or sequential vision loss has also been reported with fellow eye involvement occurring days to decades later. The relapse rate for CRION is highly variable. Like all optic neuritis, pain and/or headache herald the condition and can be sleep disruptive. Visual loss at onset is variable from none to complete. The optic disc may be normal, swollen, or atrophic. Findings on visual field testing are variable, similar to MS-ON. Interestingly, uveitis has been reported conjointly in approximately 7% of cases [60]. Therefore, extensive blood testing should be performed to rule out systemic diseases. Notably, CSF analysis is typically normal. MRI of the orbits findings vary from normal to the presence swelling at the optic nerve head and contrast enhancement, T2 hyperintensity, and/or optic atrophy. Brain MRI is typically normal [59]. Recently, a number of patients with CRION or RION were found to have the MOG antibody [43,51,61,62]. In summary, the most notable distinguishing characteristics between RION and CRION available to date remains the progression in between relapses and steroid responsiveness, recognizing the lack of agreement on the definition. Furthermore, isolated monophasic and recurrent optic neuritis seem to exist as stand-alone entities. Lastly, time will tell whether all CRION cases are MOG-antibody associated optic neuritis or not.

4.2. Multiple Sclerosis-Associated Optic Neuritis (MS-ON)

MS-ON is a common presentation of MS in approximately 20% of patients [63]. The lifetime prevalence of MS-ON is 50–66% [64–67]. Clinical presentation of retro-orbital, peri-orbital, or oculomotor pain followed by subacute and varying degrees of visual loss in a young person are hallmarks of optic neuritis. Confounding factors leading to delayed diagnosis include minimal to no visual loss with decreased color sensitivity only on examination, the absence of pain, and the presence of positive visual symptoms in a person with or without a prior history of migraines. While painful visual loss appears to be the hallmark of optic neuritis, pain has also been reported in 12% of patients with anterior ischemic optic neuropathy [68]. Further, much information on inflammatory optic neuropathy results from the optic neuritis treatment trial [69]. In a 1992 study by Beck et al., 457 patients were randomly assigned to placebo, oral prednisone at a dose of 1 mg/kg/d for two weeks or high-dose intravenous methylprednisolone (IVMP) for three days followed by an oral taper. The patients were monitored for six months; 77.2% of the subjects were women. Pain, which was present in more than 92% of the cases, preceded the visual symptoms and was unrelated to the presence or absence of optic disc swelling. Visual acuity loss was almost equally distributed between mild ($\leq$20/40), moderate (20/50–20/190), and severe ($\geq$20/200). Complete visual loss was present in 10% of patients. Noticeably, even when visual acuity was 20/20 or better, many patients had other abnormalities, such as decreased contrast sensitivity and/or abnormal color vision or unusual visual field tests. Fellow eye involvement was observed with decreased visual acuity and contrast sensitivity in 14% and 15% of patients, respectively, and abnormal color vision and visual defects were observed in 20% and 48% of patients, respectively. Disc swelling was noted in 25–40% of patients, depending on the time of the exam following the onset of symptoms, <5 days or $\geq$5 days. Positive visual symptoms, such as photopsias, were observed in 30% of patients. Notably, the pattern of painful visual loss and photopsias can be easily confused with migraines. The prolonged duration of visual loss and photopsias should alert the clinician to an alternative diagnosis considering that it is uncommon for a migraine aura to last for days and persists beyond the pain [70]. Visual field cut in patients with optic neuritis was not only centrocecal and central, but also paracentral, altitudinal, quadrantic, hemianopic, peripheral, arcuate or double arcuate, enlarged blind spot, nasal, and vertical step. Visual recovery, including acuity, contrast sensitivity, color vision and field-testing, was faster in the IVMP group. At six months, visual acuity was the same in the three groups, but the difference in low-contrast sensitivity, color vision, and field-testing persisted between the IVMP and the other groups. Only 5–7% of the patients in all groups had visual acuity of 20/50 or worse. Furthermore, only one out of 457 cases had a compressive optic neuropathy due to a pituitary tumor diagnosed by MRI. Based on these findings, the authors of the optic neuritis treatment trial did not recommend an MRI of the brain to diagnose optic neuritis unless an atypical course alerts the clinician for further imaging study [71].

4.3. Neuromyelitis Optica Spectrum Disorder-Associated Optic Neuritis

Optic neuritis is a presenting sign of NMOSD more than 50% of the time. However, the difference between MS-ON and NMOSD-ON might not be evident with the first attack of optic neuritis if the brain MRI is negative. Fortunately, an orbital MRI performed early and prior to any treatment can facilitate diagnosis and is commonly abnormal in NMOSD-ON cases [53]. Orbital MRI, however, was not performed in the optic neuritis treatment trial [72], although the technology was available [73]. A 2016 study found that the MRI of the orbits was less likely to be performed in MS-suspected optic neuritis when a neurologist, rather than an ophthalmologist, saw the patient [74]. MRI of the spinal cord might foretell the diagnosis by further disseminating the patient in space (i.e., DIS). However, if no orbital MRI is completed, and there are no further signs of DIS, the diagnosis is obscured a priori. It has been suggested that antibody testing for AQP4-IgG be reserved to patients with severe visual loss, poor visual recovery, bilateral or sequential visual involvement, recurrent optic neuritis [75] and unique findings on the MRI of the orbits. The efficiency of MRI of the brain and anterior visual pathways for differentiating NMOSD-ON from MS-ON was retrospectively examined in a study [76]. Brain MRI was included to examine the ability of DIS per the 2010 McDonald criteria [9] to differentiate between the two conditions. The absence of brain DIS, longer optic nerve lesions, an increased number of segments involved, and optic chiasma and tracts involvement were suggestive of NMOSD-ON. Further, Buch et al. examined the sensitivity and specificity for a combination of these criteria. Specifically, bilateral optic nerve, chiasma or optic tracts, and three or more segment involvement in the absence of MS-like lesions with DIS were suggestive of NMOSD-ON, with a sensitivity of 69% and a specificity of 97% [76]. More specifically, a longitudinally extensive optic neuritis (LEON) with chiasma and optic tract involvement was found to be strongly suggestive of AQP4-ON [77]. Not all LEON, however, are associated with AQP4 antibody. The test sensitivity of AQP4 antibody in the serum is about 50–80%, and MOG antibodies have been described in 25% of cases of AQP4 antibody seronegative NMOSD and cases of CRION [55]. In a study of optic neuritis by Akaishi et al., the cross-sectional prevalence of AQP4-ON, MOG-ON, MS-ON, and RION was 30–35%, 25–30%, 25–30%, and 10–15%, respectively [74]. Here, Akaishi et al. reported a comparative study of the clinical, laboratory, and imaging findings of MS-ON, AQP4-ON, MOG-ON, and RION. Female dominance was overwhelming in the AQP4 group, at 98% vs 80% in the MS group and 50% in the MOG group, which is different from other reports [43,51]. While the mean age of onset of MS-ON was less than 50 years for the entire cohort, patients with AQP4-ON had greater mean age of onset. A broader age distribution existed in the MOG-ON group, with optic neuritis diagnosed in children and the elderly similar to MOG-NMOSD. The number of optic nerve segments involved during the acute phase of optic neuritis in all groups was assessed. To clarify, the optic nerve has been divided into the following six segments anteriorly to posteriorly: (1) pre-orbital, (2) retro-orbital, (3) canalicular, (4) intracranial, (5) chiasmatic, and (6) retrochiasmatic or optic tract portion [74,77]. On MRI of the orbits, MOG-ON showed longitudinally extensive contrast enhancement, with severe swelling and a twisted running. The inflammation was anterior, with 70–80% intraorbital perineurial contrast enhancement. An example of MOG-ON is shown in Figure 2.

AQP4-ON was longitudinally extensive, with greater posterior involvement including the canalicular, chiasmatic, and retrochiasmatic segments, but with milder swelling and rare twisting. The MS-ON contrast enhancement was less extensive, with a median of only a couple of segments involved [74], as confirmed by others [77]. Optic nerve head swelling has been observed clinically with MOG-ON [78]. Significant decrease in visual acuity was associated with AQP4-ON followed by MOG-ON and MS-ON. In both the MOG-ON and the MS-ON groups, visual acuity loss during a relapse and the long-term outcome past one year were similar, but they were worse in AQP4-ON group. The less severe prognosis of MOG-ON was confirmed in a study by Matsuda et al., who also showed that the residual deficit was commonly present due to an increased number of relapses per year [62]. A 2017 study by Stiebel-Kalish et al. compared the visual acuity, field defect, and thickness of the retinal nerve fiber layer over time between a group of MOG-ON and AQP4-ON. In the MOG-ON group, the final visual acuity, mean visual field defect, and retinal nerve fiber layers were preserved, while adjusting for the number of relapses [79]. In a separate study by Havla et al., the optical coherence tomography analysis of eight patients with MOG-ON demonstrated a reduced papillary retinal nerve fiber layer compared to MS-ON; this study also revealed the presence of microcytic macular edema in six patients with MOG-ON and in two patients with AQP4-ON. Fellow eye was also affected in MOG-ON [80]. The favorable long-term prognosis of MOG-ON was not replicated in one of the largest cohort of patients ($n = 50$) with MOG-NMOSD [81]. Although short-term visual acuity was improved in patients with MOG-ON, this long-term outcome was not confirmed compared to AQP4-ON, a discrepancy that was explained by the increased number of relapses and the lack of corticosteroid use in some of these patients with MOG-ON. Regardless, a study by Piccolo et al. found severe visual acuity loss at onset and at last follow-up in five-eighths of patients with MOG-ON [55]. Recently, worse vision-related quality of life in both AQP4- and MOG-NMOSD than in MS patients was reported, steered by patients with bilateral and severe ON in the NMOSD group. Additionally, OCT, visual function and vision-related QOL parameters were similar in AQP4- and MOG-NMOSD groups [82]. Overall, there is convergence of data that visual outcome from MOG-ON is not as favorable as MS-ON, but nevertheless the visual outcome from MOG-ON is more favorable than AQP4-ON. Similar findings were recently reported in 12 Chinese Han patients [83]. The exquisite steroid sensitivity of MOG-ON, reminiscent of that observed with CRION, raises the possibility that CRION could be a manifestation of the MOG-inflammatory demyelinating disease spectrum. This observation was indeed confirmed in more than one study where patients with a clinical diagnosis of CRION or RION were found to have the MOG antibody [43,51,61,62]. The potential course of demyelinating optic neuropathy is summarized in Figure 1. Additionally, a summary of pattern recognition of ON in MS and AQP4- and MOG-NMOSD on orbital MRI is provided in Table 1.

Table 1. Comparative chart of optic nerve (ON) MRI in inflammatory demyelinating diseases (IDDs) of the CNS. AQP4-NMOSD: Aquaporin 4-antibody Associated neuromyelitis optica spectrum disorder; LEON: longitudinally extensive optic neuritis; MOG-ON: myelin oligodendrocyte glycoprotein antibody associated optic neuritis.

MRI	MS	AQP4-NMOSD	MOG-NMOSD
Optic nerve (Range) [74,76,77]	• *Unilateral* • Short segment 13 mm (8–36 mm) • Median # segments: 1 (1–4 mm)	• *Bilateral* • LEON, median length 26 mm (range 14–46 mm) • Three segments (1–8) • Posterior predominance with chiasma and optic tract involvement • Milder swelling than MOG-ON • Rare twisting	• *Bilateral* • LEON • Anterior predominance • Perineural sheath swelling • Tilting & twisting of the optic nerve

5. Transverse Myelitis Pattern Recognition: From Clinically Isolated Syndrome to MS, NMOSD and Others

5.1. Multiple Sclerosis-Associated Transverse Myelitis (MS-TM) and Myelopathy

5.1.1. Acute Complete Transverse Myelitis (ACTM) versus Acute Partial Transverse Myelitis (APTM)

The distinction between complete and partial/incomplete transverse myelitis was highlighted a quarter of a century ago [84]. Acute *partial* transverse myelitis is characterized by an asymmetric or mild loss of function, which is in contrast to the involvement of all modalities in acute *complete* transverse myelitis and severe neurologic deficit. In 2002, the ATM Working Group proposed a series of laboratory tests to try to differentiate idiopathic from post-infectious or inflammatory transverse myelitis [85]. Although this list is not fully comprehensive, it provides a framework for the workup of ATM, and can represent a work in progress that will require occasional refinement to include new knowledge. Spine MRI findings differ between ACTM and APTM. Lesions in the former involve the whole cross section of the spinal cord or at least its center. Lesions in the latter are dorsolateral. The potential for APTM to evolve into MS was recognized early. Lesions longer than three vertebrae in length exist in MS but typically affect the dorsolateral tracts, an important differentiating factor from the longitudinally extensive transverse myelitis associated with NMOSD. While an abnormal MRI of the brain predicts the future conversion into MS [86–88], an estimated 20–30% of patients with APTM and a negative cerebral MRI will convert into MS [89,90]. The presence of oligoclonal bands in the CSF appears to increase the conversion likelihood [89].

5.1.2. The Case of Progressive Solitary Sclerosis

A rare phenotype of demyelination reminiscent of primary progressive MS, i.e., progressive solitary sclerosis [91], consists of a solitary demyelinating CNS lesion most commonly located within the cervical spinal cord or cervico-medullary junction. Less commonly affected areas include the thoracic spinal cord, subcortical white matter, and ponto-mesencephalic junction. This spinal cord lesion is typically less than three vertebrae segments in length. Bilateral lesions involving the medullary pyramids or cervicomedullary junction present with quadriparesis but no brainstem symptomatology [92]. Cerebrospinal fluid analysis characteristics of MS are present in 50% of cases. Originally described in 7 patients [91], similar clinical, CSF and imaging findings were reported in 10 more patients [93–96]. Time will tell whether solitary sclerosis should belong to the MS disease or not.

5.2. Neuromyelitis Optica Spectrum Disorder-Associated Transverse Myelitis (NMOSD-TM) and Longitudinally Extensive Transverse Myelitis (LETM)

5.2.1. LETM versus Spinal Cord Infarct versus Spondylotic Myelopathy

In up to 40% of cases, transverse myelitis can be the presenting manifestation of NMOSD [97,98] The presence of a LETM almost always evokes the diagnosis of AQP4-NMOSD. However, LETM has been associated with other inflammatory diseases of the CNS such as ADEM, MS, overlap syndromes (e.g., Sjogren's and NMO), sarcoidosis, antiphospholipid syndrome, vasculitis [99], Behcet's disease, and paraneoplastic syndrome, in addition to non-inflammatory etiologies such as intramedullary tumors, dural arteriovenous fistula, Alexander's disease, metabolic and compressive myelopathies and spinal cord infarction [100,101]. An extraordinary challenge in the differential diagnosis of LETM is spinal cord infarction. A study by Kister et al. analyzed the clinical, demographic, and MRI characteristics of 11 cases with spinal cord infarction and 13 cases with LETM. More commonly associated with LETM were the female gender, non-White ethnicity, bright spotty lesions on MRI of the spinal cord described below, location within 7 cm of the cervicomedullary or cervicothoracic junction, extension to the pial surface, and contrast enhancement. Interestingly, patient age, lesion length and cross-sectional area, and cord expansion did not differentiate the two conditions [102]. Another challenging diagnosis is spondylotic myelopathy that might be confused or sometimes associated with myelitis. Flanagan et al. compiled the findings of 56 patients with the condition. A peculiar pattern of "transverse pancake gadolinium enhancement" is described caudal to the site of maximal stenosis and at the craniocaudal midpoint of a spindle-shaped T2 hyperintense lesion. On axial cuts, a complete or incomplete circumferential pattern of enhancement with gray matter sparing is observed. Distinctively, spondylotic myelopathy is associated with a prolonged contrast enhancement resolution that might extend for a year, post-surgical decompression [103].

The following sections focus on factors differentiating idiopathic LETM, AQP4- and MOG-antibodies-associated LETM.

5.2.2. Seropositive Versus Seronegative LETM: Does Seronegative LETM Truly Exist?

The answer to this question is a matter of debate in the literature. In an effort to define truly idiopathic and AQP4-antibody-associated LETM, a 2015 study by Hyun et al. enrolled 108 patients with first-ever LETM (mean follow-up periods between seropositive and negative groups were 5.4 ± 2.6 years vs. 7.0 ± 4.4 years). To determine the true seropositive and seronegative statuses, the AQP4 antibody status was repetitively confirmed by three different validated methodologies, discussed below [104]. CSF glial fibrillary acid protein (GFAP) levels were measured to investigate astrocytic damage. Of the 108 patients, 55 were positive for AQP4 antibodies (i.e., P-LETM) and 53 were consistently negative (i.e., N-LETM). Seven out of 53 N-LETM were later diagnosed with seronegative NMO (49%), and four were positive for MOG antibodies (8.2%). The remaining 42 patients (N-LETM) showed several features distinct from P-LETM, including male predominance, older age of onset, milder clinical presentation with partial transverse myelitis features, less frequent relapses, spinal cord confinement with shorter segments, and the absence of combined autoimmunity. While CSF GFAP levels were markedly elevated in P-LETM, they were not increased in N-LETM. In the group of N-LETM, 39% were true seronegative or idiopathic [104], consistent with an Italian study, which reported 41% idiopathic N-LETM among 37 first-ever LETM [105]. Interestingly, idiopathic LETM was not necessarily monophasic, although relapse rate was less than P-LETM [104]. Fewer patients were treated with immunosuppressants, most likely due to the misconception that idiopathic transverse myelitis is monophasic. The increased frequency of recurrent N-LETM in the study by Hyun et al. compared to previous studies may be due to the longer duration of follow-up. Disease heterogeneity was noted in the N-LETM group where severe cases were present [104]. A study by Kitley et al. compared P-LETM and N-LETM, and produced discrepant results. In this cohort of 76 patients presenting with LETM, 58% (*n* = 44) had the AQP4 antibody and 42% (*n* = 32) were negative. The two groups were followed for a median of 61.35 months (with a range of 2.3–260.2 months) and 25.04 months (with a range of 1.9–169.4 months), for AQP4-antibody-positive and AQP4-antibody-negative respectively. In this series, however, most of the AQP4-antibody-negative group had an identifiable etiology unlike the above two series. Six of 32 had the MOG antibody (18.75%), five had ADEM, and the rest had vasculitis, leptomeningeal syndrome, infections, paraneoplastic disease, and spinal cord infarction. The final rate of true idiopathic N-LETM was 6.5% and true N-LETM could not be clinically differentiated from P-LETM [106].

Albeit less common than AQP4-LETM, MOG-LETM is turning out to be an important differential diagnosis of N-LETM, and clinical features are crucial in differentiating these two conditions. However, the MOG antibody assay is not available commercially, and the prevalence of MOG-LETM is variable depending on the series studied. In a 2016 study by Cobo-Galvo et al., 13 cases of MOG-LETM were compared to 43 cases of N-LETM [107]. Distinctive clinical features in the MOG-seropositive group included the following: younger age at onset, increased predisposition to optic neuritis relapses, and improved prognosis. A total of 23% of patients who presented with a first episode of N-LETM tested positive for the MOG antibody [107] vs. 18.75% in another study [106].These frequencies, which are greater than previously described (8.2%; [104]), may be explained by discrepancies in the definition of LETM, which was undefined in the Hyun et al. study, as well as a genetic predisposition and unintentional selection bias. The Cobo-Calvo study had younger N-LETM patients with a more homogeneous ethnic background and followed an acknowledged definition for LETM (≥three vertebral segments). A large and comprehensive workup was also performed to rule out alternative diagnoses. Equal involvement of male gender and steroid sensitivity were noted similar to that observed with MOG-ON [74]. The clinical course of MOG-LETM patients was less severe compared to AQP4-antibody seropositive or truly seronegative forms of NMOSD, despite the similar frequency of severe episodes at onset and the increased relapse rate during the follow-up. Similar to AQP4-LETM cases, a spinal cord lesion evanescence by MRI was observed in a significant proportion of cases. A possible explanation for this recovery is the effect of the MOG antibody itself. Indeed, the intracerebral injection of the human MOG antibody in mice causes few and transient myelin changes, alteration of axonal protein expression without leukocyte infiltration, and recovery within two weeks [107]. The German Study, described in detail later, reported the findings on spinal cord MRI in MOG-NMOSD [22]. The median length of the LETM and the short segments transverse myelitis (SSTM) was 4 and 1.5 vertebral segments, respectively. Swelling and contrast enhancement were commonly present in 70.4% and 67.9% of transverse myelitis cases, respectively. The cervical spinal cord was most commonly affected, followed by the thoracolumbar areas. Other reports, however, emphasized the thoracolumbar involvement [108], particularly the conus [109]. Cord lesions were equally distributed centrally and peripherally. Asymptomatic spinal cord lesions were also present. In summary, the clinical and MRI phenotypes of MOG-LETM are reminiscent of AQP4-LETM, but a male prevalence and a less aggressive course are differentiating factors. Additionally, it appears that MOG-LETM accounts for about 20% of previously reported seronegative LETM. Furthermore, the existence of idiopathic LETM remains arguable. Further refining its definition and the serological assays, longer follow-ups, and the identification of more antibodies will likely decrease this group. In addition to its dubious nature, there needs to be a close follow-up of patients affected by N-LETM to determine long-term management, considering that the diagnosis of idiopathic LETM is associated with less potential for long-term treatment.

5.2.3. Short Segment Transverse Myelitis (SSTM) in Neuromyelitis Optica Spectrum Disorder

The presence of a short segment (<3 vertebral segments) transverse myelitis (i.e., SSTM) was reported in 14% of patients with AQP4-NMOSD [110] and most recently in patients with MOG-NMOSD [22,111]. SSTM can occur early in NMOSD, with immunosuppressive treatment and due to MRI timing [13]. Particularly, an early MRI might detect a lesion at its beginning, and a late MRI might capture the lesion following improvement. Not surprisingly, a delay in the diagnosis and treatment of the SSTM associated with NMOSD in comparison to LETM is common. However, 92% of the SSTMs were followed by a LETM. Aside from the presence of the AQP4 antibody, which is confirmatory for the SSTM disease, there are several clinical features that suggest its diagnosis, including the following: non-White ethnicity, advanced age, personal history of autoimmunity and tonic spasms, prior history of severe and bilateral optic neuritis with limited recovery, prior episode of uncontrollable nausea and vomiting, and lastly, the absence of oligoclonal bands in the CSF. Additionally, radiological features that suggest the diagnosis of SSTM include a central lesion associated with T1 hypointensity and the absence of typical MS brain lesions [110]. Interestingly, the frequency of SSTM at initial presentation was recently reported in 14.5% of 76 patients subsequently diagnosed with AQP4-NMOSD [112]. Thus, SSTM is not a rare event in NMOSD, and clinical, paraclinical, and imaging features suggestive of NMOSD are key to the diagnosis.

5.2.4. Imaging Patterns of Neuromyelitis Optica Spectrum Disorder-Associated Transverse Myelitis (NMOSD-TM)

Linear Lesions in NMOSD

Linear lesions are defined as limited ependymal inflammation in the medulla, which is due to weakness of the fluid-BBB, spinal cord, or both. While studying the relationship between linear lesions and LETM, most patients with NMOSD show linear lesions preceding LETM [113]. This raises the possibility that linear lesions are precursors to LETM. Further, the simultaneous presence of linear lesions and LETM, or linear lesions following LETM, might reflect a more severe degree of inflammation [113].

Bright Spotty Lesions (BSLs) in Neuromyelitis Optica Spectrum Disorder

Bright spotty lesions (BSLs) on the spinal cord were observed with or without LETM and in acute and chronic disease states. BSLs were found in 54% of patients with NMSOD ($n = 24$ patients) and 3% patients with MS ($n = 34$) [114]. BSLs are best visualized on axial cuts with T1- and T2-weighted imaging, where they present as hypointense and hyperintense lesions, respectively, and are located centrally or peripherally (Figure 5). Their hypointensity on T1-weighted imaging, similar to or greater than the CSF, and their contrast enhancement might reflect a destructive damage predominantly to the gray matter and blood spinal cord barrier resulting in microcystic changes [114].

To further differentiate LETM ($\geq$three vertebral segments) associated with NMOSD, MS, and other neurological diseases of the spinal cord, the most useful MRI characteristics, as found in a study by Pekcevik et al., were the presence of punctate or large cavities BSL, T1 "dark" lesions, and large lesions involving more than 50% of the spinal cord cross section [115]. In the two studies, the sensitivity of BSLs was 88% and 65%, and the specificity of BSLs was 97% and 89%, respectively [114,115]. There was an emphasis on T1-dark rather than T1-hypointense lesions due to inter-observer disagreement on the lesion definition for T1 hypointense. The presence of greater than 50% of cross-sectional involvement or "transversally" extensive lesions was sensitive, but had poor specificity. Other factors that could not differentiate between the three groups included the presence or absence and the pattern of contrast enhancement (well-defined homogeneous or ill-defined heterogeneous), as well as spinal cord abnormality extending into the brainstem [115]. In summary, BSLs are imaging markers of spinal cord lesions associated with NMOSD and can be present in isolation or in conjunction with LETM; again, their darkness on T1 should evoke the diagnosis considering that dark or severe hypointense lesions on T1 are rarely detected visually in transverse myelitis associated with MS [116]. The potential course of demyelinating transverse myelopathy is summarized in Figure 3. Additionally, a summary of pattern recognition of TM in MS and AQP4- and MOG-NMOSD on spinal cord MRI is provided in Table 2.

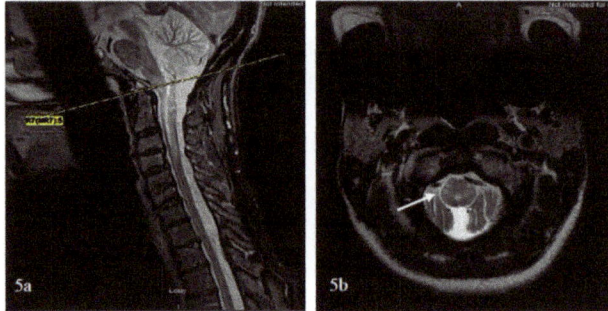

Figure 5. Added to above paragraph 55-year-old African-American female, with AQP4-NMOSD sagittal STIR cervical MRI demonstrates simultaneous linear lesion and LETM from the medulla to C4 (**5a**). An axial cut at the level of the cervicomedullary junction, axial T2 demonstrates 4 peripheral bright spotty lesions (**5b**).

Table 2. Comparative chart of optic nerve (ON) and spinal cord imaging in inflammatory demyelinating diseases (IDDs) of the CNS. LEON: longitudinally extensive optic neuritis; LETM: longitudinally extensive transverse myelitis; MOG-ON: myelin oligodendrocyte glycoprotein antibody associated optic neuritis; SSTM: short segment transverse myelitis; #: number.

MRI	MS	AQP4-NMOSD	MOG-NMOSD
Spinal Cord (Range) [113,115,117]	• SSTM • LETM less frequent • Dorsolateral lesions (SSTM & LETM)	• LETM • SSTM less frequent • Central cord lesion • Complete resolution of lesion • Bright spotty lesions • Linear lesions	• LETM • SSTM less frequent • Central cord lesion • Complete resolution of lesion • Bright spotty lesions? • Linear lesions

6. Brainstem and Cerebellar Pattern Recognition: From Clinically Isolated Syndrome to MS, NMOSD and Others

6.1. Multiple Sclerosis-Associated Brainstem and Cerebellar Symptoms

At onset, optic neuritis, transverse myelitis and vertigo, diplopia, and ataxia in a person between the ages of 20 and 50 years are quite suggestive of MS. However, although uncommon, isolated brainstem presentation can be misleading and includes oculomotor nerve palsy, trigeminal neuralgia, facial nerve palsy of the "peripheral type" and hemifacial spasms, which require a thorough workup to rule out life-threatening conditions such as an aneurysm or a brainstem tumor [118]. Asymptomatic brainstem involvement is not unusual either. Like optic neuritis and transverse myelitis, acute demyelinating brainstem syndrome can remain isolated or can evolve into MS, NMOSD, or recurrent brainstem encephalitis [119]. On imaging, acute posterior fossa lesions in MS present with T2 hyperintensity and contrast enhancement. Chronic lesions might show continuous T2 hyperintensity, but it is not unusual for them to also show focal atrophy. The presence of focal lesions and atrophy, however, is unspecific and is seen in a number of pathologies [120].

6.1.1. Trigeminal Neuralgia or Facial Sensory Loss

Trigeminal neuralgia at onset of MS is rare, and accounts for <1% of initial MS presentations [121]. However, using routine MRI of the brain in patients with MS (pwMS), trigeminal nerve enhancement was reported in 24 of 851 (2.8%) patients, with bilaterality in two-thirds of the patients and extension into Meckel's cavum in 19 patients [122]. The nerve enhancement was partial or complete, involving the nerve across its length, from its pontine exit zone (i.e., root entry zone myelinated by oligodendrocytes), passing by the central-peripheral transitional zone, and up to the Meckel's cavum (i.e., myelinated by Schwann cells). This indicates that involvement of peripheral myelin occurs in MS in addition to the central portion in relation to the dorsal root entry zone, which is supported by pathological studies [122,123]. A different study conducted by da Silva et al. found a similar frequency of trigeminal nerve enhancement (2.9% of a cohort of 275 MS patients) with bilateral involvement (75% of the cohort) [124]. Trigeminal nerve enhancement visualized by MRI is occasionally associated with sensory symptoms of pain, anesthesia, or paresthesias [124]. A third study by Mills et al. reported an increased prevalence of trigeminal nerve involvement in 11 of 47 patients (23%) using 3T MRI of the brain, with 1 mm slices through the posterior fossa [123]. Specifically, the intracranial trigeminal nerve pathway was mapped and showed T2 hyperintensity in the trigeminal root entry zone and intrapontine tract, with potential extension in either direction to the trans-cisternal portion of the nerve and what was thought to be the trigeminal nuclei (both ascending and descending). The changes were often bilateral (50% cases) and symmetrical. In this study, all the patients were asymptomatic [123]. In a study by Swinnen et al. involving 43 pwMS or clinically isolated syndrome referred for trigeminal nerve symptoms, the MRI of the brain demonstrated a linear plaque involving the intra-pontine fascicular portion of the nerve and lesions involving the spinal nucleus and tract in 48.8% and 53.4% of the patients, respectively. Lesions of the principal sensory nucleus and mesencephalic nucleus of the trigeminal nerve were less common (12–33%). In this study, however, lesions were most often unilateral (80% of the cases) [125]. In summary, uni- and bilateral trigeminal nerve enhancement is not unusually observed on the MRI of pwMS patients, albeit asymptomatic clinically, and is an example of central and peripheral myelin involvement in MS.

6.1.2. Oculomotor Abnormalities

Unilateral or bilateral internuclear ophthalmoplegia, the most common oculomotor abnormality in MS and a hallmark of the disease in a person aged 20–50 years, incites practitioners to actively follow the patient even when brain imaging is normal. [118]. However, isolated sixth nerve palsy is very rare and was reported in three out 600 pwMS seen at a neuro-ophthalmology clinic. Usually a brainstem lesion that affects the sixth nerve nucleus results in additional deficits due to the intimate relationship of the fascicular fibers to other pontine structures [126]. Isolated fourth nerve palsy is also very rare. Besides the difficulty in diagnosing superior oblique palsy, the condition is rare because the fascicular course of the trochlear nerve is exposed to little myelin [127]. Other combinations of oculomotor abnormalities occur in MS both acutely and chronically, including as one-and-a-half syndrome and walled-eyes bilateral internuclear ophthalmoplegia (WEBINO). A 2016 report detailed a list of oculomotor abnormalities and ocular instabilities observed in pwMS, including a systematic approach to their diagnosis [128]. Periaqueductal lesions, commonly seen in NMOSD, were described in 19.4% of pwMS patients and were associated with oculomotor abnormalities and higher brainstem disability scores. Some of these lesions were wedge-shaped (42%), and others had an abnormally hyperintense broad peri-aqueductal gray rim; a third group had both characteristics, meaning severe involvement. Contrast enhancement was absent. Notably, a three-dimensional direct inversion recovery technique is optimal in allowing for a strong contrast between periaqueductal gray and surrounding tissue, due to a suppression of the CSF and white matter. The pathophysiology of these lesions in the periaqueductal area is likely to involve inflammation around the subependymal veins, similar to the areas around the lateral ventricles. Moreover, the close vicinity to the CSF and potential direct gliotoxic effects from the CSF might be an additional mechanism for the formation of periacqueductal lesion in MS [129].

6.1.3. Peripheral Type Facial Nerve Palsy

The frequency of facial nerve palsy at the onset of MS varies from 1.4–4.8% [130]. In peripheral seventh nerve palsy, the lesion of the nerve usually occurs at the level of the geniculate ganglion (located in the facial canal) and therefore outside of the CNS. Peripheral facial palsy, however, can also result from a central lesion at the level of the ipsilateral facial nucleus or facial nerve at the pons [131].

6.1.4. Cerebellar Symptoms

Clinically isolated cerebellar syndrome is rare in MS, but cerebellar involvement is very common in advanced disease states and in pathological studies, even when brain imaging antemortem does not show any cerebellar findings [132]. Brainstem lesions frequently affect the cerebellum with its afferent and efferent tracts [133]. The clinical manifestations of cerebellar pathology depend on the lesion site and include truncal and appendicular ataxia, eye movement abnormalities, cognitive impairment, and tremors, which are the most common symptom. Common lesion locations include middle cerebellar peduncles and cerebellar hemispheric white matter [120].

6.2. Acute Brainstem Syndrome Associated with Neuromyelitis Optica Spectrum Disorder

NMOSD-associated acute brainstem syndrome might be difficult to diagnose, particularly when the brainstem syndrome is a precursor to NMOSD. Notably, brainstem syndromes in NMOSD have a peculiar pattern that is likely easy to diagnose by a neurologist who is familiar with the presentation, but might represent a challenge in a gastroenterology clinic. In the latest clinical criteria for NMOSD [42], the importance of acute brainstem syndrome was highlighted by making it one of the core clinical criteria for diagnosing seropositive NMOSD; the dorsal medullary or area postrema syndrome was one of two very specific criteria required for diagnosing seronegative NMOSD [42].

6.2.1. Intractable Hiccups, Nausea and Vomiting

In NMOSD, intractable hiccups and nausea preceded (54% of cases) or accompanied (29% of cases) neurological symptoms such as optic neuritis and transverse myelitis. Occasionally, a significant increase was reported in AQP4-IgG titers. Medullary involvement based on MRI, in addition to short or long segment spinal cord lesions, were present in about 50% of the cases [134]. Similarly, the initial symptom for 12 patients with NMOSD was intractable vomiting for three months prior to the onset of optic neuritis or transverse myelitis. The clinical and neuroimaging observations were consistent with area postrema involvement, a circumventricular organ that lacks the BBB thus allowing diffusion of stimulating IgG into the CNS [135]. Both intractable hiccups and vomiting were completely resolved with corticosteroids [134,135]. Heralding brainstem symptoms in demyelinating diseases are not uncommon. In a study by Cheng et al. involving 352 patients with CNS demyelinating diseases, 31 patients (8.8%) presented with an acute brainstem syndrome. The AQP4 antibody was present in only 14 of these 31 patients (45%). Intractable hiccups, nausea, and vomiting occurred more often in the positive group. Also in the positive group, five out of 14 patients had recurrent brainstem symptoms before optic neuritis or transverse myelitis vs. one out of 17 in the negative group. Dorsal medullary lesions were more often present in the positive rather than the negative group, but midbrain and pons were equally affected in the two groups. None of the 31 patients with acute brainstem syndrome had spinal cord lesions at onset, although LETM was commonly found in the positive group during follow-up. Over two years, 100% of the positive group and 17.65% of the negative group converted to NMOSD (i.e., 17 of 31 of the total group). Furthermore, seven of the 31 converted to MS, and the remaining 7 had no further neurological events. While the Expanded Disability Status Scale (EDSS) was similar at baseline, the positive group had increased EDSS at last follow-up, underlining the importance of AQP4-antibody testing for diagnosis and prognosis [136]. Not unexpectedly, in a cohort of Chinese patients with NMOSD, medullary involvement was associated with an increased annual relapse rate, worse medullary symptoms and disability, increased incidences of brain lesions and LETM, and was frequently associated with thyroid diseases [137]. Interestingly, patients who had medullary involvement more often had headaches, neuropathic pain, and a movement disorder compared with other NMOSD patients without medullary involvement [137].

6.2.2. Oculomotor Abnormalities

A number of oculomotor manifestations, similar to the ones described in MS, have been observed with both AQP4- and MOG-antibody associated NMOSD including (1) walled-eyes bilateral internuclear ophthalmoplegia (WEBINO) associated with a midbrain tegmentum lesion adjacent to the aqueduct on brain MRI [138,139], (2) ocular oscillations, including up-beating, down-beating, central vestibular nystagmus, and opsoclonus myoclonus syndrome, [140], (3) nuclear [141] and bilateral trochlear nerve palsy [142], and (4) central Horner syndrome [143], which has occasionally been described in MS in relation to brainstem lesions [144–146]. Thus, with overlay in brainstem symptomatology between MS and NMOSD, MRI of the brain and spinal cord, serology, and most importantly, high index of suspicion are expected to lead to the final diagnosis.

6.2.3. Other Atypical Brainstem Presentations

Excessive yawning unrelated to sleep deprivation or fatigue was reported in nine patients with the MOG antibody; five out of nine patients had yawning as a presentation of the illness in association with nausea, vomiting, and hiccups. The duration of this excessive yawning lasted two to 16 weeks. The MRI results were abnormal in all patients with brainstem and hypothalamic lesions [147].

Encephalopathy, albeit not a classical symptomatology of brainstem disease and NMOSD, has been associated with diencephalic and brainstem involvement and confused with Wernicke's encephalopathy. The confusion between the two entities extends to the histological level, particularly considering that the hallmarks of Wernicke's encephalopathy are periventricular involvement of thiamin-metabolism-rich areas with cytotoxic edema of astrocytes and neurons and hemorrhage [148]. While a new onset encephalopathy with focal symptoms and demyelination on CNS imaging is evocative of ADEM or Susac's syndrome [149,150], encephalopathy presentation in an established case of NMOSD should trigger the search for posterior reversible encephalopathy syndrome, a treatment complication and more recently overlap syndrome or NMOSD-encephalitis complex that will be discuss later in this review. Subsequent relapses can hint at this diagnosis.

7. Tumefactive Demyelinating Lesion Pattern Recognition: From Clinically Isolated Syndrome to MS, NMOSD and Others

7.1. Multiple Sclerosis-Associated Tumefactive Demyelinating Lesions (MS-TDLs)

TDL, defined as solitary lesions ≥2 cm, might herald symptoms of MS and represent a diagnostic challenge when occurring in an isolated manner. Given et al. reported a pictorial essay that summarized the MRI appearance of TDLs [151]. TDLs tend to be well delineated with minimal mass effects and edema. TDLs typically occur at the supratentorial level, centered in the white matter, with or without extension into the cortical gray matter. Fifty percent of TDLs typically enhance in an incomplete ring pattern, with the open side facing the cortex. Several studies reported a centrally dilated vein and decreased perfusion in comparison to tumors and normal-appearing white matter [152,153]. The presence of centrally dilated veins within TDLs was again confirmed using ultrahigh field 7T MRI of the brain [154,155]. There have been steady attempts to differentiate TDLs from brain neoplasms through locating a novel combination of imaging techniques that allow clinical rather than surgical diagnosis. For example, Mabray et al. demonstrated that TDLs can be diagnosed with a high degree of specificity and differentiated from high-grade gliomas and primary CNS lymphoma on preoperative MRI by using a combination of criteria including incomplete rim enhancement, the presence of multiple lesions, and high minimal apparent diffusion coefficient values on brain MRI [156]. Other authors used conventional and non-conventional imaging techniques to differentiate brain tumors from TDL including ^{11}C-methionine positron emission tomography (MET-PET) [157], magnetic resonance spectroscopy, and conventional angiography. In addition to vessel-like structures on TDLs, multiple venous dilatations around TDLs based on angiography can be useful for the diagnosis of large TDLs [158]. Others have attempted to differentiate TDLs from high-grade glioma using cerebral blood volume (CBV) and flow (CBF), calculated from dynamic contrast enhanced perfusion MRI. Perfusion MRI of regional CBV and CBF were reduced among demyelinating patients [159]. An additional challenge of TDL is the possible association of TDL(s) and tumors. This is illustrated by a case of a tumefactive demyelinating MS and an anaplastic oligodendroglioma where the MRI of a patient's brain fulfilled Barkhof's criteria, and the CSF study was abnormal with the presence of oligoclonal bands. An ^{18}F-FDG-PET scan was performed that demonstrated increased tracer uptake, as expected with a brain tumor and brain biopsy showed an anaplastic oligodendroglioma [160].

Another challenging scenario is the association of primary CNS lymphoma and TDLs both demonstrating the same location predilection and steroid responsiveness. Primary CNS lymphoma manifests as a uniformly contrast-enhancing mass with predilection to periventricular and superficial locations, often contacting ventricular and meningeal surfaces. The lesions are hypo- or isointense on T2-weighted imaging and have prominent perilesional edema. The presence of a mixed iso- and hyperintense lesion on T2, the lack of cortical involvement, and mass effect are in favor of TDL. A computed tomography (CT) scan of the brain demonstrates hypoattenuation in TDL and hyperattenuation in lymphoma, underlining the importance of combining imaging modality with CT and MRI. In both pathologies, magnetic resonance spectroscopy demonstrates increased lipid, choline/creatinine, and myoinositol, and decreased N-acetylaspartate peaks, but elevated glutamate/glutamine peaks favor TDL. Serial MRIs have shown continuous evolution with TDL and stability of the content of the neoplasm [161]. Long-term evolution of an isolated TDL is unknown and limited by the duration of the follow-up. However, like any clinically isolated syndrome, a group will get disseminated in time and space evolving into MS or NMOSD; a second will remain stable for the duration of follow-up, and a third might evolve into a different diagnosis [162]. Lastly, TDLs have been reported [163] with fingolimod use in MS and inadvertently in NMOSD, fingolimod discontinuation [164–166] or de-escalation from natalizumab [167–169]. These situations might pose a diagnostic challenge in case of lack of familiarity with these scenarios. Albeit an uncommon problem, TDL might represent an investigation challenge prior to, and following the diagnosis of MS [170]. Again, ultrahigh field 7T brain MRI might be promising in tumefactive demyelinating from non-demyelinating lesions [171].

7.2. Neuromyelitis Optica Spectrum Disorder-Associated Tumefactive Demyelinating Lesions (NMOSD-TDL) and Hemispheric Presentations

Extensive hemispheric lesions in areas that are not enriched with AQP4 is a pattern described in NMOSD [24,172,173]. A priori, the term tumefactive demyelinating lesion (TDL) evokes the diagnosis of MS. However, a Korean study followed 31 patients with at least one TDL over a mean period of 37.6 months. During this observation period, 11 patients remained idiopathic (six had a single event, and five had recurrent demyelinating disease inconsistent with MS or NMOSD), 11 patients developed AQP4-NMOSD, seven evolved into MS, and two had an alternative diagnosis. The increased conversion of TDL to NMOSD in this cohort could be due to the ethnicity of the studied population, but prior reports on TDL did not systematically test for the AQP4 antibody [162]. A common MRI pattern of TDL-associated NMOSD includes T2-high and T1-iso-to-hypointense lesions, increased diffusivity on apparent diffusion coefficient map, and hypo- or isointensity on diffusion-weighted images or hyperintensity, probably due to T2 shine-through. Contrast-enhancement is typically absent or faint, an indication of the integrity of the BBB [173]. However, in the absence of other clinical, paraclinical, and imaging findings, the presence of TDL with partial ring enhancement could be easily confused with tumefactive MS [174]. Magnetic resonance spectroscopy of six TDLs in three patients with NMOSD showed increased Cho/Cr and decreased N-acetylaspartate peaks/Cr ratios in all of the patients and a lactate peak in two [175]. Posterior reversible encephalopathy syndrome with supratentorial and asymmetric hemispheric presentation has been reported with NMOSD [172,176]. The clinical presentation of TDL-associated NMOSD and posterior reversible encephalopathy syndrome-associated NMOSD is somewhat similar, with a variable degree of encephalopathy and focal symptoms such homonymous hemianopia [172,176].

7.3. The relationship of Balo's Concentric Sclerosis to TDL, MS and NMOSD

Balo's concentric sclerosis lesion, which is not the focus of our review, falls under the category of atypical demyelination and is characterized radiologically and pathologically by concentric rings of demyelination and remyelination. Pathologically, the concentric configuration of the lesion is explained by the presence of radially oriented cytokines gradient that provide Balo's lesion at the edge with some preconditionning to ischemia and less demyelination. This is supported by autopsy studies confirming upregulation of hypoxia-inducible proteins [177]. BCS lesions can be confused with TDL becasuse of their large size, particularly when the layering is not easily discernible, but multiple Balo's lesions can coalesce to form a TDL radiographically. The evolution of a TDL into a BCS has also been reported in the literature [178]. However, the relationship between BCS, MS and NMOSD has not been clearly defined. Knowing that demyelination is the common denominator between these 4 entities, there are unique characteristics for TDL and BCS that differentiate them from MS and NMOSD [179]. Like TDL, BCS can evolve into MS or NMOSD; conversely, lesions of the Balo's type can be seen in MS and NMOSD [180,181]. Ultrahigh field 7T MRI of the brain holds promise in potentially differentiating MS from NMOSD, TDL from non-demyelinating ones, and prognosticating which CIS might evolve into MS versus not based on the visualization of the central vein sign [154,171]. The interdependent relationship between BCS, TDL, MS & NMOSD is summarized in Figure 6.

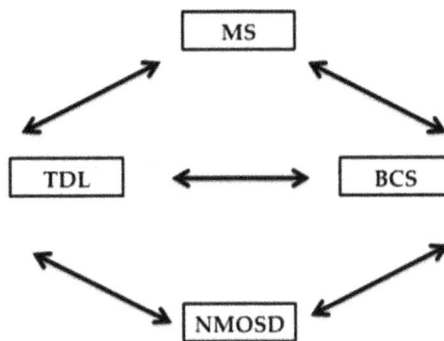

Figure 6. Diagram summarizing the relationship between TDL, BCS, MS and NMOSD; TDL: Tumefactive demyelinating lesion; BCS: Balo's concentric sclerosis; MS: Multiple sclerosis; NMOSD: Neuromyelitis optica spectrum disorder.

8. Clinical Spectrum of MOG-Antibody-Associated-Inflammatory Demyelinating Disorders

8.1. Neuromyelitis Optica Spectrum Disorder-Associated Myelin Oligodendrocyte Glycoprotein Antibody (MOG-NMOSD)

Until recently, MOG-associated NMOSD has been occluded by its grouping with the AQP4-antibody seronegative group. Currently, the diagnosis of seronegative NMOSD is made with at least two clinical core criteria meeting certain specific requirements, as recently reported [42]. With overlap in clinical and MRI presentation of AQP4-associated NMOSD and MOG-associated

NMOSD, it is necessary and important to determine a pattern to differentiate between the two conditions. Moreover, the condition of MOG-NMOSD is not yet widely recognized due to the absence of commercially available tests. Additionally, as of today, there is no gold standard for the optimal MOG-antibody assay. The largest two series of MOG-NMOSD studies were reported around the same time by the following two groups: one from Germany (n = 50 patients), referred to as the "German series", by the NEuroMyelitis Optica Study group (NEMOS) and one from Spain (n = 56 patients) referred to as the "Spanish series" [22,43]. The findings from these two studies are generally representative of what is reported in the literature to date on MOG-NMOSD and are comparatively summarized in Table 3. The two studies share some striking similarities in the mean/median age group, with wide age range, concordant with other studies including Chinese Han patients [83,108]. Females are affected more often, as observed for MS and by other studies [182], but different from the 1:1 or even male gender dominance reported by others [108]. Unlike AQP4-NMOSD, there was no female gender skewing. Almost all of the patients were White in both series. The median duration of follow-up was longer in the German study, but the range was similar in both studies. The number of relapses was greater in the German study. In both series, optic neuritis was the most common initial symptom in 60–64% of the patients in isolation and 70–74% in combination, reflecting the increased expression of MOG in the optic nerve compared to other CNS areas [108,183]. Importantly, the presentation of bilateral simultaneous optic neuritis or optic neuritis with transverse myelitis is common in MOG-NMOSD and greater than that seen for AQP4-NMOSD, which invites a careful search of the diagnostic criteria, as previously reported [108,184]. The German group reported the time duration for the second attack to be five months, but importantly, presentation of the second attack was very similar to the first attack. Indeed, 72% of patients who presented with optic neuritis had recurrent optic neuritis, and 76% who presented with transverse myelitis had recurrent transverse myelitis. A total of 71% and 81% of the patients in both Spanish and German series, respectively, had a relapsing course, which was greater than that reported by other groups who had a shorter duration of follow-up [55,108]. The clinical presentation at last follow-up was not presented similarly in both studies. At last follow-up, the disease was disseminated in 32% of patients (18 of 56) in the Spain series and 44% of patients (22 of 50) in the German series. The difference is likely due to the longer disease duration in the German series (75 months vs. 43 months in the Spanish series). While the absolute number of relapses was higher in the German group, the annual relapse rate was within range (equaling 0.8 for the isolated optic neuritis group and highest at 1.17 for the ON-TM combination group). The final outcome in both series differed, and was worse in the German series. A total of 37% of patients (14 of 38) were functionally blind or had severe visual loss in at least one eye, defined as 20/100 < visual acuity ≤ 20/40. Weakness or ataxia was the etiology of gait impairment in 25% of the patients. In the Spanish series, 19% of those affected by optic neuritis had severe vision loss, defined as visual acuity < 0.1%, and 11% had a moderately severe-to-advanced disability (Expanded Disability Status Scale (EDSS) 4–7). There was one case of death in the German series due to brainstem encephalitis with secondary respiratory failure. Explanatory factors for these discrepancies might include differences in ascertainment of the cases and definition of deficit, and genetic and environmental factors that are known to somewhat modify the risk for autoimmune diseases (such as sun exposure, vitamin D level, diet, etc.). Also likely at play are longer disease duration with a cumulative effect on the CNS from repetitive relapses. Despite the worse outcomes in the German group, comparison of recovery rates from optic neuritis and/or transverse myelitis attacks to previously published rates for AQP4-NMOSD showed an improved recovery rate (more complete and less partial/no recovery) [22], but less than observed with MS [109,185]. Both studies reported serum antibody titers; however, the German group provided more information on the profile of the antibody as a function of the clinical presentation, relapsing or remitting status, and in response to treatment. For example, in the German study, antibody titers were variable from 1/160 to 1/20,480, with the highest titers found for the combination of optic neuritis-transverse myelitis and the transverse myelitis groups at last follow-up. This trend was also seen in the Spain series. Serum antibodies sampled within few days to three months of disease onset were positive in all patients and persistent in the serum up to 10 years for available samples of this duration. The greatest antibody titers were observed within 14 days of the attack onset and occasionally during disease remission. During disease exacerbation, the titers were variable intra- and inter-individually. Serum antibody titers equally decreased during disease remission, following relapse treatment with corticosteroids and plasma exchange, and maintenance treatment with immunosuppressants, a decrease reflective of the natural remission rather than treatment [51]. Decreased MOG antibody titers were observed during rituximab therapy, and MOG antibody titers increased with B-cells recurrence. MOG antibodies in the CSF were present in 12 of 15 patients tested during disease exacerbation and decreased during remission [51]. Greatest CSF MOG antibody titers were observed for transverse myelitis. To determine antibody class and sub-class, the serum and CSF of eight seropositive patients during exacerbation and six in remission were analyzed. All of the patients had positive MOG-IgG1 in the serum and in the CSF [51]. Two out of the 20 patients tested from the MOG antibody group were positive for IgM in the serum only; no CSF or serum tested positive for MOG-IgA [22]. Overall, the CSF profile of MOG-NMOSD mirrors the profile of AQP4-NMOSD. Like AQP4-NMOSD, neutrophils and granulocytes were present in the CSF in about two-thirds of the cases. Consequently, the absence of oligoclonal bands and/or the presence of granulocytes and neutrophils in the CSF should challenge the diagnosis of MS. MRI of the brain, brainstem, spinal cord, and orbits in MOG-NMOSD are reminiscent of that seen in AQP4-NMOSD, except for the differences noted in Table 3. For the brain MRI, supratentorial abnormalities were seen at onset in 35.4% of patients, and infratentorial was seen in 14.6% of the patients [51]. Supratentorial lesions were periventricular, involving the corpus callosum in a confluent manner in the frontal, parietal, temporoparietal, and occipital deep white matter. Some lesions were subcortical or juxtacortical, including the insular cortex. Leptomeningeal enhancement and basal ganglia involvement were observed in 1 in 50 patients each. Infratentorial lesions at onset involved the cerebral peduncles, pons, medulla (particularly the area postrema), the periaqueductal gray matter, and the cerebellar hemisphere and peduncles. Corpus callosum lesions were longitudinally extensive as seen with patients with AQP4 antibodies [23]. Most importantly, 50% of the patients fulfilled the McDonald 2010 criteria by MRI, and 62.3% fulfilled the 2006 NMO criteria. Barkhof's criteria for DIS were fulfilled in 15% of the cases that were positive MOG antibody and had a history of transverse myelitis and/or optic neuritis, and in 26.9% of the cases that were positive for the MOG antibody and had a history of brain lesions [51]. Visual evoked potentials demonstrated subclinical optic nerve damage, and asymptomatic lesions were present in the brain, cerebellum, or spinal cord, which is a pattern observed in MS but not in AQP4-NMOSD. Importantly, the presentation of bilateral simultaneous optic neuritis or optic neuritis with transverse myelitis is common and invites a careful search for the correct diagnosis, as reported by others [108,184]. Typically, optic neuritis patients had improved recovery compared to transverse myelitis patients. Deterioration was observed after corticosteroid withdrawal, as described by others [61,186]. Treatment with plasma exchange and immunoadsorption, with or without steroids, has been associated with complete recovery in about one-third of the cases, and partial recovery has been observed in the majority of the cases. There were 14 untreated attacks, two without recovery, with one being fatal, three with partial recovery, and nine with complete, or almost complete, recovery [22].

Table 3. Comparative chart of the two largest reported series on MOG-NMOSD. * Functional blindness
is defined as a person who has to use many alternative techniques to perform tasks that are ordinarily
performed with sight that his/her pattern of daily living is substantially altered. Such alternative
techniques might include reading a newspaper by listening to it over the telephone or using Braille
to read a book. ** EDSS score includes visual functional score system (VFSS); *** BMRC: British
Medical Research Council score; a score ≤2, refers to severe paresis; ADEM: acute disseminated
encephalomyelitis; CBA: cell-based assay; FU: follow-up; IVMP: intravenous methylprednisolone;
LETM: longitudinally extensive transverse myelitis; MOG: myelin oligodendrocyte glycoprotein; ON:
optic neuritis; VA: visual acuity; TM: transverse myelitis; +/−: with or without.

	Spanish Study [76] Sepulveda et al., 2016; *n* = 56	German Study [60] Jarius et al., 2016; *n* = 50
MOG assay	CBA	CBA
Gender (Female:Male)	1.67:1	2.8:1
Race	54 Caucasian; 2 others	49 Caucasian; 1 Asian
Median age disease onset (Range)	37 years (18–70)	31 years (6–70)
Clinical syndrome at onset (N)	• ON (34) • LETM (12)/TM (1) • ON + TM (5) • ADEM (3) • Brainstem syndrome (1)	• ON (32) • TM (9) • ON + transverse myelitis (5) • ADEM (3) • Brainstem syndrome (1)
Median time to second attack	No data	5 month
Clinical syndrome at last FU (N)	• ON (27) • Recurrent (21); CRION (3/21); BL simultaneous (1/21) • Monophasic (6); BL simultaneous (5/6) • LETM (7)/TM (3) • LETM (7); Relapsing (1) • SSTM (3); Relapsing (3) • ON + TM (1) • ADEM (2): Monophasic • MDEM (1): Relapse at 3 and 4 years • Relapsing brainstem syndrome (1) • NMOSD (14) • Relapsing (12): ON, TM, brainstem relapse, ON + TM (5) • Monophasic (2)	• ON (22) • Recurrent (14) • Monophasic (8) • TM (6) • Multifocal: ON+ transverse myelitis +/− brain +/− brainstem (22)
Relapsing% versus Monophasic %	71% versus 29%	80% versus 20%
Median follow-up (range), months	43 (4–554) months	75 (1–507) months
MOG titers, serum (range) MOG titers, CSF	1/960 (1/160–1/10240) No data	1/160–1/20480 1/2–1/64
CSF • Cells • OCB	Mean= 41 (SD = 70) 3/53	Median = 33 (IQR 13–125) 6/45
Relapse number	125 • 85 ON • 29 TM • 2 ON+TM • 5 brain • 4 brainstem	• 205 ON • 73 TM • 3 cerebellum • 9 brain • 20 brainstem

Table 3. *Cont.*

	Spanish Study [76] Sepulveda et al., 2016; *n* = 56	German Study [60] Jarius et al., 2016; *n* = 50
Annualized relapse rate	1.11	0.83
Outcome	**Data on VA in 46 patients** • VA < 20/100 in 19% (8/46) ON patients **Last EDSS 2 (0–7)** • 0–2.5: 71% • 3–3.5: 18% • 4–5.5: 7% • ≥6: 4%	**Data on VA in 38 patients** • VA ≤ 20/100; 10/38 (26.3%) • Functional blindness* in at least one eye • Severe visual loss at least one eye 20/100 < VA ≤ 20/40; 4/38 (10.5%) • Moderate visual loss at least one eye • 20/40 < VA < 20/25; 2/38 (5.3%) • Mild visual loss at least one eye • 20/25 < VA < 20/20; 5/38 (13.2%) **Last EDSS** (1–10) • Median 3 (*n* = 40 relapsing cohort) • Median 2.5 (*n* =47 total cohort) **Data on motor deficit available for 28 TM patients** • Severe paresis (BMRC*** ≤ 2): 1/28 or 3.6% • Moderate paresis: 4/28 or 14.3% • Mild paresis: 6/28 or 21.4% • Impaired ambulation (paresis + ataxia): 25%
Concomitant autoimmune antibodies Concomitant autoimmune disorders	No data *n* = 7 (13%)	*n* = 19/45 (42.2%) *n* = 4/47 (8.5%)
Relapse management	• No detailed data • Corticosteroids	• 122 relapses treated with IVMP • Almost-complete recovery: 50% • Partial recovery: 44.3% • Almost no recovery: 5.7%
Maintenance Therapy Medications	46% No data	Monoclonal B-cell therapies No data

8.2. Pediatric Acute Disseminated Encephalomyelitis (ADEM)

Few studies demonstrated an increase in serum anti-MOG antibodies in patients with MS compared to controls. The prevalence and titers of MOG antibodies in the studies were low and had no predictive or prognostic significance in MS [187]. Subsequently, anti-MOG antibodies were identified in pediatric demyelinating diseases, including ADEM and clinically isolated syndromes. The presence of MOG antibodies correlated with a younger age of onset and the initial clinical presentation of ADEM. Additionally, increased titers correlated with an initial presentation of ADEM rather than clinically isolated syndrome. Most children with MOG antibody-associated ADEM (i.e., MOG-ADEM) had a rapid decline in MOG antibody titers with recovery and persistence with incomplete recovery. Based on the MRI of the brain and spinal cord, these children had more fluffy lesions and LETM with complete resolution and improved outcome [49,50].

An interesting syndrome associated with MOG-ADEM in children is the ADEM- optic neuritis (ADEM-ON or ADEMON) or multiphasic disseminated encephalomyelitis (MDEM-ON) complex. Table 4 summarizes the 2 cases-series reported in the literature. The natural history and prognosis for ADEM/MDEM-ON remains unknown considering that the median follow-up years was 4–6 years in the 2-case series [188]. Another case of ADEM followed by optic neuritis 71 days later and positive MOG antibody in the serum was reported in a five-year-old Japanese girl [189]. Adding to the list of cases of MDEM-ON is the case of a 4.5-year-old Malyasian girl with MDEM followed by optic neuritis and positive MOG antibodies with relapses triggered by viral infections and a gluten rich diet [190].

Extending the spectrum of MOG-antibody inflammatory demyelinating diseases, Baumann et al. reported a case series involving eight children, six White and two Asian, with MDEM and who were positive for the MOG antibody. These children were followed for four years. The mean number of relapses was three, (ranging from 2 to 4), and the mean inter-attack interval was four months (ranging from one to 48 months). All children had at least two attacks separated by at least three months. Initial multifocal deficit with and without encephalopathy were followed by optic neuritis in some children or MDEM-ON. MOG antibodies were present in all children, but none had the AQP4 antibody. Initial and repeat CSF studies were similar to ADEM-ON and NMOSD. MRI of the brain was characterized as hazy and with TDLs that improved over time, along with cortical gray matter involvement in seven patients, an area rarely involved in AQP4-NMOSD [23]. Asymptomatic white matter lesions were seen in one child. Two LETM and two SSTM were present in four children [191].

Table 4. Summary of the 2-case series on ADEM/MDEM followed by optic neuritis. ADEM: Acute disseminated encephalomyelitis; AQP4: Aquaporin4; GA: Glatiramer acetate; INF-beta: interferon beta; IVIG: Intravenous immunoglobulins; MDEM: Multiphasic disseminated encephalomyelitis; MOG: Myelin oligodendrocyte; ND: No data; OCB: Oligoclonal bands; ON: Optic neuritis; PLEX: Plasma exchange; TDL: Tumefactive demyelinating lesions.

	Hupke et al., 2012 [188]	Baumann et al., 2016 [191]
N of patients	7	8
Median Age (range) years	6 (4–8)	3 (1–7)
Gender (F/M)	6/1	5/3
Clinical presentation • ADEM • MDEM • ON	$n = 7$ $n = 3/7$ Unilateral in $n = 6$; Bilateral in $n = 1$ Always following ADEM/MDEM N of attacks: 1–7 Inter-attack intervals: 3 weeks-2 years	$n = 7$ Unilateral/Bilateral: ND; $n = 2$
Median inter-attack interval	Minimum 4 weeks for ON	4 months for MDEM
Preceding febrile illness; N; (weeks prior)	ND	Yes; $n = 4$; 4 weeks prior
Median follow up (range) years	6 for $n = 4/7$	4 (1–8)
Autoantibody	ADEM stage: MOG-antibody (+) 3/7 and ND 4/7 MOG- antibody (+) with ON AQP4-antibody (-)	MDEM stage: MOG- antibody (+) 8/8 AQP4-antibody (-)
CSF	Pleocytosis Negative OCB	Pleocytosis Negative OCB in $n = 7/8$
MRI Brain MRI spinal cord	Classical ADEM findings New lesions with MDEM No new brain MRI lesions during ON ND	Classical ADEM findings with TDL and cortical GM lesions. New lesions with MDEM LETM $n = 2$ SSTM $n = 2$
Treatment	Improvement with corticosteroids Azathioprine: Partial effectiveness IFN-beta and GA: Not effective	Corticosteroids during attacks ($n = 8$) IVIG ($n = 1$) during attack PLEX ($n = 1$) during attack IVIG Monthly ($n = 4$)
Outcome	Minimal to no relapses, $n = 2$ Continuous relapses, $n = 2$ Mild vision loss, $n = 4$	Normal $n = 4$ Mild-moderate deficit $n = 4$ (psychomotor and/or seizures)

8.3. Overlap Syndrome or Complex Neuromyelitis Optica Spectrum Disorder Encephalitis

In this review, Table 5 summarizes all of the cases of NMOSD associated with encephalitis ($n = 46$) reported in the literature from 2010 onward [192–200]. While the authors of these case reports or series aimed to report clinical cases with "overlap syndrome", the series by Hacohen et al. was a systemic evaluation of CNS autoantibodies in pediatric demyelination syndrome including ADEM, optic neuritis, LETM, and NMO [194]. The age range for the entire group as shown in Table 5 is broad, from 16-months- to 65-years-old. The clinical presentation was of encephalitis, abnormal movement disorders, unilateral or bilateral optic neuritis, transverse myelitis, seizures, or a combination of the above symptoms. Abnormalities on imaging were variable from normal MRI brain to multifocal white matter lesions and LETM, multiple cranial nerve enhancement, and gray matter involvement (e.g., cortical and basal ganglia). Some of these cases were associated with positive serum AQP4-IgG, others were associated with the MOG antibody, and in the majority of the cases, whenever done, CSF NMDA-R antibody was positive. Serum NMDA titers were not available for the majority of cases, except in one study. Glycine receptor and voltage-gated potassium channel antibodies were positive in one and three cases, respectively [194]. In some case, antibodies for both NMDA-R and MOG were exclusively present in the CSF. CT scan of the chest, abdomen and pelvis were performed in a number of cases and were negative, except in one case of ovarian teratoma [199]. Pleocytosis in the CSF was frequently present with protein elevation. Oligoclonal bands were present in 13 cases, absent in 18 cases, and not available for the rest. Full recovery was reported in 10 cases, incomplete recovery in 21 cases, unchanged in two patients, death in one case due to rapidly evolving pneumocystis pneumonia, and data was not available for 12 cases. Acute management included multimodality treatment with corticosteroids, plasma exchange, and intravenous immunoglobulin in different combinations. Immunosuppressants, such as rituximab [197,198,200], mycophenolate mofetil [197,198], and azathioprine [199], were used mostly on acute basis, but for an unknown period of time. We recently had a referral for diagnosis and treatment of a 21-year-old woman with history of bilateral LEON associated with perineural sheath thickening and optic nerve twisting on orbital MRI at the age of 9 (case not published). Throughout the years, she had three more episodes of cerebral encephalitis (see Figure 7) associated with cortical swelling on brain MRI and seizures at the age of 13, 17, and 19. A MOG-antibody associated disease was suspected. Extensive serum and CSF autoantibodies testing was positive for serum NMDA-R, thyroid peroxidase, voltage gated potassium channel autoantibodies at low titers and for MOG-antibody.

Upon literature review, we found that she had an overlap syndrome reported as benign, unilateral, cerebral cortical encephalitis with epilepsy associated with the MOG-antibody, recently reported in four men, aged from 23 to 39 years. All of these men had seizures, and three had associated encephalopathy and/or psychosis. Unilateral optic neuritis (two cases) and seizures (one case) were observed at seven months and 35 months prior to encephalitis. Encephalitis, seizure, and optic neuritis were coincident in one case [192]. Our case was different, as following an initial episode of bilateral LEON, she had three more episodes of encephalitis and seizures associated with waxing and waning of cortical T2 hyperintensity over the course of 10 years. She was very recently initiated on Rituximab, and remains on lamotrigine for seizure management, although she has not had any seizures in isolation of encephalitis. In summary, the number of overlap syndrome cases might be underestimated. With the lack of familiarity with this novel entity, it is perceivable that cases of NMOSD-encephalitis complex are misdiagnosed as ADEM, paraneoplastic syndromes or toxic metabolic encephalopathy, to cite a few.

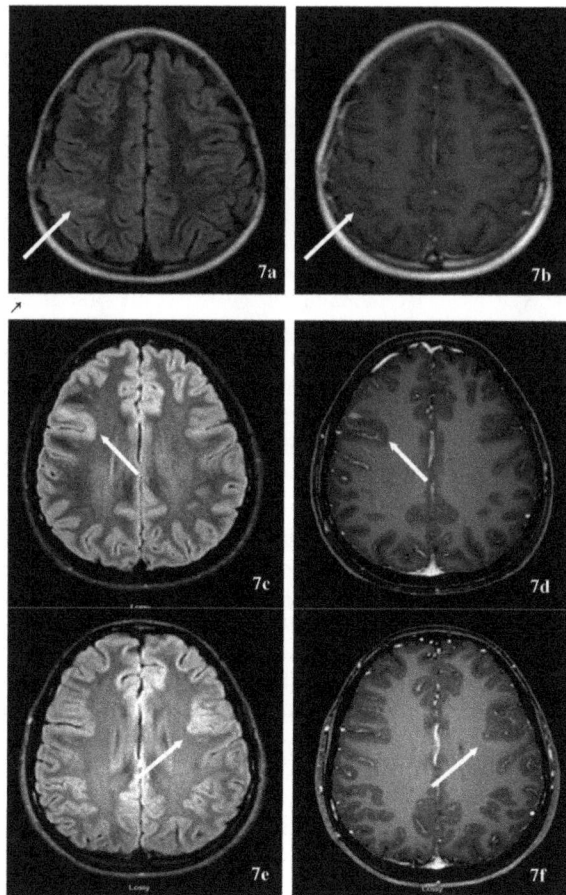

Figure 7. This is the case of a 21-year-old woman who presented with a history of bilateral optic neuritis at the age of 10. She subsequently had recurrent encephalitis with cortical swelling and seizures at the age of 13, 17 and 19. Axial FLAIR shows a right parietal cortical T2 hyperintensity with swelling (**7a**) associated with minimal T1 hypointensity without contrast enhancement (**7b**). Recurrent encephalitis was associated with a new right frontal cortical T2 hyperintensity with swelling (**7c**) associated with minimal T1 hypointensity with contrast enhancement. (**7d**). A third episode demonstrated a new left frontal cortical T2 hyperintensity with swelling (**7e**) associated with minimal T1 hypointensity with contrast enhancement (**7f**). Serum MOG-antibody was positive.

Table 5. Summary of Overlap Syndrome or NMOSD-Encephalitis complex reported in the literature. Ab: Antibodies; ADEM: acute disseminated encephalomyelitis; CEL, contrast-enhancing lesions; CSF: cerebral spinal fluid; HSV: herpes simplex virus; EDSS: expanded disability status scale; INO: internuclear opthalmoplegia; IV: intravenous; IVIG: intravenous immunoglobulins; LETM: longitudinally extensive transverse myelitis; MOG: myelin oligodendrocyte glycoprotein; MRI: Magnetic resonance imaging; ND: no data; NL: normal; NMDAR antibody: N-Methyl-D-Asparate receptor antibody: NMO: neuromyelitis optica; OCB: oligoclonal bands; ON: optic neuritis; PLEX: plasma exchange; PRES: posterior reversible encephalopathy syndrome; TM: transverse myelitis; VGKC: Voltage-gated potassium channel; − : negative; + : positive; (): number of patients; * Multi-modality therapy: combination therapy of IVIG, steroids, and plasma exchange * CT scan includes imaging of the abdomen, chest, and pelvis to rule out malignancy.

	Demographics	Presentation	MRI	Treatment/Response	Serum Ab (Number of Patients)	CSF/an Cillary Testing (Number of Patients)	Final Outcome
Kruer et al., 2010 [193]	n = 1 F, 15 years	Seizures, encephalopathy, movement disorder, myelopathy, unilateral ON	Multifocal CEL, LETM	Prednisone: improvement. Interferon-B therapy, cyclophosphamide: failed therapy Multi-modality therapy *: drastic improvement	AQP4: -	Pleocytosis: + Protein: 2x NL Glucose: low OCB: + NMDAR: + Pan CT scan: -	Asymptomatic
Lekoubou et al., 2012 [198]	n = 1 F, 34 years	Encephalopathy, movement disorder, myelopathy	Widespread, multifocal white matter lesions, right frontal contrast enhancing lesion, LETM	IVIG, steroids: failed therapy Immunosuppressant: dramatic improvement	AQP4: -	Pleocytosis: + Protein: elevation Glucose: low OCB: + NMDAR: + Viral panel: - Pan CT scan: -	Minor residual cognitive impairment
Zoccarato et al., 2013 [199]	n = 1 F, 50 years	Encephalopathy, movement disorder, myelopathy, unilateral ON	T2 hyperintense lesions in cortical medial temporal lobe, pons, hypothalamus, medulla, cervical & dorsal spine	Hystero-adnexectomy, oral steroids: improvement PLEX, steroids: improvement Immunosuppressant: stable disease	AQP4: + NMDAR: +	Pleocytosis: ND Protein: ND Glucose: ND OCB: + NMDAR: + EEG: - Pan CT scan: ovarian teratoma	Stable disease
Outteryck et al., 2013 [200]	n = 1 F, 65 years	Encephalopathy, movement disorder, myelopathy, subclinical unilateral ON	LETM with gadolinium enhancement, T2 hyperintensities in insular regions, medial temporal lobes & thalamus, gadolinium enhancement in meninges & ventricles	IV steroids: paraparesis worsened PLEX, immunosuppressant, oral steroids: significant improvement	AQP4: - NMDAR: +	Pleocytosis: + Protein: elevation Glucose: ND OCB: + NMDAR: + Pan CT scan: -	Death: due to rapidly evolving pneumocystis pneumonia

Table 5. *Cont.*

Demographics	Presentation	MRI	Treatment/Response	Serum Ab (Number of Patients)	CSF/an Cillary Testing (Number of Patients)	Final Outcome
Hacohen et al., 2014 [194] $n = 15$ M (6), 3–15 years F (9), 2–15 years	Seizures, encephalopathy, movement disorder, myelopathy, unilateral/bilateral ON	Normal, optic nerve signal changes, multiple white matter abnormalities, spinal cord involvement, periventricular/juxtacortical, brainstem lesions	ND	AQP4: +(3) NMDAR: +(2) MOG: +(7) VGKC: +(3) GlyR: +(1)	Pleocytosis: +(3) Protein: ND Glucose: ND OCB: +(3)/-(7)/ND (5) NMDAR: ND Viral panel: ND MOG: ND EEG: ND Pan CT scan: ND	Full recovery (4), EDSS 1 (6), EDSS 3 (2), EDSS 4 (2), EDSS 7 (1), seizures (2), Visual loss (1)
Hacohen et al., 2014 [195] $n = 10$ M (5), 2–18 years F (5), 1.33–11 years	Brainstem encephalitis HSV encephalitis Unilateral/bilateral ON ADEM Recurrent hyperventilation, dizziness, double vision	Normal (2), brainstem signal change (2), white matter changes (7), multiple cranial nerve enhancement (1), gray matter changes (1)	No treatment: (1) Multimodality treatment with IV, oral steroids, IVIG and immunosuppressants	AQP4: ND NMDAR: +(10) MOG: ND VGKC: ND GlyR: ND	Pleocytosis: ND Protein: ND Glucose: ND OCB: +(1)/-(2)/ ND (7) NMDAR: +(4)/ ND (6) MOG: +(3)/ND (7)	Full recovery (4) Incomplete (5) Unchanged (1)
Titulaer et al., 2014 [197] $n = 12$ M (5), 10–38 years F (7), 8–55 years	Seizures, encephalopathy, movement disorder, myelopathy (LETM) unilateral/bilateral ON	ND (2), Normal (1), optic nerve signal changes (1), white matter lesions (4), T2/FLAIR abnormalities in various aspects of the brain (7), spinal cord involvement (4), Gd enhanced lesions (6), brainstem lesions (4)	No treatment: (1) Multimodality treatment with steroids, IVIG, PLEX, interferon, imunosuppressanT	AQP4: +(3), -(5), ND (4) NMDAR: +(2), -(5), ND (5) MOG: +(6), -(3), ND (3) VGKC: ND GlyR: ND	Pleocytosis: +(8)/-(4) Protein and glucose ND OCB: +(5)/-(5)/ND (2) NMDAR: +(12) MOG: +(7)/-(4)/ND (1)	ND
Splendiani et al., 2016 [196] $n = 1$ M, 17 years	Seizures, encephalopathy, movement disorder, myelopathy	T2 & FLAIR cortical-subcortical hyperintensities in right cerebellar hemisphere, ipsilateral cerebellar tonsil, with faint T1 hypointensity	Olanzapine: progression of symptoms IV steroids: gradual improvement	AQP4: ND NMDAR: + MOG: ND VGKC: ND GlyR: ND	CSF findings: ND NMDAR: + Viral panel: - EEG: - Pan CT scan: -	Asymptomatic
Ogawa et al., 2017 [192] $n = 4$ patients M, 23–39 years	Seizures, encephalopathy, ON	T2 & FLAIR unilateral cortical hyperintensities; no ADEM-like lesions.	IV steroids with significant improvement 2–3 antiepileptics/patient	AQP4: - MOG: +(3) NMDAR: -(4) AMPA: -(4) LGI1: -(4) CASPR2: -(4) GABAb: -(4)	Pleocytosis + Protein: NL-1.5xNL OCB: ND MBP: NL (3); ND (1) MOG: +(3); ND (1)	Remain on antiepileptics Otherwise full recovery

8.4. The Myelin Oligodendrocyte Glycoprotein (MOG) Antibody and Its Association to Other Autoantibodies

The association of the MOG antibody in inflammatory demyelinating diseases to other autoantibodies (e.g., thyroid, celiac, antinuclear antibodies, Sjogren's, glomerular basement membrane) has been reported in several papers [22,43,53,55], but this association is less commonly found in comparison to AQP4-NMOSD. Double seropositivity of AQP4 and MOG antibodies in NMOSD and its limited forms was found by ELISA techniques [201], by cell-based assays [62,98,202], and in a single patient with gastric cancer and NMO [53]. Known to be rare, these cases present clinically with severe deficits because of recurrent optic neuritis or bilateral optic neuritis and simultaneous transverse myelitis. The frequency of AQP4, glycine receptor alpha 1 subunit, and MOG antibodies was determined in a cohort of patients with isolated optic neuritis; the combination was found in 45% (23 of 51) of the cases of the cases and was associated with unilateral or bilateral, severe or recurrent optic neuritis [203].

9. Other Autoantibodies, Diseases, and Biomarkers Associated with Neuromyelitis Optica Spectrum Disorder

9.1. Aquaporin 1-Antibody Associated with NMOSD (-NMOSD)

As previously noted, there are at least 12 related AQPs expressed in mammalian tissues, and a greater number of homologues are expressed in plants and lower organisms. Both aquaporin 1 (AQP1) and APQ4 are expressed in astrocytes, reflecting some redundancy that might functionally important [33]. The differential expression of AQP1 and AQP4 in different parts of the astrocytes has been demonstrated in the gray and white matter [204]. In an attempt to determine if AQP4-seronegative NMOSD had other associated antibodies, a group of researchers from Greece analyzed the sera of patients with inflammatory demyelinating diseases, other neuroimmune diseases and healthy control for the presence of AQP1 antibodies. A total of 348 sera referred for AQP4-IgG testing were analyzed. Antibodies to AQP1 and AQP4 were analyzed using radioimmunoprecipitation assay (RIPA) and two-steps RIPA, a more sensitive technique. Aquaporin 1 antibodies could not be detected by cell-based assays. A total of 42 out of 348 sera (12%) tested positive for the AQP4 antibody using RIPA, and 44 out of 306 remaining sera (14%) tested positive for the AQP1 antibody. A total of 14 out of 42 sera (33%) were double positive for the AQP4 and AQP1 antibodies, meaning there were a total of 58 out of 306 patients (19%) who had the AQP1 antibody. None of the MS, neuroimmune diseases or healthy control samples had the AQP1 antibody. The female-to-male ratio was 2:1 for the AQP1 antibody and 10:1 for the AQP4 antibody. Although both AQPs share amino acid sequence identity (51%), antibodies in the double-positive sera did not cross-react with the other antigen. Most AQP1 antibodies were more often bound to the extracellular portion of the AQP1 loop A compared to loop C. Loop A is the more antigenic portion of AQP, has reduced identity, and shows homology to the corresponding AQP4 sequence. Most APQ1 antibodies were IgG1. The double-positive sera, however, had high antibody titers, indicating an interdependent immune system triggering mechanism. Epitope spreading across the AQPs in the double-positive group is also a possibility. The clinical and imaging data of a cohort of 22 patients, 17 with NMOSD and five with MS were reviewed. A total of 16 out of these 22 patients with imaging data had LETM, one had transverse myelitis only, and two had MS with a dominant spinal load. Five out 16 patients had LETM and optic neuritis. Three out of the 22 patients had classic MS and were positive for the AQP1 antibody. However, the antibodies in the MS sera were different and could not bind the extracellular portion or the whole AQP1 antigen. Interestingly, these three patients had neoplasms, thus raising the possibility that the antibodies were triggered by some neoplastic antigens. The authors concluded that the role of AQP1 in NMOSD remains unknown [205]. Findings in this study were not replicated by two more studies [206,207]. AQP1 antibody was tested in a group of 249 patients with different types of inflammatory demyelinating diseases, using CBA with Triton X-100. There were 98 AQP4-antibody-positive and 151 antibody-negative serum. A total of 73 out of 98 serums (74.5%) were positive for AQP1, meaning 73 were double positive. A total of 49 out of 151 AQP4-antibody-negative cases turned to be AQP1-antibody-positive cases, and these had relapsing optic neuritis, LETM, MS, or NMO. The authors concluded that there was some limited value in using the AQP1 antibody when the AQP4 antibody was negative. Interestingly, and in keeping with the prior group's findings [205], CBA without the use of Triton X-100 had a low efficiency for detecting the AQP1 antibody. However, adding Triton X-100 resulted in a dramatic increase in AQP1 antibody detection/determination. Triton-100 is a detergent commonly used experimentally to permeabilize the membrane of living cells and increase antigen retrieval [207]. A third group, Sanchez Gomar et al., attempted to detect AQP1 antibodies using CBA and ELISA. The AQP1 antibody was undetectable using CBA and detected with low titers by ELISA in few samples. The conclusion was that the study did not allow a sustained detection of anti-AQP1 in serum of NMOSD patients analyzed by CBA or ELISA [206]. This was further confirmed by the absence of AQP1-antibody using a live CBA in a cohort of patients with NMOSD, MS and controls [208]. Thus, AQP1 antibody does not convincingly appear to be a biomarker for NMOSD. Nevertheless, the quest for more autoantibodies to diagnose or prognosticate inflammatory demyelinating diseases will continue. The story of AQP1 antibody is not unique and the inward rectifying potassium channel 4.1 (KIR 4.1) is another antibody where different groups across the world had conflicting results [209–214]. KIR 4.1 is a membrane protein expressed by oligodendrocytes, a subset of astrocytes and various tissues such as kidney. An editorial by Hemmer provides a comprehensive analysis of the problem of isolating autoantibodies in human diseases. Detection of autoantibodies by protein based assays remains the winner but even then, posttranslational modification of the protein needs to be monitored during the assay [215].

9.2. Neuromyelitis Optica Spectrum Disorder as a Paraneoplastic Syndrome

Incidental malignancies were found in 31 patients among 180,000 patients evaluated for paraneoplastic antibodies and included breast, lung, thymic, uterine cervix, B-cell lymphoma, monoclonal gammopathy, thyroid Hurthle cell, carcinoid and pituitary somatotropinoma. These malignancies preceded or ensued NMOSD [216]. Treatment-resistant AQP4-LETM significantly improved, and autoantibody response reverted following breast cancer treatment. Screening for malignancy was thus suggested in treatment-resistant demyelinating disorders [217]. A case of invasive, poorly differentiated breast ductal carcinoma was reported in a 29-year-old woman who presented with brainstem symptomatology. Because of a prior history of cardiac surgery, brain MRI was not possible, but a PET scan demonstrated increased tracer uptake in the brainstem, medulla, pons, dorsal midbrain, and left medial temporal lobe, which was consistent with brainstem and limbic encephalitis. Serum and CSF were positive for the AQP4 antibody. The patient had a modified radical mastectomy followed by plasma exchange (PLEX) and rituximab with significant recovery at three months [218]. Paraneoplastic NMOSD has been also reported in association to a metastatic carcinoid tumor to the liver. While the tumor antedated the neurological manifestation by six years, hepatic metastases were coincident. Interestingly, AQP4 cells were present interspersed between neuroendocrine tumor cells on pathological examination of the metastases [219]. Similar findings of thyroid cancer expressing AQP4 were reported in a patient with thyroid cancer, predating the onset of NMOSD [220]. Negative AQP4-antibody NMO-like disorders have been associated with collapsing response-mediated protein 5 (CRMP5) and different malignancies such as small cell lung, prostate, thyroid papillary, and renal cancer, and thymoma [221] as well as anti-amphiphysin antibody and invasive poorly differentiated breast adenocarcinoma [222]. The association of the MOG antibody inflammatory demyelinating diseases to tumors was reported in the MOG-NMOSD German series; there was one case of mature cystic teratoma and a ganglioneuroma in one patient. Otherwise, there were no other cases of malignancy [22].

9.3. Other Biomarkers Associated with Neuromyelitis Optica Spectrum Disorder

As diagnostic biomarkers, AQP4 and MOG antibodies allow for improved tailoring of maintenance therapy. In a sense, they are also prognostic biomarkers as MOG-inflammatory demyelinating disease appears to be less aggressive than AQP4-antibody-associated diseases. The following section addresses several available biomarkers relevant to the diagnosis and prognosis of MOG-inflammatory demyelinating diseases and AQP4-antibody-associated diseases and allow differentiation of the two conditions. This is by no means an exhaustive list, and the reader is directed to the following update on NMO biomarkers [223].

9.3.1. Cerebrospinal Fluid Myelin Basic Protein (CSF MBP)

Ikeda et al. [224] analyzed MBP and GFAP levels in patients with MOG-NMOSD patients with the MOG antibody [224]. During an acute attack, CSF MBP was increased by 10-fold, but CSF and GFAP was undetectable. Similar work was reproduced in a larger multicenter international collaborative study where MOG and AQP4 antibodies and markers of damage myelin (i.e., MBP) and astrocytes (i.e., GFAP) were evaluated in the CSF of patients with NMOSD and MS. A third of the NMOSD patient sera were positive for the MOG antibody and two-third were positive for the AQP4 antibody. Three-fourths of the patients with NMOSD had either MOG- or APQ4-antibody in the CSF; again, one-third of the patients were positive for the MOG antibody, and two-thirds were positive for the AQP4 antibody. CSF GFAP was elevated in the AQP4-antibody-positive patients but not in the MOG-antibody-positive and MS cases. Elevated CSF MBP was observed in both MOG- and AQP4-positive antibodies, and both groups exhibited higher MBP levels compared with the MS group. Myelin damage in AQP4-antibody-positive patients with NMOSD is recognized to be secondary to antibody-dependent and complement-dependent astrocytic injury. The direct binding of MOG antibodies to MOG causes myelin damage and/or oligodendrocyte dysfunction. Pro-inflammatory cytokines released during attacks may induce the recruitment of immune cells (MOG-specific reactive T-cells and B-cells, macrophages, etc.) and the release or synthesis of other cofactors promoting demyelination [225].

9.3.2. Cerebrospinal Fluid (CSF) Glial Fibrillary Acid Protein (GFAP) and S100

Glial fibrillary acid protein (GFAP) is a monomeric intermediate filament protein in the astrocytes. A marked increase in GFAP in the CSF of AQP4-antibody-positive NMO is supported by the work of Misu et al. [226]. CSF-GFAP values in the AQP4-antibody-positive group were increased by 10,000-fold compared with the MS and ADEM groups and by 20-fold in the spinal cord infarct group. Treatment of three NMO patients with corticosteroid remarkably decreased CSF-GFAP close to normal levels. The significant increase of CSF-GFAP is supported by pathological studies showing a loss of GFAP and AQP4 reactivity in the acute perivascular lesions [31,227]. In MS, mild CSF-GFAP elevation likely results from reactive astrogliosis in chronic lesions, whereby GFAP is released into the CSF. In NMO, the CSF-GFAP levels strongly correlated with EDSS and spinal lesion length ($r > 0.9$ for both correlations). The CSF-GFAP levels were reportedly high in other destructive pathologies such as stroke and herpes encephalitis, but the levels were still 100-fold less than NMO. The CSF levels of S100-B, expressed by astrocytes ensheathing blood vessels and by glia (microglia and oligodendrocyte), were also analyzed. During acute NMO exacerbations, CSF-S100B levels were 100-fold greater than MS, ADEM and spinal cord infarction. Like CSF-GFAP, the values decreased after corticosteroid therapy, and the levels strongly correlated with the length of spinal cord lesions ($r > 0.9$). Thus, measuring astrocytic markers, especially CSF-GFAP, would be useful for assessing astrocytopathy and clinical severity of NMO, as well as to discriminate between AQP4-antibody-positve NMOSD and MS [226,228].

9.3.3. Interleukin 6 (IL-6) in Neuromyelitis Optica Spectrum Disorder (NMOSD)

Between all the cytokines and chemokines described with NMOSD, interleukin 6 (IL-6) appears to be the most relevant clinically, as it has shown the strongest correlation with clinical variables, including CSF GFAP, cell counts, and AQP4 antibody titers [229,230] and EDSS [230]. IL-6 has immunologic and non-immunologic roles in the CNS [231]. Parallel to its increase in the CSF, increased serum IL-6 was observed in patients with NMO [229] Further, IL-6 induces the differentiation and maturation of B-lymphocytes into plasmablasts further increasing the production of AQP4 antibodies, and the differentiation of naïve T-cells into TH17 cells. Likewise, TH17 cells produce IL-17 that further increases the production of IL-6. During initial [229] and recurrent NMOSD exacerbations, CSF-IL6 levels increase to the same extent, independent of lesion location. It is not specifically known if IL-6 is released by astrocytes or responsible for their injury, but an autocrine function of IL-6 has been reported [232]. The increase in serum and CSF-IL-6 has been therapeutically exploited, allowing for treatment of NMOSD with tocilizumab, a humanized monoclonal antibody inhibitor of IL-6 that is described below [229,233,234].

10. Tips on Management of Inflammatory Demyelinating Diseases of the Central Nervous System (CNS), with a Focus on Neuromyelitis Optica Spectrum Disorder (NMOSD)

10.1. Management of Acute Demyelinating Relapses Of The Central Nervous System

Acute management of demyelinating events is similar across the entire demyelination spectrum and consists of high doses of intravenous methylprednisolone (IVMP) and plasma exchange (PLEX) for severe demyelinating attacks of the CNS [235]. High-dose IVMP did not change the six-month visual acuity in the treated vs. untreated groups during an optic neuritis treatment trial [72]. However, corticosteroids are the cornerstone of CRION management, starting with high-dose pulse methylprednisolone followed by a prolonged oral prednisone taper [59]. Timely institution of corticosteroids leads to prompt recovery from pain, but rapid weaning can lead to irreversible damage. This deleterious response to steroid withdrawal should be an admonition to the diagnosis of CRION [59]. Likewise, exquisite steroid sensitivity is observed with granulomatous optic neuropathy and sarcoidosis and both conditions should always be ruled out [57]. Furthermore, corticosteroids are crucial in the management of NMOSD-ON, specifically MOG-ON. Ultrahigh doses of IVMP (2–5 g) have been used in severe cases [236]. Furthermore, while oral prednisone taper following IVMP failed to improve disability from relapses in MS [237], a prolonged prednisone taper is recommended with MOG-ON and CRION considering that a rapid withdrawal might lead to abrupt visual loss [22]. This prolonged oral taper is an intermediate therapy providing segue into chronic management. Additionally, the benefits of PLEX in treating severe demyelinating events have been reported [235]. PLEX is a standard of care for steroid-refractory demyelinating relapses per the European guidelines [238] and level B recommendation by the American Academy of Neurology [239]. Due to loss of proteins during plasma exchange, replacement with albumin or other plasma proteins is required. Immunoadsorption (IA) bypasses this problem and allows for a more selective adsorption of antibodies using a selective adsorbent to the antibody. Immunoadsorption using tryptophan or sepharose-conjugated sheep antibodies to human Ig is not infrequently used in Europe for management of steroid refractory NMOSD relapses [240]. The German Neuromyelitis Optica Study Group (NEMOS) recently conducted a retrospective analysis of 871 attacks and 1153 treatment courses in NMO/NMOSD [236]. While steroids were used with the first round of relapse management in 83.6% cases, PLEX was used in 100% of cases at the fifth round. The frequency of attacks treated with a second, third, fourth and fifth treatment modality was 28.2%, 7.1%, 1.4%, and 0.5%, respectively. Reportedly, the percentage of complete and partial recovery increased and the no recovery decreased with successive treatment courses [236]. Other treatment modalities included 54 identified combination therapies, the most common being IVMP followed by PLEX. Predictors of complete recovery included younger age of onset, PLEX/IA as a first treatment course and complete recovery from a prior attack. On the other hand, transverse myelitis attack was inversely associated with complete recovery. Characteristics such as gender, AQP4 antibody status, time from disease onset to attack, and time since previous attack were not predictors of complete recovery. Importantly, time from attack onset to therapy showed a trend that appeared significant. The NEMOS group retrospectively reviewed different therapies used in the management of MOG-NMOSD. High-dose corticosteroids were partially effective, resulting in ultrahigh doses of IVMP use. Oral prednisone taper and additional PLEX were recommended. Variable response to PLEX was observed, perhaps due to timing of the procedure, MOG antibody titers, number of sessions, intensity of the relapse, the extension and site of the demyelinating event, and optic neuritis vs. transverse myelitis vs. other locations. The exchange sessions varied from three to 11, with relapses observed when the number of PLEX sessions was less than seven. The authors recommended using PLEX as a substitute for ultrahigh-dose IV method prednisolone because of the risk of cerebral venous thrombosis that was reported in one case [22].

10.2. Maintenance Therapy of Neuromyelitis Optica Spectrum Disorder (NMOSD)

A detailed overview of maintenance therapy in NMOSD and MS is beyond the scope of this review (Table 6). Today, both AQP4- and NMOSD are similarly treated. Regarding maintenance therapies for MOG-NMOSD, azathioprine was ineffective if used without steroids the first six months of treatment. Methotrexate caused a reduced relapse rate in some but not all patients. Attacks were observed during rituximab initiation, likely due to B-cell activating factor (BAFF) increase [241,242]. It is in the present author's clinical practice to use 1 g of IVMP a week following the first dose of Rituximab (1 g) as worsening inflammation has been observed first hand clinically and by imaging (personal observation). A rapid relapse rate with B-cell repopulation was noticed underlining the importance of monitoring CD19 cells. Like NMOSD-AQP4, interferon beta caused an increase relapsed rate and glatiramer acetate was not effective in relapse prevention [22]. Fingolimod exacerbated optic neuritis symptoms in a case of MOG-ON within three weeks of medication initiation [53]. With the absence of disease definition and serological markers for NMOSD in the past and the persistence of clinical manifestations overlap to date between MS and NMOSD, empirical clinical experience, resulted inadvertently, in morbidity and mortality using MS disease modifying therapies such as interferon beta, glatiramer acetate, natalizumab and very lately alemtuzumab and dimethylfumarate to treat NMOSD [243–249]. Promising therapies, however, are emerging. Tocilizumab, an IL6- receptor antagonist, was shown to reduce relapse rate in 8 patients with NMOSD by about 90%. A pragmatic use of the infusion at the dose of 8 mg/kg administered every 4 weeks might improve its performance [250]. A humanized IL-6R neutralizing monoclonal antibody, SA237, designed by applying recycling antibody technology to tocilizumab, is currently being tested in 2 clinical trials [251,252]. Recycling allows SA237 to bind to IL-6 receptor multiple times and slows medication clearance from plasma [253].

Table 6. Maintenance therapy for NMOSD [233,234]. GI: Gastrointestinal; UTI: urinary tract infection; URI: upper respiratory infection; PML: progressive multifocal leukoencephalopathy; DVT: deep venous thrombosis; TB: tuberculosis.

Medication Name	Mechanism of Action (MOA)	Dosage	Treatment Response	Side Effects
Azathioprine	Thiopurine antagonist of endogenous purines in DNA and RNA, interferes with lymphocyte proliferation	Initial: 2–3 mg/kg/day with concomitant prednisone (5–60 mg daily) for 6–12 months Maintenance: 2–3 mg/kg/day	Approximately 50/50 chance of preventing additional relapse	Nausea, diarrhea, rash, recurrent infections, leukopenia, transaminase elevation, increased risk of lymphoma
Cyclophosphamide	Cytotoxic alkylating agent, inhibits mitosis	Initial: 1000 mg every 2 months with associated steroid Maintenance: same as initial dosing	Specific treatment response unavailable—only recommended when other immunosuppressive therapies fail or are not available due to contradictory preliminary findings.	GI symptoms, hyponatremia, heart block, pancytopenia, opportunistic infections
Eculizumab	Binds to the complement protein C5 specifically, inhibiting its cleavage to C5a and C5b and subsequent generation of the terminal complement complex C5b-9	Standard dose: IV 600 mg weekly for four weeks, then IV 900 mg every two weeks	Specific treatment response unavailable at this time	Headache, increased risk of infection with encapsulated organisms, especially meningococcal infections
Methotrexate	Folic acid antagonist	Initial: start with 7.5 mg weekly with upward titration and concomitant prednisone (5–60 mg daily) Maintenance: 7.5–15 mg weekly with concurrent prednisone (5–10 mg daily for at least sixmonths)	Remission rates in up to 2/3 of subjects when used as monotherapy or in conjunction with corticosteroids	Pneumonitis, GI upset, cytopenia, hepatotoxicity
Mitoxatrone	Causes DNA cross-linking and strand breaks, interferes with DNA repair	Initial: 12 mg/m² for 3–6 months Maintenance: 6–12 mg/m² every 3 months	Remission in up to 70% of subjects when dosed appropriately	Nausea, transaminase elevation, leukopenia, hair loss, amenorrhea, minor infections including UTI and URI, rarely heart failure and acute leukemia
Mycophenolate mofetil	Inhibits inosine monophosphate dehydrogenase, impairs B- and T-cell synthesis	Initial: 1000–2000 mg daily with concurrent prednisone (5–60 mg daily) Maintenance: 1000–2000 mg	Approximately 60–75% achieve remission with fewer side effects and adverse effects	Photosensitivity, recurrent infections, headache, constipation, abdominal pain, leukopenia, PML is rare
Rituximab	Removal of B cells as antigen presenting cells and reduction in the CD20+ early plasmablast population generating anti-aquaporin-4 antibodies	Initial: 1000 mg weekly for two weeks or 375 mg/m² weekly for four weeks Maintenance: 375 mg/m² or 1000 mg weekly for 2 weeks when CD19 count >1% on flow cytometry	Remission rates up to 83% were achieved with persistent B cell depletion	Sepsis, infections (Herpes zoster, UTIs, URIs), leukopenia, transaminase elevation, PML is rare
Tocilizumab	Directed against the IL-6 receptor reducing plasmablast survival, inhibiting AQP4 antibody production	Standard dose: 8 mg/kg every four weeks	Specific treatment response unavailable at this time	GI disturbance, fatigue, UTIs, neutropenia, leukopenia, elevation of cholesterol, transient mild transaminase elevation, DVT, TB reactivation

11. Discussion and Conclusions

A number of practical learning points emerge in this review, which is geared toward the pattern recognition of optic neuritis, transverse myelitis, brainstem/cerebellar and hemispheric TDL-associated MS, AQP4-antibody and MOG-antibody NMOSD, overlap syndrome, and some yet-to-be-defined/classified demyelinating disease all unspecifically labeled under *MS syndrome*. In the case of demyelinating syndrome occurring past the age of 50 and as a broad suggestion, one should suspect AQP4-NMOSD and MOG-NMOSD, although both diseases, particularly MOG-NMOSD, are seen in the younger population. Whenever a man presents with optic neuritis or other demyelinating symptom, the most likely root disease is MS or MOG-NMOSD, particularly considering the overrepresentation of women in relapsing AQP4-NMOSD with a women-to-men ratio of 9:1 [111,254]. Whenever optic neuritis presents in an individual younger than 20-year-old, it is recommended that the MOG antibody and AQP4 antibody levels be checked [74]. The classical teaching has been that corticosteroids speed the recovery of optic neuritis but do not necessarily alter the visual acuity outcome at six months. However, severe visual deficit has been associated with AQP4-ON, MOG-ON, and CRION, thus calling for prompt treatment with high-dose steroids in any new patients with optic neuritis particularly when immediate imaging of the brain is not available. With the broadening spectrum of inflammatory demyelinating optic neuritis, MRI of the orbit is crucial; the positive findings have diagnostic and prognostic implications, and can help tailor treatment. Because the long-term management of these conditions differs significantly, testing for both antibodies in a patient with his or her first optic neuritis relapse and normal MRI of the brain is recommended. The spectrum of MOG-inflammatory demyelinating diseases underlines that female gender is not always overrepresented in autoimmune disorders, and the MOG antibody autoimmune phenotype appears in a considerable proportion of male patients compared to other demyelinating diseases. Although the McDonald criteria should be reserved for patients suspected to have MS and are supposed to be of prognostication value, their use to diagnosing MS is common particularly when the clinical presentation of demyelinating disease is in question. By applying these criteria in clinical practice to cases of NMOSD, there appears to be a significant phenotypic overlap with the 2006 NMO, 2015 NMOSD, and the 2010 McDonald diagnostic criteria. Whether AQP4-inflammatory demyelinating diseases and MOG-inflammatory demyelinating diseases should be broadly categorized under the umbrella of NMOSD remains a subject of debate [3,255]. Clinically, the two conditions share striking similarities with the presence of longitudinally extensive optic neuritis and transverse myelitis, the presence of neutrophils and eosinophils in the CSF, and the absence of oligoclonal bands, despite a different underlying pathophysiology, the first being an astrocytopathy and the second a demyelinating disease. A peculiar pattern for MOG-NMOSD emerges, including clinically, the striking presentation of bilateral simultaneous/sequential optic neuritis with transverse myelitis; a dominant optic neuritis phenotype, with a relapse rate higher than MS or NMOSD, nevertheless an intermediate prognosis between MS-ON and AQP4-ON; a wide age distribution with the pediatric and geriatric population on both ends of the spectrum; a versatile clinical phenotypes that has optic neuritis as a component but does not fulfill MS or NMOSD diagnostic criteria such ADEM-ON [188], MDEM-ON [190,191], benign unilateral cortical encephalitis with epilepsy [192]. Pattern recognition extends to imaging with the presence of cortical lesions on MRI; a LEON with anterior involvement, perineural sheath swelling and optic nerve twisting. The course is less aggressive than AQP4-NMOSD due to the difference in the substrate of attack by the antibody. In MOG-optic neuritis, the antibody attacks the MOG antigen, leading to demyelination. In AQP4-optic neuritis, the antibody is pathogenic against the astrocyte, leading to direct neuronal and oligodendroglia damage. The similarity in clinical phenotype, however, is intriguing, and raises the possibility of a downstream common pathway for damage. Another intriguing finding is systemic autoimmunity, which is less often associated with MOG-NMOSD compared to AQP4-NMOSD. Overlap syndromes or NMOSD-encephalitis complex associated with NMOSD and neuronal antibodies (NMDA-R, VGKC, glycine receptor alpha 1 subunit antibodies) have been reported with both conditions but seem to be more common with NMOSD-MOG.

These findings parallel the auto-antigens present in AQP4- and MOG-NMOSD, the first being present in the central and peripheral nervous system and other organs, the latter being restricted to the nervous system. Whether the location of the antigen has a bearing on the presence or absence of systemic autoimmunity remains to be determined. Whether the proper classification of NMOSD should be based on clinical or biological phenotype with the identification of new target autoantigens remains unanswered. However, the nosology of NMOSD might need to be revised. With more antibodies being unraveled, the seropositive/seronegative terminology should be abandoned or modified, as it will become a source of confusion once the MOG antibody testing becomes widely available. We propose using AQP4-NMOSD and MOG-NMOSD for true seropositive NMOSD and undefined NMOSD when an antibody is unknown/not present. This terminology will allow the incorporation of future antibodies in the classification of NMOSD. Lastly, based on a recent article showing the MOG antibody against native MOG in patients with MS [256], a question remains unanswered: whether or not MOG-NMOSD is the nebulous borderland between MS and NMOSD.

Supplementary Materials: The following are available online at www.mdpi.com/link/2076-3425/7/10/138/s1, Table S1: Evolution of MRI diagnostic criteria for dissemination in time (DIT) and dissemination in space (DIS). CEL: Contrast-enhancing lesion, Table S2: Brain Imaging Findings in Neuromyelitis Optica Spectrum Disorder, Figure S1: 50-year-old female, with seronegative NMO and cloudlike enhancement on axial T1 with contrast enhancement (**1a**). 55-year-old African American female, with AQP4-NMOSD; presence of an ovoid right frontal juxtacortical/subcortical T2 hyperintensity (**1b**) with cloud like enhancement on axial T1 with contrast (**1c**). Repeat MRI of the brain 6 months later showed a significant improvement in T2 hyperintensity (**1d**) underlining the evanescent nature of NMOSD lesions, Figure S2: Axial FLAIR cuts (2a and 2c) demonstrate a right middle cerebellar peduncle and midbrain lesions with leptomeningeal contrast enhancement on T1 with contrast (2b and 2d) in a 50-year-old female with seronegative NMOSD, Figure S3: Sagittal and axial FLAIR MRI of the brain demonstrate diffuse involvement and swelling of the corpus callosum (3a and 3c) with high intensity rim and lower intensity core. Axial T1 with contrast demonstrates heterogeneous contrast enhancement. Repeat brain MRI (3b), 8 months later, demonstrates resolved edematous state with some callosal atrophy. Figure S4: 40-year-old Caucasian women presenting with intractable hiccups, nausea and vomiting and a dorsal brainstem lesion with a linear component involving the medulla and cervicomedullary junction seen on sagittal STIR (4a) and enhancing with contrast on T1. Aquaporin 4 antibody was positive. Figure S5: 43-year-old female, with AQP4-NMOSD; axial FLAIR demonstrates non-specific white matter lesions, (5a) a periventricular lesion around the posterior horn of the left lateral ventricle (5b), confirmed on sagittal FLAIR (5c).

Acknowledgments: We would like to thank Melody Montgomery for technical and language editing and proofreading and Crystal Upshaw for general administrative support.

Author Contributions: R.K.Z., R.S. and K.H. contributed equally to this review.

Conflicts of Interest: Rana K. Zabad has been a site investigator or site PI for clinical trials funded by Biogen, Genentech, Novartis, Sunpharma. In the last 2 years, she has served as a consultant for Bayer, Genzyme, TEVA Neuroscience and TG therapeutics and has given unbranded lectures sponsored by TEVA and is a member of the Adjudication Committee for a clinical trial of biotin in primary and secondary progressive multiple sclerosis sponsored by PAREXEL and medDay pharmaceutical. Renee Stewart has nothing to disclose. Kathleen Healey has received funding in the past for investigation-initiated research from the Multiple Sclerosis Foundation. In relation to this review specifically, the authors declare no conflict of interest.

Abbreviations

The following abbreviations are used in this manuscript:

ADEM	Acute Disseminated Encephalomyelitis
ADEM-ON	Acute Disseminated Encephalomyelitis-Optic Neuritis
ATM	Acute Transverse Myelitis
APTM	Acute Partial Transverse Myelitis
ACTM	Acute complete Transverse Myelitis
CRION	Chronic Relapsing Inflammatory Neuropathy
GFAP	Glial Fibrillary Acid Protein
LETM	Longitudinally Extensive Transverse Myelitis
MS	Multiple Sclerosis
MS-ON	MS-Associated Optic Neuritis
MS-TM	MS-Asociated Transverse Myelitis

MS-BS	MS-Asociated Brainstem Syndrome
MOG	Myelin Oligodendrocyte Glycoprotein
MOG-ON	Myelin Oligodendrocyte Glycoprotein- Associated Optic Neuritis
MOG-TM	Myelin Oligodendrocyte Glycoprotein- Associated Transverse Myelitis
MDEM-ON	Multiphasic Disseminated Encephalomyelitis-Optic Neuritis
MDEM-ON	Multiphasic Disseminated Encephalomyelitis-Optic Neuritis
APQ4	aquaporin 4
NMOSD	Neuromyelitis Optica Spectrum Disorder
NMOSD-ON	Neuromyelitis Optica Spectrum Disorder-Associated Optic Neuritis
NMOSD-TM	Neuromyelitis Optica Spectrum Disorder-Associated Transverse Myelitis Optic Neuritis
NMOSD-BS	Neuromyelitis Optica Spectrum Disorder-Associated Brainstem Syndrome
NMDA-R	N-Methyl-D-Aspartate Receptor
ON	Optic Neuritis
SION	Single Inflammatory or Isolated Optic Neuritis
RION	Recurrent Inflammatory or Isolated Optic Neuritis
SSTM	Short Segment Transverse Myelitis

References

1. Fang, B.; McKeon, A.; Hinson, S.R.; Kryzer, T.J.; Pittock, S.J.; Aksamit, A.J.; Lennon, V.A. Autoimmune Glial Fibrillary Acidic Protein Astrocytopathy: A Novel Meningoencephalomyelitis. *JAMA Neurol.* **2016**, *73*, 1297–1307. [CrossRef] [PubMed]

2. Holtje, M.; Mertens, R.; Schou, M.B.; Saether, S.G.; Kochova, E.; Jarius, S.; Pruss, H.; Komorowski, L.; Probst, C.; Paul, F.; et al. Synapsin-antibodies in psychiatric and neurological disorders: Prevalence and clinical findings. *Brain Behav. Immun.* **2017**, *66*, 125–134. [CrossRef] [PubMed]

3. Zamvil, S.S.; Slavin, A.J. Does MOG Ig-positive AQP4-seronegative opticospinal inflammatory disease justify a diagnosis of NMO spectrum disorder? *Neurol. Neuroimmunol. Neuroinflamm.* **2015**, *2*, e62. [CrossRef] [PubMed]

4. Zekeridou, A.; Lennon, V.A. Aquaporin-4 autoimmunity. *Neurol. Neuroimmunol. Neuroinflamm.* **2015**, *2*, e110. [CrossRef] [PubMed]

5. Poser, C.M.; Brinar, V.V. Diagnostic criteria for multiple sclerosis: An historical review. *Clin. Neurol. Neurosurg.* **2004**, *106*, 147–158. [CrossRef] [PubMed]

6. Schumacher, G.A.; Beebe, G.; Kibler, R.F.; Kurland, L.T.; Kurtzke, J.F.; Mcdowell, F.; Nagler, B.; Sibley, W.A.; Tourtellotte, W.W.; Willmon, T.L. Problems of Experimental Trials of Therapy in Multiple Sclerosis: Report by the Panel on the Evaluation of Experimental Trials of Therapy in Multiple Sclerosis. *Ann. N. Y. Acad. Sci.* **1965**, *122*, 552–568. [CrossRef] [PubMed]

7. Poser, C.M.; Paty, D.W.; Scheinberg, L.; McDonald, W.I.; Davis, F.A.; Ebers, G.C.; Johnson, K.P.; Sibley, W.A.; Silberberg, D.H.; Tourtellotte, W.W. New diagnostic criteria for multiple sclerosis: Guidelines for research protocols. *Ann. Neurol.* **1983**, *13*, 227–231. [CrossRef] [PubMed]

8. McDonald, W.I.; Compston, A.; Edan, G.; Goodkin, D.; Hartung, H.P.; Lublin, F.D.; McFarland, H.F.; Paty, D.W.; Polman, C.H.; Reingold, S.C.; et al. Recommended diagnostic criteria for multiple sclerosis: Guidelines from the International Panel on the diagnosis of multiple sclerosis. *Ann. Neurol.* **2001**, *50*, 121–127. [CrossRef] [PubMed]

9. Polman, C.H.; Reingold, S.C.; Banwell, B.; Clanet, M.; Cohen, J.A.; Filippi, M.; Fujihara, K.; Havrdova, E.; Hutchinson, M.; Kappos, L.; et al. Diagnostic criteria for multiple sclerosis: 2010 revisions to the McDonald criteria. *Ann. Neurol.* **2011**, *69*, 292–302. [CrossRef] [PubMed]

10. Polman, C.H.; Reingold, S.C.; Edan, G.; Filippi, M.; Hartung, H.P.; Kappos, L.; Lublin, F.D.; Metz, L.M.; McFarland, H.F.; O'Connor, P.W.; et al. Diagnostic criteria for multiple sclerosis: 2005 revisions to the "McDonald Criteria". *Ann. Neurol.* **2005**, *58*, 840–846. [CrossRef] [PubMed]

11. Rovira, A.; Wattjes, M.P.; Tintore, M.; Tur, C.; Yousry, T.A.; Sormani, M.P.; De Stefano, N.; Filippi, M.; Auger, C.; Rocca, M.A.; et al. Evidence-based guidelines: MAGNIMS consensus guidelines on the use of MRI in multiple sclerosis-clinical implementation in the diagnostic process. *Nat. Rev. Neurol.* **2015**, *11*, 471–482. [CrossRef] [PubMed]

12. Jarius, S.; Wildemann, B.; Paul, F. Neuromyelitis optica: Clinical features, immunopathogenesis and treatment. *Clin. Exp. Immunol.* **2014**, *176*, 149–164. [CrossRef] [PubMed]

13. Asgari, N.; Skejoe, H.P.; Lennon, V.A. Evolution of longitudinally extensive transverse myelitis in an aquaporin-4 IgG-positive patient. *Neurology* **2013**, *81*, 95–96. [CrossRef] [PubMed]

14. Ho, M.L.; Moonis, G.; Ginat, D.T.; Eisenberg, R.L. Lesions of the corpus callosum. *AJR Am. J. Roentgenol.* **2013**, *200*, W1–W16. [CrossRef] [PubMed]

15. Kazi, A.Z.; Joshi, P.C.; Kelkar, A.B.; Mahajan, M.S.; Ghawate, A.S. MRI evaluation of pathologies affecting the corpus callosum: A pictorial essay. *Indian J. Radiol. Imaging* **2013**, *23*, 321–332. [CrossRef] [PubMed]

16. Bourekas, E.C.; Varakis, K.; Bruns, D.; Christoforidis, G.A.; Baujan, M.; Slone, H.W.; Kehagias, D. Lesions of the corpus callosum: MR imaging and differential considerations in adults and children. *AJR Am. J. Roentgenol.* **2002**, *179*, 251–257. [CrossRef] [PubMed]

17. Jurynczyk, M.; Weinshenker, B.; Akman-Demir, G.; Asgari, N.; Barnes, D.; Boggild, M.; Chaudhuri, A.; D'hooghe, M.; Evangelou, N.; Geraldes, R.; et al. Status of diagnostic approaches to AQP4-IgG seronegative NMO and NMO/MS overlap syndromes. *J. Neurol.* **2016**, *263*, 140–149. [CrossRef] [PubMed]

18. Matthews, L.; Marasco, R.; Jenkinson, M.; Kuker, W.; Luppe, S.; Leite, M.I.; Giorgio, A.; De Stefano, N.; Robertson, N.; Johansen-Berg, H.; et al. Distinction of seropositive NMO spectrum disorder and MS brain lesion distribution. *Neurology* **2013**, *80*, 1330–1337. [CrossRef] [PubMed]

19. Chan, K.H.; Tse, C.T.; Chung, C.P.; Lee, R.L.; Kwan, J.S.; Ho, P.W.; Ho, J.W. Brain involvement in neuromyelitis optica spectrum disorders. *Arch. Neurol.* **2011**, *68*, 1432–1439. [CrossRef] [PubMed]

20. Wang, F.; Liu, Y.; Duan, Y.; Li, K. Brain MRI abnormalities in neuromyelitis optica. *Eur. J. Radiol.* **2011**, *80*, 445–449. [CrossRef] [PubMed]

21. Asgari, N.; Lillevang, S.T.; Skejoe, H.P.; Falah, M.; Stenager, E.; Kyvik, K.O. A population-based study of neuromyelitis optica in Caucasians. *Neurology* **2011**, *76*, 1589–1595. [CrossRef] [PubMed]

22. Jarius, S.; Ruprecht, K.; Kleiter, I.; Borisow, N.; Asgari, N.; Pitarokoili, K.; Pache, F.; Stich, O.; Beume, L.A.; Hummert, M.W.; et al. MOG-IgG in NMO and related disorders: A multicenter study of 50 patients. Part 2: Epidemiology, clinical presentation, radiological and laboratory features, treatment responses, and long-term outcome. *J. Neuroinflamm.* **2016**, *13*, 280. [CrossRef] [PubMed]

23. Kim, W.; Kim, S.H.; Huh, S.Y.; Kim, H.J. Brain abnormalities in neuromyelitis optica spectrum disorder. *Mult. Scler. Int.* **2012**, *2012*, 735486. [CrossRef] [PubMed]

24. Kim, H.J.; Paul, F.; Lana-Peixoto, M.A.; Tenembaum, S.; Asgari, N.; Palace, J.; Klawiter, E.C.; Sato, D.K.; de Seze, J.; Wuerfel, J.; et al. MRI characteristics of neuromyelitis optica spectrum disorder: An international update. *Neurology* **2015**, *84*, 1165–1173. [CrossRef] [PubMed]

25. Pekcevik, Y.; Orman, G.; Lee, I.H.; Mealy, M.A.; Levy, M.; Izbudak, I. What do we know about brain contrast enhancement patterns in neuromyelitis optica? *Clin. Imaging* **2016**, *40*, 573–580. [CrossRef] [PubMed]

26. Asgari, N.; Flanagan, E.P.; Fujihara, K.; Kim, H.J.; Skejoe, H.P.; Wuerfel, J.; Kuroda, H.; Kim, S.H.; Maillart, E.; Marignier, R.; et al. Disruption of the leptomeningeal blood barrier in neuromyelitis optica spectrum disorder. *Neurol. Neuroimmunol. Neuroinflamm.* **2017**, *4*, e343. [CrossRef] [PubMed]

27. Tahara, M.; Ito, R.; Tanaka, K.; Tanaka, M. Cortical and leptomeningeal involvement in three cases of neuromyelitis optica. *Eur. J. Neurol.* **2012**, *19*, e47–e48. [CrossRef] [PubMed]

28. Pittock, S.J.; Weinshenker, B.G.; Lucchinetti, C.F.; Wingerchuk, D.M.; Corboy, J.R.; Lennon, V.A. Neuromyelitis optica brain lesions localized at sites of high aquaporin 4 expression. *Arch. Neurol.* **2006**, *63*, 964–968. [CrossRef] [PubMed]

29. Jarius, S.; Kleiter, I.; Ruprecht, K.; Asgari, N.; Pitarokoili, K.; Borisow, N.; Hummert, M.W.; Trebst, C.; Pache, F.; Winkelmann, A.; et al. MOG-IgG in NMO and related disorders: A multicenter study of 50 patients. Part 3: Brainstem involvement—frequency, presentation and outcome. *J. Neuroinflamm.* **2016**, *13*, 281. [CrossRef] [PubMed]

30. Wingerchuk, D.M.; Hogancamp, W.F.; O'Brien, P.C.; Weinshenker, B.G. The clinical course of neuromyelitis optica (Devic's syndrome). *Neurology* **1999**, *53*, 1107–1114. [CrossRef] [PubMed]

31. Lennon, V.A.; Kryzer, T.J.; Pittock, S.J.; Verkman, A.S.; Hinson, S.R. IgG marker of optic-spinal multiple sclerosis binds to the aquaporin-4 water channel. *J. Exp. Med.* **2005**, *202*, 473–477. [CrossRef] [PubMed]

32. Papadopoulos, M.C.; Verkman, A.S. Aquaporin 4 and neuromyelitis optica. *Lancet Neurol.* **2012**, *11*, 535–544. [CrossRef]

33. Gomes, D.; Agasse, A.; Thiebaud, P.; Delrot, S.; Geros, H.; Chaumont, F. Aquaporins are multifunctional water and solute transporters highly divergent in living organisms. *Biochim. Biophys. Acta* **2009**, *1788*, 1213–1228. [CrossRef] [PubMed]

34. Verkman, A.S.; Phuan, P.W.; Asavapanumas, N.; Tradtrantip, L. Biology of AQP4 and anti-AQP4 antibody: Therapeutic implications for NMO. *Brain Pathol.* **2013**, *23*, 684–695. [CrossRef] [PubMed]

35. Jarius, S.; Paul, F.; Franciotta, D.; Waters, P.; Zipp, F.; Hohlfeld, R.; Vincent, A.; Wildemann, B. Mechanisms of disease: Aquaporin-4 antibodies in neuromyelitis optica. *Nat. Clin. Pract. Neurol.* **2008**, *4*, 202–214. [CrossRef] [PubMed]

36. Phuan, P.W.; Ratelade, J.; Rossi, A.; Tradtrantip, L.; Verkman, A.S. Complement-dependent cytotoxicity in neuromyelitis optica requires aquaporin-4 protein assembly in orthogonal arrays. *J. Biol. Chem.* **2012**, *287*, 13829–13839. [CrossRef] [PubMed]

37. Hoshi, A.; Yamamoto, T.; Kikuchi, S.; Soeda, T.; Shimizu, K.; Ugawa, Y. Aquaporin-4 expression in distal myopathy with rimmed vacuoles. *BMC Neurol.* **2012**, *12*. [CrossRef] [PubMed]

38. Guo, Y.; Lennon, V.A.; Popescu, B.F.; Grouse, C.K.; Topel, J.; Milone, M.; Lassmann, H.; Parisi, J.E.; Pittock, S.J.; Stefoski, D.; et al. Autoimmune aquaporin-4 myopathy in neuromyelitis optica spectrum. *JAMA Neurol.* **2014**, *71*, 1025–1029. [CrossRef] [PubMed]

39. He, D.; Li, Y.; Dai, Q.; Zhang, Y.; Xu, Z.; Li, Y.; Cai, G.; Chu, L. Myopathy associated with neuromyelitis optica spectrum disorders. *Int. J. Neurosci.* **2016**, *126*, 863–866. [CrossRef] [PubMed]

40. Ratelade, J.; Zhang, H.; Saadoun, S.; Bennett, J.L.; Papadopoulos, M.C.; Verkman, A.S. Neuromyelitis optica IgG and natural killer cells produce NMO lesions in mice without myelin loss. *Acta Neuropathol.* **2012**, *123*, 861–872. [CrossRef] [PubMed]

41. Ratelade, J.; Bennett, J.L.; Verkman, A.S. Evidence against cellular internalization in vivo of NMO-IgG, aquaporin-4, and excitatory amino acid transporter 2 in neuromyelitis optica. *J. Biol. Chem.* **2011**, *286*, 45156–45164. [CrossRef] [PubMed]

42. Wingerchuk, D.M.; Banwell, B.; Bennett, J.L.; Cabre, P.; Carroll, W.; Chitnis, T.; de Seze, J.; Fujihara, K.; Greenberg, B.; Jacob, A.; et al. International consensus diagnostic criteria for neuromyelitis optica spectrum disorders. *Neurology* **2015**, *85*, 177–189. [CrossRef] [PubMed]

43. Sepulveda, M.; Armangue, T.; Martinez-Hernandez, E.; Arrambide, G.; Sola-Valls, N.; Sabater, L.; Tellez, N.; Midaglia, L.; Arino, H.; Peschl, P.; et al. Clinical spectrum associated with MOG autoimmunity in adults: Significance of sharing rodent MOG epitopes. *J. Neurol.* **2016**, *263*, 1349–1360. [CrossRef] [PubMed]

44. Boyle, L.H.; Traherne, J.A.; Plotnek, G.; Ward, R.; Trowsdale, J. Splice variation in the cytoplasmic domains of myelin oligodendrocyte glycoprotein affects its cellular localisation and transport. *J. Neurochem.* **2007**, *102*, 1853–1862. [CrossRef] [PubMed]

45. Berger, T.; Rubner, P.; Schautzer, F.; Egg, R.; Ulmer, H.; Mayringer, I.; Dilitz, E.; Deisenhammer, F.; Reindl, M. Antimyelin antibodies as a predictor of clinically definite multiple sclerosis after a first demyelinating event. *N. Engl. J. Med.* **2003**, *349*, 139–145. [CrossRef] [PubMed]

46. Kuhle, J.; Pohl, C.; Mehling, M.; Edan, G.; Freedman, M.S.; Hartung, H.P.; Polman, C.H.; Miller, D.H.; Montalban, X.; Barkhof, F.; et al. Lack of association between antimyelin antibodies and progression to multiple sclerosis. *N. Engl. J. Med.* **2007**, *356*, 371–378. [CrossRef] [PubMed]

47. Reindl, M.; Di Pauli, F.; Rostasy, K.; Berger, T. The spectrum of MOG autoantibody-associated demyelinating diseases. *Nat. Rev. Neurol.* **2013**, *9*, 455–461. [CrossRef] [PubMed]

48. Waters, P.; Woodhall, M.; O'Connor, K.C.; Reindl, M.; Lang, B.; Sato, D.K.; Jurynczyk, M.; Tackley, G.; Rocha, J.; Takahashi, T.; et al. MOG cell-based assay detects non-MS patients with inflammatory neurologic disease. *Neurol. Neuroimmunol. Neuroinflamm.* **2015**, *2*, e89. [CrossRef] [PubMed]

49. Brilot, F.; Dale, R.C.; Selter, R.C.; Grummel, V.; Kalluri, S.R.; Aslam, M.; Busch, V.; Zhou, D.; Cepok, S.; Hemmer, B. Antibodies to native myelin oligodendrocyte glycoprotein in children with inflammatory demyelinating central nervous system disease. *Ann. Neurol.* **2009**, *66*, 833–842. [CrossRef] [PubMed]

50. Di Pauli, F.; Mader, S.; Rostasy, K.; Schanda, K.; Bajer-Kornek, B.; Ehling, R.; Deisenhammer, F.; Reindl, M.; Berger, T. Temporal dynamics of anti-MOG antibodies in CNS demyelinating diseases. *Clin. Immunol.* **2011**, *138*, 247–254. [CrossRef] [PubMed]

51. Jarius, S.; Ruprecht, K.; Kleiter, I.; Borisow, N.; Asgari, N.; Pitarokoili, K.; Pache, F.; Stich, O.; Beume, L.A.; Hummert, M.W.; et al. MOG-IgG in NMO and related disorders: A multicenter study of 50 patients. Part 1: Frequency, syndrome specificity, influence of disease activity, long-term course, association with AQP4-IgG, and origin. *J. Neuroinflamm.* **2016**, *13*, 279. [CrossRef] [PubMed]

52. Kortvelyessy, P.; Breu, M.; Pawlitzki, M.; Metz, I.; Heinze, H.J.; Matzke, M.; Mawrin, C.; Rommer, P.; Kovacs, G.G.; Mitter, C.; et al. ADEM-like presentation, anti-MOG antibodies, and MS pathology: TWO case reports. *Neurol. Neuroimmunol. Neuroinflamm.* **2017**, *4*, e335. [CrossRef] [PubMed]

53. Nakajima, H.; Motomura, M.; Tanaka, K.; Fujikawa, A.; Nakata, R.; Maeda, Y.; Shima, T.; Mukaino, A.; Yoshimura, S.; Miyazaki, T.; et al. Antibodies to myelin oligodendrocyte glycoprotein in idiopathic optic neuritis. *BMJ Open* **2015**, *5*, e007766. [CrossRef] [PubMed]

54. De Seze, J. Inflammatory Optic Neuritis: From Multiple Sclerosis to Neuromyelitis Optica. *Neuroophthalmology* **2013**, *37*, 141–145. [CrossRef] [PubMed]

55. Piccolo, L.; Woodhall, M.; Tackley, G.; Jurynczyk, M.; Kong, Y.; Domingos, J.; Gore, R.; Vincent, A.; Waters, P.; Leite, M.I.; et al. Isolated new onset 'atypical' optic neuritis in the NMO clinic: Serum antibodies, prognoses and diagnoses at follow-up. *J. Neurol.* **2016**, *263*, 370–379. [CrossRef] [PubMed]

56. Benoilid, A.; Tilikete, C.; Collongues, N.; Arndt, C.; Vighetto, A.; Vignal, C.; de Seze, J. Relapsing optic neuritis: A multicentre study of 62 patients. *Mult. Scler.* **2014**, *20*, 848–853. [CrossRef] [PubMed]

57. Pirko, I.; Blauwet, L.A.; Lesnick, T.G.; Weinshenker, B.G. The natural history of recurrent optic neuritis. *Arch. Neurol.* **2004**, *61*, 1401–1405. [CrossRef] [PubMed]

58. Kidd, D.; Burton, B.; Plant, G.T.; Graham, E.M. Chronic relapsing inflammatory optic neuropathy (CRION). *Brain* **2003**, *126*, 276–284. [CrossRef] [PubMed]

59. Petzold, A.; Plant, G.T. Chronic relapsing inflammatory optic neuropathy: A systematic review of 122 cases reported. *J. Neurol.* **2014**, *261*, 17–26. [CrossRef] [PubMed]

60. Thorne, J.E.; Wittenberg, S.; Kedhar, S.R.; Dunn, J.P.; Jabs, D.A. Optic neuropathy complicating multifocal choroiditis and panuveitis. *Am. J. Ophthalmol.* **2007**, *143*, 721–723. [CrossRef] [PubMed]

61. Chalmoukou, K.; Alexopoulos, H.; Akrivou, S.; Stathopoulos, P.; Reindl, M.; Dalakas, M.C. Anti-MOG antibodies are frequently associated with steroid-sensitive recurrent optic neuritis. *Neurol. Neuroimmunol. Neuroinflamm.* **2015**, *2*, e131. [CrossRef] [PubMed]

62. Matsuda, R.; Kezuka, T.; Umazume, A.; Okunuki, Y.; Goto, H.; Tanaka, K. Clinical Profile of Anti-Myelin Oligodendrocyte Glycoprotein Antibody Seropositive Cases of Optic Neuritis. *Neuroophthalmology* **2015**, *39*, 213–219. [CrossRef] [PubMed]

63. Confavreux, C.; Vukusic, S.; Moreau, T.; Adeleine, P. Relapses and progression of disability in multiple sclerosis. *N. Engl. J. Med.* **2000**, *343*, 1430–1438. [CrossRef] [PubMed]

64. Brass, S.D.; Zivadinov, R.; Bakshi, R. Acute demyelinating optic neuritis: A review. *Front. Biosci.* **2008**, *13*, 2376–2390. [CrossRef] [PubMed]

65. Rodriguez, M.; Siva, A.; Cross, S.A.; O'Brien, P.C.; Kurland, L.T. Optic neuritis: A population-based study in Olmsted County, Minnesota. *Neurology* **1995**, *45*, 244–250. [CrossRef] [PubMed]

66. Galetta, S.L.; Villoslada, P.; Levin, N.; Shindler, K.; Ishikawa, H.; Parr, E.; Cadavid, D.; Balcer, L.J. Acute optic neuritis: Unmet clinical needs and model for new therapies. *Neurol. Neuroimmunol. Neuroinflamm.* **2015**, *2*, e135. [CrossRef] [PubMed]

67. Petzold, A.; Wattjes, M.P.; Costello, F.; Flores-Rivera, J.; Fraser, C.L.; Fujihara, K.; Leavitt, J.; Marignier, R.; Paul, F.; Schippling, S.; et al. The investigation of acute optic neuritis: A review and proposed protocol. *Nat. Rev. Neurol.* **2014**, *10*, 447–458. [CrossRef] [PubMed]

68. Swartz, N.G.; Beck, R.W.; Savino, P.J.; Sergott, R.C.; Bosley, T.M.; Lam, B.L.; Drucker, M.; Katz, B. Pain in anterior ischemic optic neuropathy. *J. Neuroophthalmol.* **1995**, *15*, 9–10. [CrossRef] [PubMed]

69. Beck, R.W.; Cleary, P.A.; Anderson, M.M., Jr.; Keltner, J.L.; Shults, W.T.; Kaufman, D.I.; Buckley, E.G.; Corbett, J.J.; Kupersmith, M.J.; Miller, N.R. A randomized, controlled trial of corticosteroids in the treatment of acute optic neuritis. The Optic Neuritis Study Group. *N. Engl. J. Med.* **1992**, *326*, 581–588. [CrossRef] [PubMed]

70. Viana, M.; Sprenger, T.; Andelova, M.; Goadsby, P.J. The typical duration of migraine aura: A systematic review. *Cephalalgia* **2013**, *33*, 483–490. [CrossRef] [PubMed]

71. The clinical profile of optic neuritis. Experience of the Optic Neuritis Treatment Trial. Optic Neuritis Study Group. *Arch. Ophthalmol.* **1991**, *109*, 1673–1678.

72. Beck, R.W. The optic neuritis treatment trial. Implications for clinical practice. Optic Neuritis Study Group. *Arch. Ophthalmol.* **1992**, *110*, 331–332. [CrossRef] [PubMed]

73. Abràm, M.D.; Baker, M.S. *New Concepts in Orbital Imaging*; Karcioglu, Z., Ed.; Orbital Tumors; Springer: New York, NY, USA, 2014; pp. 111–121.

74. Akaishi, T.; Nakashima, I.; Takeshita, T.; Mugikura, S.; Sato, D.K.; Takahashi, T.; Nishiyama, S.; Kurosawa, K.; Misu, T.; Nakazawa, T.; et al. Lesion length of optic neuritis impacts visual prognosis in neuromyelitis optica. *J. Neuroimmunol.* **2016**, *293*, 28–33. [CrossRef] [PubMed]

75. Matiello, M.; Lennon, V.A.; Jacob, A.; Pittock, S.J.; Lucchinetti, C.F.; Wingerchuk, D.M.; Weinshenker, B.G. NMO-IgG predicts the outcome of recurrent optic neuritis. *Neurology* **2008**, *70*, 2197–2200. [CrossRef] [PubMed]

76. Buch, D.; Savatovsky, J.; Gout, O.; Vignal, C.; Deschamps, R. Combined brain and anterior visual pathways' MRIs assist in early identification of neuromyelitis optica spectrum disorder at onset of optic neuritis. *Acta Neurol. Belg.* **2017**, *117*, 67–74. [CrossRef] [PubMed]

77. Storoni, M.; Davagnanam, I.; Radon, M.; Siddiqui, A.; Plant, G.T. Distinguishing optic neuritis in neuromyelitis optica spectrum disease from multiple sclerosis: A novel magnetic resonance imaging scoring system. *J. Neuroophthalmol.* **2013**, *33*, 123–127. [CrossRef] [PubMed]

78. Ramanathan, S.; Prelog, K.; Barnes, E.H.; Tantsis, E.M.; Reddel, S.W.; Henderson, A.P.; Vucic, S.; Gorman, M.P.; Benson, L.A.; Alper, G.; et al. Radiological differentiation of optic neuritis with myelin oligodendrocyte glycoprotein antibodies, aquaporin-4 antibodies, and multiple sclerosis. *Mult. Scler.* **2016**, *22*, 470–482. [CrossRef] [PubMed]

79. Stiebel-Kalish, H.; Lotan, I.; Brody, J.; Chodick, G.; Bialer, O.; Marignier, R.; Bach, M.; Hellmann, M.A. Retinal Nerve Fiber Layer May Be Better Preserved in MOG-IgG versus AQP4-IgG Optic Neuritis: A Cohort Study. *PLoS ONE* **2017**, *12*, e0170847. [CrossRef] [PubMed]

80. Havla, J.; Kumpfel, T.; Schinner, R.; Spadaro, M.; Schuh, E.; Meinl, E.; Hohlfeld, R.; Outteryck, O. Myelin-oligodendrocyte-glycoprotein (MOG) autoantibodies as potential markers of severe optic neuritis and subclinical retinal axonal degeneration. *J. Neurol.* **2017**, *264*, 139–151. [CrossRef] [PubMed]

81. Pache, F.; Zimmermann, H.; Mikolajczak, J.; Schumacher, S.; Lacheta, A.; Oertel, F.C.; Bellmann-Strobl, J.; Jarius, S.; Wildemann, B.; Reindl, M.; et al. MOG-IgG in NMO and related disorders: A multicenter study of 50 patients. Part 4: Afferent visual system damage after optic neuritis in MOG-IgG-seropositive versus AQP4-IgG-seropositive patients. *J. Neuroinflamm.* **2016**, *13*, 282. [CrossRef] [PubMed]

82. Schmidt, F.; Zimmermann, H.; Mikolajczak, J.; Oertel, F.C.; Pache, F.; Weinhold, M.; Schinzel, J.; Bellmann-Strobl, J.; Ruprecht, K.; Paul, F.; et al. Severe structural and functional visual system damage leads to profound loss of vision-related quality of life in patients with neuromyelitis optica spectrum disorders. *Mult. Scler. Relat. Disord.* **2017**, *11*, 45–50. [CrossRef] [PubMed]

83. Zhou, L.; Huang, Y.; Li, H.; Fan, J.; Zhangbao, J.; Yu, H.; Li, Y.; Lu, J.; Zhao, C.; Lu, C.; Wang, M.; Quan, C. MOG-antibody associated demyelinating disease of the CNS: A clinical and pathological study in Chinese Han patients. *J. Neuroimmunol.* **2017**, *305*, 19–28. [CrossRef] [PubMed]

84. Ford, B.; Tampieri, D.; Francis, G. Long-term follow-up of acute partial transverse myelopathy. *Neurology* **1992**, *42*, 250–252. [CrossRef] [PubMed]

85. Transverse Myelitis Consortium Working Group. Proposed diagnostic criteria and nosology of acute transverse myelitis. *Neurology* **2002**, *59*, 499–505.

86. Morrissey, S.P.; Miller, D.H.; Kendall, B.E.; Kingsley, D.P.; Kelly, M.A.; Francis, D.A.; MacManus, D.G.; McDonald, W.I. The significance of brain magnetic resonance imaging abnormalities at presentation with clinically isolated syndromes suggestive of multiple sclerosis. A 5-year follow-up study. *Brain* **1993**, *116 Pt 1*, 135–146. [CrossRef] [PubMed]

87. Brex, P.A.; Ciccarelli, O.; O'Riordan, J.I.; Sailer, M.; Thompson, A.J.; Miller, D.H. A longitudinal study of abnormalities on MRI and disability from multiple sclerosis. *N. Engl. J. Med.* **2002**, *346*, 158–164. [CrossRef] [PubMed]

88. Tintore, M.; Rovira, A.; Hernandez, D.; Rio, J.; Marzo, M.E.; Montalban, X. Optic neuritis, brain stem syndromes and myelitis: Rapid conversion to multiple sclerosis. *Med. Clin.* **1999**, *112*, 693–694.

89. Perumal, J.; Zabad, R.; Caon, C.; MacKenzie, M.; Tselis, A.; Bao, F.; Latif, Z.; Zak, I.; Lisak, R.; Khan, O. Acute transverse myelitis with normal brain MRI: Long-term risk of MS. *J. Neurol.* **2008**, *255*, 89–93. [CrossRef] [PubMed]

90. Scott, T.F.; Kassab, S.L.; Singh, S. Acute partial transverse myelitis with normal cerebral magnetic resonance imaging: Transition rate to clinically definite multiple sclerosis. *Mult. Scler.* **2005**, *11*, 373–377. [CrossRef] [PubMed]

91. Schmalstieg, W.F.; Keegan, B.M.; Weinshenker, B.G. Solitary sclerosis: Progressive myelopathy from solitary demyelinating lesion. *Neurology* **2012**, *78*, 540–544. [CrossRef] [PubMed]

92. Keegan, B.M.; Kaufmann, T.J.; Weinshenker, B.G.; Kantarci, O.H.; Schmalstieg, W.F.; Paz Soldan, M.M.; Flanagan, E.P. Progressive solitary sclerosis: Gradual motor impairment from a single CNS demyelinating lesion. *Neurology* **2016**, *87*, 1713–1719. [CrossRef] [PubMed]

93. Ayrignac, X.; Carra-Dalliere, C.; Homeyer, P.; Labauge, P. Solitary sclerosis: Progressive myelopathy from solitary demyelinating lesion. A new entity? *Acta Neurol. Belg.* **2013**, *113*, 533–534. [CrossRef] [PubMed]

94. Lebrun, C.; Cohen, M.; Mondot, L.; Ayrignac, X.; Labauge, P. A Case Report of Solitary Sclerosis: This is Really Multiple Sclerosis. *Neurol. Ther.* **2017**. [CrossRef] [PubMed]

95. Rathnasabapathi, D.; Elsone, L.; Krishnan, A.; Young, C.; Larner, A.; Jacob, A. Solitary sclerosis: Progressive neurological deficit from a spatially isolated demyelinating lesion: A further report. *J. Spinal Cord Med.* **2015**, *38*, 551–555. [CrossRef] [PubMed]

96. Lattanzi, S.; Logullo, F.; Di Bella, P.; Silvestrini, M.; Provinciali, L. Multiple sclerosis, solitary sclerosis or something else? *Mult. Scler.* **2014**, *20*, 1819–1824. [CrossRef] [PubMed]

97. Awad, A.; Stuve, O. Idiopathic transverse myelitis and neuromyelitis optica: Clinical profiles, pathophysiology and therapeutic choices. *Curr. Neuropharmacol.* **2011**, *9*, 417–428. [CrossRef] [PubMed]

98. Sepulveda, M.; Armangue, T.; Sola-Valls, N.; Arrambide, G.; Meca-Lallana, J.E.; Oreja-Guevara, C.; Mendibe, M.; Alvarez de Arcaya, A.; Aladro, Y.; Casanova, B.; et al. Neuromyelitis optica spectrum disorders: Comparison according to the phenotype and serostatus. *Neurol. Neuroimmunol. Neuroinflamm.* **2016**, *3*, e225. [CrossRef] [PubMed]

99. Scott, T.F. Nosology of idiopathic transverse myelitis syndromes. *Acta Neurol. Scand.* **2007**, *115*, 371–376. [CrossRef] [PubMed]

100. Tobin, W.O.; Weinshenker, B.G.; Lucchinetti, C.F. Longitudinally extensive transverse myelitis. *Curr. Opin. Neurol.* **2014**, *27*, 279–289. [CrossRef] [PubMed]

101. Trebst, C.; Raab, P.; Voss, E.V.; Rommer, P.; Abu-Mugheisib, M.; Zettl, U.K.; Stangel, M. Longitudinal extensive transverse myelitis–it's not all neuromyelitis optica. *Nat. Rev. Neurol.* **2011**, *7*, 688–698. [CrossRef] [PubMed]

102. Kister, I.; Johnson, E.; Raz, E.; Babb, J.; Loh, J.; Shepherd, T.M. Specific MRI findings help distinguish acute transverse myelitis of Neuromyelitis Optica from spinal cord infarction. *Mult. Scler. Relat. Disord.* **2016**, *9*, 62–67. [CrossRef] [PubMed]

103. Flanagan, E.P.; Krecke, K.N.; Marsh, R.W.; Giannini, C.; Keegan, B.M.; Weinshenker, B.G. Specific pattern of gadolinium enhancement in spondylotic myelopathy. *Ann. Neurol.* **2014**, *76*, 54–65. [CrossRef] [PubMed]

104. Hyun, J.W.; Kim, S.H.; Huh, S.Y.; Kim, W.; Yun, J.; Joung, A.; Sato, D.K.; Fujihara, K.; Kim, H.J. Idiopathic aquaporin-4 antibody negative longitudinally extensive transverse myelitis. *Mult. Scler.* **2015**, *21*, 710–717. [CrossRef] [PubMed]

105. Iorio, R.; Damato, V.; Mirabella, M.; Evoli, A.; Marti, A.; Plantone, D.; Frisullo, G.; Batocchi, A.P. Distinctive clinical and neuroimaging characteristics of longitudinally extensive transverse myelitis associated with aquaporin-4 autoantibodies. *J. Neurol.* **2013**, *260*, 2396–2402. [CrossRef] [PubMed]

106. Kitley, J.; Leite, M.I.; Kuker, W.; Quaghebeur, G.; George, J.; Waters, P.; Woodhall, M.; Vincent, A.; Palace, J. Longitudinally extensive transverse myelitis with and without aquaporin 4 antibodies. *JAMA Neurol.* **2013**, *70*, 1375–1381. [CrossRef] [PubMed]

107. Cobo-Calvo, A.; Sepulveda, M.; Bernard-Valnet, R.; Ruiz, A.; Brassat, D.; Martinez-Yelamos, S.; Saiz, A.; Marignier, R. Antibodies to myelin oligodendrocyte glycoprotein in aquaporin 4 antibody seronegative longitudinally extensive transverse myelitis: Clinical and prognostic implications. *Mult. Scler.* **2016**, *22*, 312–319. [CrossRef] [PubMed]

108. Sato, D.K.; Callegaro, D.; Lana-Peixoto, M.A.; Waters, P.J.; de Haidar Jorge, F.M.; Takahashi, T.; Nakashima, I.; Apostolos-Pereira, S.L.; Talim, N.; Simm, R.F.; et al. Distinction between MOG antibody-positive and AQP4 antibody-positive NMO spectrum disorders. *Neurology* **2014**, *82*, 474–481. [CrossRef] [PubMed]

109. Kitley, J.; Waters, P.; Woodhall, M.; Leite, M.I.; Murchison, A.; George, J.; Kuker, W.; Chandratre, S.; Vincent, A.; Palace, J. Neuromyelitis optica spectrum disorders with aquaporin-4 and myelin-oligodendrocyte glycoprotein antibodies: A comparative study. *JAMA Neurol.* **2014**, *71*, 276–283. [CrossRef] [PubMed]

110. Flanagan, E.P.; Weinshenker, B.G.; Krecke, K.N.; Lennon, V.A.; Lucchinetti, C.F.; McKeon, A.; Wingerchuk, D.M.; Shuster, E.A.; Jiao, Y.; Horta, E.S.; et al. Short myelitis lesions in aquaporin-4-IgG-positive neuromyelitis optica spectrum disorders. *JAMA Neurol.* **2015**, *72*, 81–87. [CrossRef] [PubMed]

111. Jarius, S.; Ruprecht, K.; Wildemann, B.; Kuempfel, T.; Ringelstein, M.; Geis, C.; Kleiter, I.; Kleinschnitz, C.; Berthele, A.; Brettschneider, J.; et al. Contrasting disease patterns in seropositive and seronegative neuromyelitis optica: A multicentre study of 175 patients. *J. Neuroinflamm.* **2012**, *9*, 14. [CrossRef] [PubMed]

112. Huh, S.Y.; Kim, S.H.; Hyun, J.W.; Jeong, I.H.; Park, M.S.; Lee, S.H.; Kim, H.J. Short segment myelitis as a first manifestation of neuromyelitis optica spectrum disorders. *Mult. Scler.* **2017**, *23*, 413–419. [CrossRef] [PubMed]

113. Cai, W.; Tan, S.; Zhang, L.; Shan, Y.; Wang, Y.; Lin, Y.; Zhou, F.; Zhang, B.; Chen, X.; Zhou, L.; et al. Linear lesions may assist early diagnosis of neuromyelitis optica and longitudinally extensive transverse myelitis, two subtypes of NMOSD. *J. Neurol. Sci.* **2016**, *360*, 88–93. [CrossRef] [PubMed]

114. Yonezu, T.; Ito, S.; Mori, M.; Ogawa, Y.; Makino, T.; Uzawa, A.; Kuwabara, S. "Bright spotty lesions" on spinal magnetic resonance imaging differentiate neuromyelitis optica from multiple sclerosis. *Mult. Scler.* **2014**, *20*, 331–337. [CrossRef] [PubMed]

115. Pekcevik, Y.; Mitchell, C.H.; Mealy, M.A.; Orman, G.; Lee, I.H.; Newsome, S.D.; Thompson, C.B.; Pardo, C.A.; Calabresi, P.A.; Levy, M.; et al. Differentiating neuromyelitis optica from other causes of longitudinally extensive transverse myelitis on spinal magnetic resonance imaging. *Mult. Scler.* **2016**, *22*, 302–311. [CrossRef] [PubMed]

116. Gass, A.; Filippi, M.; Rodegher, M.E.; Schwartz, A.; Comi, G.; Hennerici, M.G. Characteristics of chronic MS lesions in the cerebrum, brainstem, spinal cord, and optic nerve on T1-weighted MRI. *Neurology* **1998**, *50*, 548–550. [CrossRef] [PubMed]

117. Matthews, L.A.; Palace, J.A. The role of imaging in diagnosing neuromyelitis optica spectrum disorder. *Mult. Scler. Relat. Disord.* **2014**, *3*, 284–293. [CrossRef] [PubMed]

118. Zaffaroni, M.; Baldini, S.M.; Ghezzi, A. Cranial nerve, brainstem and cerebellar syndromes in the differential diagnosis of multiple sclerosis. *Neurol. Sci.* **2001**, *22* (Suppl. 2), S74–S78. [CrossRef] [PubMed]

119. Renaud, M.; Aupy, J.; Camuset, G.; Collongues, N.; Chanson, J.B.; de Seze, J.; Blanc, F. Chronic Bickerstaff's encephalitis with cognitive impairment, a reality? *BMC Neurol.* **2014**, *14*, 99. [CrossRef] [PubMed]

120. Gass, A.; Filippi, M.; Grossman, R.I. The contribution of MRI in the differential diagnosis of posterior fossa damage. *J. Neurol. Sci.* **2000**, *172* (Suppl. 1), S43–S49. [CrossRef]

121. Hooge, J.P.; Redekop, W.K. Trigeminal neuralgia in multiple sclerosis. *Neurology* **1995**, *45*, 1294–1296. [CrossRef] [PubMed]

122. Van der Meijs, A.H.; Tan, I.L.; Barkhof, F. Incidence of enhancement of the trigeminal nerve on MRI in patients with multiple sclerosis. *Mult. Scler.* **2002**, *8*, 64–67. [CrossRef] [PubMed]

123. Mills, R.J.; Young, C.A.; Smith, E.T. Central trigeminal involvement in multiple sclerosis using high-resolution MRI at 3 T. *Br. J. Radiol.* **2010**, *83*, 493–498. [CrossRef] [PubMed]

124. Da Silva, C.J.; da Rocha, A.J.; Mendes, M.F.; Maia, A.C., Jr.; Braga, F.T.; Tilbery, C.P. Trigeminal involvement in multiple sclerosis: Magnetic resonance imaging findings with clinical correlation in a series of patients. *Mult. Scler.* **2005**, *11*, 282–285. [CrossRef] [PubMed]

125. Swinnen, C.; Lunskens, S.; Deryck, O.; Casselman, J.; Vanopdenbosch, L. MRI characteristics of trigeminal nerve involvement in patients with multiple sclerosis. *Mult. Scler. Relat. Disord.* **2013**, *2*, 200–203. [CrossRef] [PubMed]

126. Barr, D.; Kupersmith, M.J.; Turbin, R.; Bose, S.; Roth, R. Isolated sixth nerve palsy: An uncommon presenting sign of multiple sclerosis. *J. Neurol.* **2000**, *247*, 701–704. [CrossRef] [PubMed]

127. Jacobson, D.M.; Moster, M.L.; Eggenberger, E.R.; Galetta, S.L.; Liu, G.T. Isolated trochlear nerve palsy in patients with multiple sclerosis. *Neurology* **1999**, *53*, 877–879. [CrossRef] [PubMed]

128. Nerrant, E.; Tilikete, C. Ocular Motor Manifestations of Multiple Sclerosis. *J. Neuroophthalmol.* **2017**, *33*, 332–340. [CrossRef] [PubMed]

129. Papadopoulou, A.; Naegelin, Y.; Weier, K.; Amann, M.; Hirsch, J.; von Felten, S.; Yaldizli, O.; Sprenger, T.; Radue, E.W.; Kappos, L.; et al. MRI characteristics of periaqueductal lesions in multiple sclerosis. *Mult. Scler. Relat. Disord.* **2014**, *3*, 542–551. [CrossRef] [PubMed]

130. Fukazawa, T.; Moriwaka, F.; Hamada, K.; Hamada, T.; Tashiro, K. Facial palsy in multiple sclerosis. *J. Neurol.* **1997**, *244*, 631–633. [CrossRef] [PubMed]

131. Saleh, C.; Patsi, O.; Mataigne, F.; Beyenburg, S. Peripheral (Seventh) Nerve Palsy and Multiple Sclerosis: A Diagnostic Dilemma—A Case Report. *Case Rep. Neurol.* **2016**, *8*, 27–33. [CrossRef] [PubMed]

132. Kutzelnigg, A.; Faber-Rod, J.C.; Bauer, J.; Lucchinetti, C.F.; Sorensen, P.S.; Laursen, H.; Stadelmann, C.; Bruck, W.; Rauschka, H.; Schmidbauer, M.; et al. Widespread demyelination in the cerebellar cortex in multiple sclerosis. *Brain Pathol.* **2007**, *17*, 38–44. [CrossRef] [PubMed]

133. Tornes, L.; Conway, B.; Sheremata, W. Multiple sclerosis and the cerebellum. *Neurol. Clin.* **2014**, *32*, 957–977. [CrossRef] [PubMed]

134. Takahashi, T.; Miyazawa, I.; Misu, T.; Takano, R.; Nakashima, I.; Fujihara, K.; Tobita, M.; Itoyama, Y. Intractable hiccup and nausea in neuromyelitis optica with anti-aquaporin-4 antibody: A herald of acute exacerbations. *J. Neurol. Neurosurg. Psychiatry* **2008**, *79*, 1075–1078. [CrossRef] [PubMed]

135. Apiwattanakul, M.; Popescu, B.F.; Matiello, M.; Weinshenker, B.G.; Lucchinetti, C.F.; Lennon, V.A.; McKeon, A.; Carpenter, A.F.; Miller, G.M.; Pittock, S.J. Intractable vomiting as the initial presentation of neuromyelitis optica. *Ann. Neurol.* **2010**, *68*, 757–761. [CrossRef] [PubMed]

136. Cheng, C.; Jiang, Y.; Lu, X.; Gu, F.; Kang, Z.; Dai, Y.; Lu, Z.; Hu, X. The role of anti-aquaporin 4 antibody in the conversion of acute brainstem syndrome to neuromyelitis optica. *BMC Neurol.* **2016**, *16*, 203. [CrossRef] [PubMed]

137. Wang, Y.; Zhang, L.; Zhang, B.; Dai, Y.; Kang, Z.; Lu, C.; Qiu, W.; Hu, X.; Lu, Z. Comparative clinical characteristics of neuromyelitis optica spectrum disorders with and without medulla oblongata lesions. *J. Neurol.* **2014**, *261*, 954–962. [CrossRef] [PubMed]

138. Shinoda, K.; Matsushita, T.; Furuta, K.; Isobe, N.; Yonekawa, T.; Ohyagi, Y.; Kira, J. Wall-eyed bilateral internuclear ophthalmoplegia (WEBINO) syndrome in a patient with neuromyelitis optica spectrum disorder and anti-aquaporin-4 antibody. *Mult. Scler.* **2011**, *17*, 885–887. [CrossRef] [PubMed]

139. Lee, J.; Jeong, S.H.; Park, S.M.; Sohn, E.H.; Lee, A.Y.; Kim, J.M.; Jo, H.J.; Lee, Y.H.; Kim, J.S. Anti-aquaporin-4 antibody-positive dorsal midbrain syndrome. *Mult. Scler.* **2015**, *21*, 477–480. [CrossRef] [PubMed]

140. Hage, R., Jr.; Merle, H.; Jeannin, S.; Cabre, P. Ocular oscillations in the neuromyelitis optica spectrum. *J. Neuroophthalmol.* **2011**, *31*, 255–259. [CrossRef] [PubMed]

141. Ogasawara, M.; Sakai, T.; Kono, Y.; Shikishima, K.; Tsuneoka, H. A limited form of neuromyelitis optica with a lesion of the fourth nerve nucleus. *J. Neuroophthalmol.* **2013**, *33*, 414–416. [CrossRef] [PubMed]

142. Rizek, P.; Nicolle, D.; Yeow Tay, K.; Kremenchutzky, M. Bilateral trochlear nerve palsy in a patient with neuromyelitis optica. *Mult. Scler. Relat. Disord.* **2014**, *3*, 273–275. [CrossRef] [PubMed]

143. Uludag, I.F.; Sariteke, A.; Ocek, L.; Zorlu, Y.; Sener, U.; Tokucoglu, F.; Uludag, B. Neuromyelitis optica presenting with horner syndrome: A case report and review of literature. *Mult. Scler. Relat. Disord.* **2017**, *14*, 32–34. [CrossRef] [PubMed]

144. De Seze, J.; Vukusic, S.; Viallet-Marcel, M.; Tilikete, C.; Zephir, H.; Delalande, S.; Stojkovic, T.; Defoort-Dhellemmes, S.; Confavreux, C.; Vermersch, P. Unusual ocular motor findings in multiple sclerosis. *J. Neurol. Sci.* **2006**, *243*, 91–95. [CrossRef] [PubMed]

145. Giles, C.L.; Henderson, J.W. Horner's syndrome: An analysis of 216 cases. *Am. J. Ophthalmol.* **1958**, *46*, 289–296. [CrossRef]

146. Walton, K.A.; Buono, L.M. Horner syndrome. *Curr. Opin. Ophthalmol.* **2003**, *14*, 357–363. [CrossRef] [PubMed]

147. Lana-Peixoto, M.A.; Callegaro, D.; Talim, N.; Talim, L.E.; Pereira, S.A.; Campos, G.B.; Brazilian Committee for Treatment and Research in Multiple Sclerosis. Pathologic yawning in neuromyelitis optica spectrum disorders. *Mult. Scler. Relat. Disord.* **2014**, *3*, 527–532. [CrossRef] [PubMed]

148. Shan, F.; Zhong, R.; Wu, L.; Fan, Y.; Long, Y.; Gao, C. Neuromyelitis optica spectrum disorders may be misdiagnosed as Wernicke's encephalopathy. *Int. J. Neurosci.* **2016**, *126*, 922–927. [CrossRef] [PubMed]

149. Jarius, S.; Kleffner, I.; Dorr, J.M.; Sastre-Garriga, J.; Illes, Z.; Eggenberger, E.; Chalk, C.; Ringelstein, M.; Aktas, O.; Montalban, X.; et al. Clinical, paraclinical and serological findings in Susac syndrome: An international multicenter study. *J. Neuroinflamm.* **2014**, *11*, 46. [CrossRef] [PubMed]

150. Kleffner, I.; Dorr, J.; Ringelstein, M.; Gross, C.C.; Bockenfeld, Y.; Schwindt, W.; Sundermann, B.; Lohmann, H.; Wersching, H.; Promesberger, J.; et al. Diagnostic criteria for Susac syndrome. *J. Neurol. Neurosurg. Psychiatry* **2016**, *87*, 1287–1295. [CrossRef] [PubMed]

151. Given, C.A., 2nd; Stevens, B.S.; Lee, C. The MRI appearance of tumefactive demyelinating lesions. *AJR Am. J. Roentgenol.* **2004**, *182*, 195–199. [CrossRef] [PubMed]

152. Cha, S. Update on brain tumor imaging: From anatomy to physiology. *AJNR Am. J. Neuroradiol.* **2006**, *27*, 475–487. [CrossRef] [PubMed]

153. Qi, W.; Jia, G.E.; Wang, X.; Zhang, M.; Ma, Z. Cerebral tumefactive demyelinating lesions. *Oncol. Lett.* **2015**, *10*, 1763–1768. [CrossRef] [PubMed]

154. Kister, I.; Herbert, J.; Zhou, Y.; Ge, Y. Ultrahigh-Field MR (7 T) Imaging of Brain Lesions in Neuromyelitis Optica. *Mult. Scler. Int.* **2013**, *2013*, 398259. [CrossRef] [PubMed]

155. Sinnecker, T.; Dorr, J.; Pfueller, C.F.; Harms, L.; Ruprecht, K.; Jarius, S.; Bruck, W.; Niendorf, T.; Wuerfel, J.; Paul, F. Distinct lesion morphology at 7-T MRI differentiates neuromyelitis optica from multiple sclerosis. *Neurology* **2012**, *79*, 708–714. [CrossRef] [PubMed]

156. Mabray, M.C.; Cohen, B.A.; Villanueva-Meyer, J.E.; Valles, F.E.; Barajas, R.F.; Rubenstein, J.L.; Cha, S. Performance of Apparent Diffusion Coefficient Values and Conventional MRI Features in Differentiating Tumefactive Demyelinating Lesions From Primary Brain Neoplasms. *AJR Am. J. Roentgenol.* **2015**, *205*, 1075–1085. [CrossRef] [PubMed]

157. Ninomiya, S.; Hara, M.; Morita, A.; Teramoto, H.; Momose, M.; Takahashi, T.; Kamei, S. Tumefactive Demyelinating Lesion Differentiated from a Brain Tumor Using a Combination of Magnetic Resonance Imaging and (11)C-methionine Positron Emission Tomography. *Intern. Med.* **2015**, *54*, 1411–1414. [CrossRef] [PubMed]

158. Kiriyama, T.; Kataoka, H.; Taoka, T.; Tonomura, Y.; Terashima, M.; Morikawa, M.; Tanizawa, E.; Kawahara, M.; Furiya, Y.; Sugie, K.; et al. Characteristic neuroimaging in patients with tumefactive demyelinating lesions exceeding 30 mm. *J. Neuroimaging* **2011**, *21*, e69–e77. [CrossRef] [PubMed]

159. Parks, N.E.; Bhan, V.; Shankar, J.J. Perfusion Imaging of Tumefactive Demyelinating Lesions Compared to High Grade Gliomas. *Can. J. Neurol. Sci.* **2016**, *43*, 316–318. [CrossRef] [PubMed]

160. Kebir, S.; Gaertner, F.C.; Mueller, M.; Nelles, M.; Simon, M.; Schafer, N.; Stuplich, M.; Schaub, C.; Niessen, M.; Mack, F.; et al. 18F-fluoroethyl-L-tyrosine positron emission tomography for the differential diagnosis of tumefactive multiple sclerosis versus glioma: A case report. *Oncol. Lett.* **2016**, *11*, 2195–2198. [CrossRef] [PubMed]

161. Kalus, S.; Di Muzio, B.; Gaillard, F. Demyelination preceding a diagnosis of central nervous system lymphoma. *J. Clin. Neurosci.* **2016**, *24*, 146–148. [CrossRef] [PubMed]

162. Jeong, I.H.; Kim, S.H.; Hyun, J.W.; Joung, A.; Cho, H.J.; Kim, H.J. Tumefactive demyelinating lesions as a first clinical event: Clinical, imaging, and follow-up observations. *J. Neurol. Sci.* **2015**, *358*, 118–124. [CrossRef] [PubMed]

163. Min, J.H.; Kim, B.J.; Lee, K.H. Development of extensive brain lesions following fingolimod (FTY720) treatment in a patient with neuromyelitis optica spectrum disorder. *Mult. Scler.* **2012**, *18*, 113–115. [CrossRef] [PubMed]

164. Faissner, S.; Hoepner, R.; Lukas, C.; Chan, A.; Gold, R.; Ellrichmann, G. Tumefactive multiple sclerosis lesions in two patients after cessation of fingolimod treatment. *Ther. Adv. Neurol. Disord.* **2015**, *8*, 233–238. [CrossRef] [PubMed]

165. Totaro, R.; di Carmine, C.; Carolei, A. Tumefactive demyelinating lesion in a patient with relapsing remitting multiple sclerosis treated with fingolimod. *J. Neurol. Neurophysiol.* **2014**, *S12*, 1–2.

166. Hellmann, M.A.; Lev, N.; Lotan, I.; Mosberg-Galili, R.; Inbar, E.; Luckman, J.; Fichman-Horn, S.; Yakimov, M.; Steiner, I. Tumefactive demyelination and a malignant course in an MS patient during and following fingolimod therapy. *J. Neurol. Sci.* **2014**, *344*, 193–197. [CrossRef] [PubMed]

167. Giovannoni, G.; Naismith, R.T. Natalizumab to fingolimod washout in patients at risk of PML: When good intentions yield bad outcomes. *Neurology* **2014**, *82*, 1196–1197. [CrossRef] [PubMed]

168. Rinaldi, F.; Seppi, D.; Calabrese, M.; Perini, P.; Gallo, P. Switching therapy from natalizumab to fingolimod in relapsing-remitting multiple sclerosis: Clinical and magnetic resonance imaging findings. *Mult. Scler.* **2012**, *18*, 1640–1643. [CrossRef] [PubMed]

169. Jander, S.; Turowski, B.; Kieseier, B.C.; Hartung, H.P. Emerging tumefactive multiple sclerosis after switching therapy from natalizumab to fingolimod. *Mult. Scler.* **2012**, *18*, 1650–1652. [CrossRef] [PubMed]

170. Algahtani, H.; Shirah, B.; Alassiri, A. Tumefactive demyelinating lesions: A comprehensive review. *Mult. Scler. Relat. Disord.* **2017**, *14*, 72–79. [CrossRef] [PubMed]

171. Sinnecker, T.; Kuchling, J.; Dusek, P.; Dorr, J.; Niendorf, T.; Paul, F.; Wuerfel, J. Ultrahigh field MRI in clinical neuroimmunology: A potential contribution to improved diagnostics and personalised disease management. *EPMA J.* **2015**, *6*, 16. [CrossRef] [PubMed]

172. Cheng, C.; Jiang, Y.; Chen, X.; Dai, Y.; Kang, Z.; Lu, Z.; Peng, F.; Hu, X. Clinical, radiographic characteristics and immunomodulating changes in neuromyelitis optica with extensive brain lesions. *BMC Neurol.* **2013**, *13*, 72. [CrossRef] [PubMed]

173. Saiki, S.; Ueno, Y.; Moritani, T.; Sato, T.; Sekine, T.; Kawajiri, S.; Adachi, S.; Yokoyama, K.; Tomizawa, Y.; Motoi, Y.; et al. Extensive hemispheric lesions with radiological evidence of blood-brain barrier integrity in a patient with neuromyelitis optica. *J. Neurol. Sci.* **2009**, *284*, 217–219. [CrossRef] [PubMed]

174. Roy, U.; Saini, D.S.; Pan, K.; Pandit, A.; Ganguly, G.; Panwar, A. Neuromyelitis Optica Spectrum Disorder with Tumefactive Demyelination mimicking Multiple Sclerosis: A Rare Case. *Front. Neurol.* **2016**, *7*, 73. [CrossRef] [PubMed]

175. Matsushita, T.; Isobe, N.; Matsuoka, T.; Ishizu, T.; Kawano, Y.; Yoshiura, T.; Ohyagi, Y.; Kira, J. Extensive vasogenic edema of anti-aquaporin-4 antibody-related brain lesions. *Mult. Scler.* **2009**, *15*, 1113–1117. [CrossRef] [PubMed]

176. Magana, S.M.; Matiello, M.; Pittock, S.J.; McKeon, A.; Lennon, V.A.; Rabinstein, A.A.; Shuster, E.; Kantarci, O.H.; Lucchinetti, C.F.; Weinshenker, B.G. Posterior reversible encephalopathy syndrome in neuromyelitis optica spectrum disorders. *Neurology* **2009**, *72*, 712–717. [CrossRef] [PubMed]

177. Stadelmann, C.; Ludwin, S.; Tabira, T.; Guseo, A.; Lucchinetti, C.F.; Leel-Ossy, L.; Ordinario, A.T.; Bruck, W.; Lassmann, H. Tissue preconditioning may explain concentric lesions in Balo's type of multiple sclerosis. *Brain* **2005**, *128*, 979–987. [CrossRef] [PubMed]

178. Hardy, T.A.; Corboy, J.R.; Weinshenker, B.G. Balo concentric sclerosis evolving from apparent tumefactive demyelination. *Neurology* **2017**, *88*, 2150–2152. [CrossRef] [PubMed]

179. Hardy, T.A.; Tobin, W.O.; Lucchinetti, C.F. Exploring the overlap between multiple sclerosis, tumefactive demyelination and Balo's concentric sclerosis. *Mult. Scler.* **2016**, *22*, 986–992. [CrossRef] [PubMed]

180. Graber, J.J.; Kister, I.; Geyer, H.; Khaund, M.; Herbert, J. Neuromyelitis optica and concentric rings of Balo in the brainstem. *Arch. Neurol.* **2009**, *66*, 274–275. [CrossRef] [PubMed]

181. Masuda, H.; Mori, M.; Katayama, K.; Kikkawa, Y.; Kuwabara, S. Anti-aquaporin-4 antibody-seronegative NMO spectrum disorder with Balo's concentric lesions. *Intern. Med.* **2013**, *52*, 1517–1521. [CrossRef] [PubMed]

182. Hyun, J.W.; Woodhall, M.R.; Kim, S.H.; Jeong, I.H.; Kong, B.; Kim, G.; Kim, Y.; Park, M.S.; Irani, S.R.; Waters, P.; et al. Longitudinal analysis of myelin oligodendrocyte glycoprotein antibodies in CNS inflammatory diseases. *J. Neurol. Neurosurg. Psychiatry* **2017**, *88*, 811–817. [CrossRef] [PubMed]

183. Rostasy, K.; Mader, S.; Schanda, K.; Huppke, P.; Gartner, J.; Kraus, V.; Karenfort, M.; Tibussek, D.; Blaschek, A.; Bajer-Kornek, B.; et al. Anti-myelin oligodendrocyte glycoprotein antibodies in pediatric patients with optic neuritis. *Arch. Neurol.* **2012**, *69*, 752–756. [CrossRef] [PubMed]

184. Kim, S.M.; Kim, S.J.; Lee, H.J.; Kuroda, H.; Palace, J.; Fujihara, K. Differential diagnosis of neuromyelitis optica spectrum disorders. *Ther. Adv. Neurol. Disord.* **2017**, *10*, 265–289. [CrossRef] [PubMed]

185. Probstel, A.K.; Rudolf, G.; Dornmair, K.; Collongues, N.; Chanson, J.B.; Sanderson, N.S.; Lindberg, R.L.; Kappos, L.; de Seze, J.; Derfuss, T. Anti-MOG antibodies are present in a subgroup of patients with a neuromyelitis optica phenotype. *J. Neuroinflamm.* **2015**, *12*, 46. [CrossRef] [PubMed]

186. Chang, T.; Waters, P.; Woodhall, M.; Vincent, A. Recurrent Optic Neuritis Associated With MOG Antibody Seropositivity. *Neurologist* **2017**, *22*, 101–102. [CrossRef] [PubMed]

187. Reindl, M.; Linington, C.; Brehm, U.; Egg, R.; Dilitz, E.; Deisenhammer, F.; Poewe, W.; Berger, T. Antibodies against the myelin oligodendrocyte glycoprotein and the myelin basic protein in multiple sclerosis and other neurological diseases: A comparative study. *Brain* **1999**, *122 Pt 11*, 2047–2056. [CrossRef] [PubMed]

188. Huppke, P.; Rostasy, K.; Karenfort, M.; Huppke, B.; Seidl, R.; Leiz, S.; Reindl, M.; Gartner, J. Acute disseminated encephalomyelitis followed by recurrent or monophasic optic neuritis in pediatric patients. *Mult. Scler.* **2013**, *19*, 941–946. [CrossRef] [PubMed]

189. Miyauchi, A.; Monden, Y.; Watanabe, M.; Sugie, H.; Morita, M.; Kezuka, T.; Momoi, M.; Yamagata, T. Persistent presence of the anti-myelin oligodendrocyte glycoprotein autoantibody in a pediatric case of acute disseminated encephalomyelitis followed by optic neuritis. *Neuropediatrics* **2014**, *45*, 196–199. [CrossRef] [PubMed]

190. Ching, B.H.; Mohamed, A.R.; Khoo, T.B.; Ismail, H.I. Multiphasic disseminated encephalomyelitis followed by optic neuritis in a child with gluten sensitivity. *Mult. Scler.* **2015**, *21*, 1209–1211. [CrossRef] [PubMed]

191. Baumann, M.; Hennes, E.M.; Schanda, K.; Karenfort, M.; Kornek, B.; Seidl, R.; Diepold, K.; Lauffer, H.; Marquardt, I.; Strautmanis, J.; et al. Children with multiphasic disseminated encephalomyelitis and antibodies to the myelin oligodendrocyte glycoprotein (MOG): Extending the spectrum of MOG antibody positive diseases. *Mult. Scler.* **2016**, *22*, 1821–1829. [CrossRef] [PubMed]

192. Ogawa, R.; Nakashima, I.; Takahashi, T.; Kaneko, K.; Akaishi, T.; Takai, Y.; Sato, D.K.; Nishiyama, S.; Misu, T.; Kuroda, H.; et al. MOG antibody-positive, benign, unilateral, cerebral cortical encephalitis with epilepsy. *Neurol. Neuroimmunol. Neuroinflamm.* **2017**, *4*, e322. [CrossRef] [PubMed]

193. Kruer, M.C.; Koch, T.K.; Bourdette, D.N.; Chabas, D.; Waubant, E.; Mueller, S.; Moscarello, M.A.; Dalmau, J.; Woltjer, R.L.; Adamus, G. NMDA receptor encephalitis mimicking seronegative neuromyelitis optica. *Neurology* **2010**, *74*, 1473–1475. [CrossRef] [PubMed]

194. Hacohen, Y.; Absoud, M.; Woodhall, M.; Cummins, C.; De Goede, C.G.; Hemingway, C.; Jardine, P.E.; Kneen, R.; Pike, M.G.; Whitehouse, W.P.; et al. Autoantibody biomarkers in childhood-acquired demyelinating syndromes: Results from a national surveillance cohort. *J. Neurol. Neurosurg. Psychiatry* **2014**, *85*, 456–461. [CrossRef] [PubMed]

195. Hacohen, Y.; Absoud, M.; Hemingway, C.; Jacobson, L.; Lin, J.P.; Pike, M.; Pullaperuma, S.; Siddiqui, A.; Wassmer, E.; Waters, P.; et al. NMDA receptor antibodies associated with distinct white matter syndromes. *Neurol. Neuroimmunol. Neuroinflamm.* **2014**, *1*, e2. [CrossRef] [PubMed]

196. Splendiani, A.; Felli, V.; Di Sibio, A.; Gennarelli, A.; Patriarca, L.; Stratta, P.; Di Cesare, E.; Rossi, A.; Massimo, G. Magnetic resonance imaging and magnetic resonance spectroscopy in a young male patient with anti-N-methyl-D-aspartate receptor encephalitis and uncommon cerebellar involvement: A case report with review of the literature. *Neuroradiol. J.* **2016**, *29*, 30–35. [CrossRef] [PubMed]

197. Titulaer, M.J.; Hoftberger, R.; Iizuka, T.; Leypoldt, F.; McCracken, L.; Cellucci, T.; Benson, L.A.; Shu, H.; Irioka, T.; Hirano, M.; et al. Overlapping demyelinating syndromes and anti-N-methyl-D-aspartate receptor encephalitis. *Ann. Neurol.* **2014**, *75*, 411–428. [CrossRef] [PubMed]

198. Lekoubou, A.; Viaccoz, A.; Didelot, A.; Anastasi, A.; Marignier, R.; Ducray, F.; Rogemond, V.; Honnorat, J. Anti-N-methyl-D-aspartate receptor encephalitis with acute disseminated encephalomyelitis-like MRI features. *Eur. J. Neurol.* **2012**, *19*, e16–e17. [CrossRef] [PubMed]

199. Zoccarato, M.; Saddi, M.V.; Serra, G.; Pelizza, M.F.; Rosellini, I.; Peddone, L.; Ticca, A.; Giometto, B.; Zuliani, L. Aquaporin-4 antibody neuromyelitis optica following anti-NMDA receptor encephalitis. *J. Neurol.* **2013**, *260*, 3185–3187. [CrossRef] [PubMed]

200. Outteryck, O.; Baille, G.; Hodel, J.; Giroux, M.; Lacour, A.; Honnorat, J.; Zephir, H.; Vermersch, P. Extensive myelitis associated with anti-NMDA receptor antibodies. *BMC Neurol.* **2013**, *13*, 211. [CrossRef] [PubMed]

201. Kezuka, T.; Usui, Y.; Yamakawa, N.; Matsunaga, Y.; Matsuda, R.; Masuda, M.; Utsumi, H.; Tanaka, K.; Goto, H. Relationship between NMO-antibody and anti-MOG antibody in optic neuritis. *J. Neuroophthalmol.* **2012**, *32*, 107–110. [CrossRef] [PubMed]

202. Kezuka, T.; Tanaka, K.; Matsunaga, Y.; Goto, H. Distinction between MOG antibody-positive and AQP4 antibody-positive NMO spectrum disorders. *Neurology* **2014**, *83*, 475. [CrossRef] [PubMed]

203. Martinez-Hernandez, E.; Sepulveda, M.; Rostasy, K.; Hoftberger, R.; Graus, F.; Harvey, R.J.; Saiz, A.; Dalmau, J. Antibodies to aquaporin 4, myelin-oligodendrocyte glycoprotein, and the glycine receptor alpha1 subunit in patients with isolated optic neuritis. *JAMA Neurol.* **2015**, *72*, 187–193. [CrossRef] [PubMed]

204. Popescu, B.F.; Bunyan, R.F.; Guo, Y.; Parisi, J.E.; Lennon, V.A.; Lucchinetti, C.F. Evidence of aquaporin involvement in human central pontine myelinolysis. *Acta Neuropathol. Commun.* **2013**, *1*, 40. [CrossRef] [PubMed]

205. Tzartos, J.S.; Stergiou, C.; Kilidireas, K.; Zisimopoulou, P.; Thomaidis, T.; Tzartos, S.J. Anti-aquaporin-1 autoantibodies in patients with neuromyelitis optica spectrum disorders. *PLoS ONE* **2013**, *8*, e74773. [CrossRef] [PubMed]

206. Sanchez Gomar, I.; Diaz Sanchez, M.; Ucles Sanchez, A.J.; Casado Chocan, J.L.; Suarez-Luna, N.; Ramirez-Lorca, R.; Villadiego, J.; Toledo-Aral, J.J.; Echevarria, M. Comparative Analysis for the Presence of IgG Anti-Aquaporin-1 in Patients with NMO-Spectrum Disorders. *Int. J. Mol. Sci.* **2016**, *17*, 1195. [CrossRef] [PubMed]

207. Long, Y.; Zheng, Y.; Shan, F.; Chen, M.; Fan, Y.; Zhang, B.; Gao, C.; Gao, Q.; Yang, N. Development of a cell-based assay for the detection of anti-aquaporin 1 antibodies in neuromyelitis optica spectrum disorders. *J. Neuroimmunol.* **2014**, *273*, 103–110. [CrossRef] [PubMed]

208. Schanda, K.; Waters, P.; Holzer, H.; Aboulenein-Djamshidian, F.; Leite, M.I.; Palace, J.; Vukusic, S.; Marignier, R.; Berger, T.; Reindl, M. Antibodies to aquaporin-1 are not present in neuromyelitis optica. *Neurol. Neuroimmunol. Neuroinflamm.* **2015**, *2*, e160. [CrossRef] [PubMed]

209. Watanabe, M.; Yamasaki, R.; Kawano, Y.; Imamura, Y.; Kira, J. Anti-KIR4.1 antibodies in Japanese patients with idiopathic central nervous system demyelinating diseases. *Clin. Exp. Neuroimmunol.* **2013**, *4*, 241–242. [CrossRef]

210. Srivastava, R.; Aslam, M.; Kalluri, S.R.; Schirmer, L.; Buck, D.; Tackenberg, B.; Rothhammer, V.; Chan, A.; Gold, R.; Berthele, A.; et al. Potassium channel KIR4.1 as an immune target in multiple sclerosis. *N. Engl. J. Med.* **2012**, *367*, 115–123. [CrossRef] [PubMed]

211. Kraus, V.; Srivastava, R.; Kalluri, S.R.; Seidel, U.; Schuelke, M.; Schimmel, M.; Rostasy, K.; Leiz, S.; Hosie, S.; Grummel, V.; et al. Potassium channel KIR4.1-specific antibodies in children with acquired demyelinating CNS disease. *Neurology* **2014**, *82*, 470–473. [CrossRef] [PubMed]

212. Brickshawana, A.; Hinson, S.R.; Romero, M.F.; Lucchinetti, C.F.; Guo, Y.; Buttmann, M.; McKeon, A.; Pittock, S.J.; Chang, M.H.; Chen, A.P.; et al. Investigation of the KIR4.1 potassium channel as a putative antigen in patients with multiple sclerosis: A comparative study. *Lancet Neurol.* **2014**, *13*, 795–806. [CrossRef]

213. Brill, L.; Goldberg, L.; Karni, A.; Petrou, P.; Abramsky, O.; Ovadia, H.; Ben-Hur, T.; Karussis, D.; Vaknin-Dembinsky, A. Increased anti-KIR4.1 antibodies in multiple sclerosis: Could it be a marker of disease relapse? *Mult. Scler.* **2015**, *21*, 572–579. [CrossRef] [PubMed]

214. Higuchi, O.; Nakane, S.; Sakai, W.; Maeda, Y.; Niino, M.; Takahashi, T.; Fukazawa, T.; Kikuchi, S.; Fujihara, K.; Matsuo, H. Lack of KIR4.1 autoantibodies in Japanese patients with MS and NMO. *Neurol. Neuroimmunol. Neuroinflamm.* **2016**, *3*, e263. [CrossRef] [PubMed]

215. Hemmer, B. Antibodies to the inward rectifying potassium channel 4.1 in multiple sclerosis: Different methodologies–conflicting results? *Mult. Scler.* **2015**, *21*, 537–539. [CrossRef] [PubMed]

216. Pittock, S.J.; Lennon, V.A. Aquaporin-4 autoantibodies in a paraneoplastic context. *Arch. Neurol.* **2008**, *65*, 629–632. [CrossRef] [PubMed]

217. Armagan, H.; Tuzun, E.; Icoz, S.; Birisik, O.; Ulusoy, C.; Demir, G.; Altintas, A.; Akman-Demir, G. Long extensive transverse myelitis associated with aquaporin-4 antibody and breast cancer: Favorable response to cancer treatment. *J. Spinal Cord Med.* **2012**, *35*, 267–269. [CrossRef] [PubMed]

218. Moussawi, K.; Lin, D.J.; Matiello, M.; Chew, S.; Morganstern, D.; Vaitkevicius, H. Brainstem and limbic encephalitis with paraneoplastic neuromyelitis optica. *J. Clin. Neurosci.* **2016**, *23*, 159–161. [CrossRef] [PubMed]

219. Figueroa, M.; Guo, Y.; Tselis, A.; Pittock, S.J.; Lennon, V.A.; Lucchinetti, C.F.; Lisak, R.P. Paraneoplastic neuromyelitis optica spectrum disorder associated with metastatic carcinoid expressing aquaporin-4. *JAMA Neurol.* **2014**, *71*, 495–498. [CrossRef] [PubMed]

220. Soelberg, K.; Larsen, S.R.; Moerch, M.T.; Thomassen, M.; Brusgaard, K.; Paul, F.; Smith, T.J.; Godballe, C.; Grauslund, J.; Lillevang, S.T.; et al. Aquaporin-4 IgG autoimmune syndrome and immunoreactivity associated with thyroid cancer. *Neurol. Neuroimmunol. Neuroinflamm.* **2016**, *3*, e252. [CrossRef] [PubMed]

221. Jarius, S.; Wandinger, K.P.; Borowski, K.; Stoecker, W.; Wildemann, B. Antibodies to CV2/CRMP5 in neuromyelitis optica-like disease: Case report and review of the literature. *Clin. Neurol. Neurosurg.* **2012**, *114*, 331–335. [CrossRef] [PubMed]

222. Faissner, S.; Lukas, C.; Reinacher-Schick, A.; Tannapfel, A.; Gold, R.; Kleiter, I. Amphiphysin-positive paraneoplastic myelitis and stiff-person syndrome. *Neurol. Neuroimmunol. Neuroinflamm.* **2016**, *3*, e285. [CrossRef] [PubMed]

223. Melamed, E.; Levy, M.; Waters, P.J.; Sato, D.K.; Bennett, J.L.; John, G.R.; Hooper, D.C.; Saiz, A.; Bar-Or, A.; Kim, H.J.; et al. Update on biomarkers in neuromyelitis optica. *Neurol. Neuroimmunol. Neuroinflamm.* **2015**, *2*, e134. [CrossRef] [PubMed]

224. Ikeda, K.; Kiyota, N.; Kuroda, H.; Sato, D.K.; Nishiyama, S.; Takahashi, T.; Misu, T.; Nakashima, I.; Fujihara, K.; Aoki, M. Severe demyelination but no astrocytopathy in clinically definite neuromyelitis optica with anti-myelin-oligodendrocyte glycoprotein antibody. *Mult. Scler.* **2015**, *21*, 656–659. [CrossRef] [PubMed]

225. Kaneko, K.; Sato, D.K.; Nakashima, I.; Nishiyama, S.; Tanaka, S.; Marignier, R.; Hyun, J.W.; Oliveira, L.M.; Reindl, M.; Seifert-Held, T.; et al. Myelin injury without astrocytopathy in neuroinflammatory disorders with MOG antibodies. *J. Neurol. Neurosurg. Psychiatry* **2016**, *87*, 1257–1259. [CrossRef] [PubMed]

226. Misu, T.; Takano, R.; Fujihara, K.; Takahashi, T.; Sato, S.; Itoyama, Y. Marked increase in cerebrospinal fluid glial fibrillar acidic protein in neuromyelitis optica: An astrocytic damage marker. *J. Neurol. Neurosurg. Psychiatry* **2009**, *80*, 575–577. [CrossRef] [PubMed]

227. Misu, T.; Fujihara, K.; Kakita, A.; Konno, H.; Nakamura, M.; Watanabe, S.; Takahashi, T.; Nakashima, I.; Takahashi, H.; Itoyama, Y. Loss of aquaporin 4 in lesions of neuromyelitis optica: Distinction from multiple sclerosis. *Brain* **2007**, *130*, 1224–1234. [CrossRef] [PubMed]

228. Takano, R.; Misu, T.; Takahashi, T.; Sato, S.; Fujihara, K.; Itoyama, Y. Astrocytic damage is far more severe than demyelination in NMO: A clinical CSF biomarker study. *Neurology* **2010**, *75*, 208–216. [CrossRef] [PubMed]

229. Uzawa, A.; Mori, M.; Kuwabara, S. Cytokines and chemokines in neuromyelitis optica: Pathogenetic and therapeutic implications. *Brain Pathol.* **2014**, *24*, 67–73. [CrossRef] [PubMed]

230. Icoz, S.; Tuzun, E.; Kurtuncu, M.; Durmus, H.; Mutlu, M.; Eraksoy, M.; Akman-Demir, G. Enhanced IL-6 production in aquaporin-4 antibody positive neuromyelitis optica patients. *Int. J. Neurosci.* **2010**, *120*, 71–75. [CrossRef] [PubMed]

231. Erta, M.; Quintana, A.; Hidalgo, J. Interleukin-6, a major cytokine in the central nervous system. *Int. J. Biol. Sci.* **2012**, *8*, 1254–1266. [CrossRef] [PubMed]

232. Van Wagoner, N.J.; Oh, J.W.; Repovic, P.; Benveniste, E.N. Interleukin-6 (IL-6) production by astrocytes: Autocrine regulation by IL-6 and the soluble IL-6 receptor. *J. Neurosci.* **1999**, *19*, 5236–5244. [PubMed]

233. Sherman, E.; Han, M.H. Acute and Chronic Management of Neuromyelitis Optica Spectrum Disorder. *Curr. Treat. Options Neurol.* **2015**, *17*, 48. [CrossRef] [PubMed]

234. Kessler, R.A.; Mealy, M.A.; Levy, M. Treatment of Neuromyelitis Optica Spectrum Disorder: Acute, Preventive, and Symptomatic. *Curr. Treat. Options Neurol.* **2016**, *18*, 2. [CrossRef] [PubMed]

235. Weinshenker, B.G.; O'Brien, P.C.; Petterson, T.M.; Noseworthy, J.H.; Lucchinetti, C.F.; Dodick, D.W.; Pineda, A.A.; Stevens, L.N.; Rodriguez, M. A randomized trial of plasma exchange in acute central nervous system inflammatory demyelinating disease. *Ann. Neurol.* **1999**, *46*, 878–886. [CrossRef]

236. Kleiter, I.; Gahlen, A.; Borisow, N.; Fischer, K.; Wernecke, K.D.; Wegner, B.; Hellwig, K.; Pache, F.; Ruprecht, K.; Havla, J.; et al. Neuromyelitis optica: Evaluation of 871 attacks and 1,153 treatment courses. *Ann. Neurol.* **2016**, *79*, 206–216. [CrossRef] [PubMed]

237. Perumal, J.S.; Caon, C.; Hreha, S.; Zabad, R.; Tselis, A.; Lisak, R.; Khan, O. Oral prednisone taper following intravenous steroids fails to improve disability or recovery from relapses in multiple sclerosis. *Eur. J. Neurol.* **2008**, *15*, 677–680. [CrossRef] [PubMed]

238. Multiple Sclerosis Therapy Consensus Group (MSTCG); Wiendl, H.; Toyka, K.V.; Rieckmann, P.; Gold, R.; Hartung, H.P.; Hohlfeld, R. Basic and escalating immunomodulatory treatments in multiple sclerosis: Current therapeutic recommendations. *J. Neurol.* **2008**, *255*, 1449–1463. [PubMed]

239. Cortese, I.; Chaudhry, V.; So, Y.T.; Cantor, F.; Cornblath, D.R.; Rae-Grant, A. Evidence-based guideline update: Plasmapheresis in neurologic disorders: Report of the Therapeutics and Technology Assessment Subcommittee of the American Academy of Neurology. *Neurology* **2011**, *76*, 294–300. [CrossRef] [PubMed]

240. Schneider-Gold, C.; Krenzer, M.; Klinker, E.; Mansouri-Thalegani, B.; Mullges, W.; Toyka, K.V.; Gold, R. Immunoadsorption versus plasma exchange versus combination for treatment of myasthenic deterioration. *Ther. Adv. Neurol. Disord.* **2016**, *9*, 297–303. [CrossRef] [PubMed]

241. Nakashima, I.; Takahashi, T.; Cree, B.A.; Kim, H.J.; Suzuki, C.; Genain, C.P.; Vincent, T.; Fujihara, K.; Itoyama, Y.; Bar-Or, A. Transient increases in anti-aquaporin-4 antibody titers following rituximab treatment in neuromyelitis optica, in association with elevated serum BAFF levels. *J. Clin. Neurosci.* **2011**, *18*, 997–998. [CrossRef] [PubMed]

242. Perumal, J.S.; Kister, I.; Howard, J.; Herbert, J. Disease exacerbation after rituximab induction in neuromyelitis optica. *Neurol. Neuroimmunol. Neuroinflamm.* **2015**, *2*, e61. [CrossRef] [PubMed]

243. Stellmann, J.P.; Krumbholz, M.; Friede, T.; Gahlen, A.; Borisow, N.; Fischer, K.; Hellwig, K.; Pache, F.; Ruprecht, K.; Havla, J.; et al. Immunotherapies in neuromyelitis optica spectrum disorder: Efficacy and predictors of response. *J. Neurol. Neurosurg. Psychiatry* **2017**, *88*, 639–647. [CrossRef] [PubMed]

244. Kleiter, I.; Hellwig, K.; Berthele, A.; Kumpfel, T.; Linker, R.A.; Hartung, H.P.; Paul, F.; Aktas, O.; Neuromyelitis Optica Study Group. Failure of natalizumab to prevent relapses in neuromyelitis optica. *Arch. Neurol.* **2012**, *69*, 239–245. [CrossRef] [PubMed]

245. Gelfand, J.M.; Cotter, J.; Klingman, J.; Huang, E.J.; Cree, B.A. Massive CNS monocytic infiltration at autopsy in an alemtuzumab-treated patient with NMO. *Neurol. Neuroimmunol. Neuroinflamm.* **2014**, *1*, e34. [CrossRef] [PubMed]

246. Ayzenberg, I.; Schollhammer, J.; Hoepner, R.; Hellwig, K.; Ringelstein, M.; Aktas, O.; Kumpfel, T.; Krumbholz, M.; Trebst, C.; Paul, F.; et al. Efficacy of glatiramer acetate in neuromyelitis optica spectrum disorder: A multicenter retrospective study. *J. Neurol.* **2016**, *263*, 575–582. [CrossRef] [PubMed]

247. Gahlen, A.; Trampe, A.K.; Haupeltshofer, S.; Ringelstein, M.; Aktas, O.; Berthele, A.; Wildemann, B.; Gold, R.; Jarius, S.; Kleiter, I. Aquaporin-4 antibodies in patients treated with natalizumab for suspected MS. *Neurol. Neuroimmunol. Neuroinflamm.* **2017**, *4*, e363. [CrossRef] [PubMed]

248. Kira, J.I. Unexpected exacerbations following initiation of disease-modifying drugs in neuromyelitis optica spectrum disorder: Which factor is responsible, anti-aquaporin 4 antibodies, B cells, Th1 cells, Th2 cells, Th17 cells, or others? *Mult. Scler.* **2017**, *23*, 1300–1302. [CrossRef] [PubMed]

249. Yamout, B.I.; Beaini, S.; Zeineddine, M.M.; Akkawi, N. Catastrophic relapses following initiation of dimethyl fumarate in two patients with neuromyelitis optica spectrum disorder. *Mult. Scler.* **2017**, *23*, 1297–1300. [CrossRef] [PubMed]

250. Ringelstein, M.; Ayzenberg, I.; Harmel, J.; Lauenstein, A.S.; Lensch, E.; Stogbauer, F.; Hellwig, K.; Ellrichmann, G.; Stettner, M.; Chan, A.; et al. Long-term Therapy With Interleukin 6 Receptor Blockade in Highly Active Neuromyelitis Optica Spectrum Disorder. *JAMA Neurol.* **2015**, *72*, 756–763. [CrossRef] [PubMed]

251. Chugai Pharmaceutical A Multicenter, Randomized, Double-blind, Placebo-controlled, Phase 3 Study to Evaluate the Efficacy and Safety of SA237 as Monotherapy in Patients With Neuromyelitis Optica (NMO) and NMO Spectrum Disorder (NMOSD). Available online: https://clinicaltrials.gov/ct2/show/NCT02073279?term=SA237&cond=NMO+Spectrum+Disorder&rank=1 (accessed on 16 September 2017).

252. Chugai Pharmaceutical A Multicenter, Randomized, Addition to Baseline Treatment, Double-blind, Placebo-controlled, Phase 3 Study to Evaluate the Efficacy and Safety of SA237 in Patients With Neuromyelitis Optica (NMO) and NMO Spectrum Disorder (NMOSD). Available online: https://clinicaltrials.gov/ct2/show/NCT02028884?term=SA237&cond=NMO+Spectrum+Disorder&rank=2 (accessed on 16 September 2017).

253. Kang, S.; Tanaka, T.; Kishimoto, T. Therapeutic uses of anti-interleukin-6 receptor antibody. *Int. Immunol.* **2015**, *27*, 21–29. [CrossRef] [PubMed]

254. Pandit, L.; Asgari, N.; Apiwattanakul, M.; Palace, J.; Paul, F.; Leite, M.I.; Kleiter, I.; Chitnis, T.; GJCF International Clinical Consortium & Biorepository for Neuromyelitis Optica. Demographic and clinical features of neuromyelitis optica: A review. *Mult. Scler.* **2015**, *21*, 845–853. [PubMed]

Brain Sci. **2017**, *7*, 138

255. Reindl, M.; Rostasy, K. MOG antibody-associated diseases. *Neurol. Neuroimmunol. Neuroinflamm.* **2015**, *2*, e60. [CrossRef] [PubMed]
256. Spadaro, M.; Gerdes, L.A.; Krumbholz, M.; Ertl-Wagner, B.; Thaler, F.S.; Schuh, E.; Metz, I.; Blaschek, A.; Dick, A.; Bruck, W.; et al. Autoantibodies to MOG in a distinct subgroup of adult multiple sclerosis. *Neurol. Neuroimmunol. Neuroinflamm.* **2016**, *3*, e257. [CrossRef] [PubMed]

brain
sciences

MDPI

Review

Role of Immunological Memory Cells as a Therapeutic Target in Multiple Sclerosis

Tanima Bose

Institute of Human Genetics, University of Regensburg, Franz-Josef-Strauss-Allee 11,
D-93053 Regensburg, Germany; tanimabose@gmail.com or tanima.bose@ur.de;
Tel.: +49-941-944-5449; Fax: +49-941-944-5402

Received: 31 August 2017; Accepted: 2 November 2017; Published: 7 November 2017

Abstract: Pharmacological targeting of memory cells is an attractive treatment strategy in various autoimmune diseases, such as psoriasis and rheumatoid arthritis. Multiple sclerosis is the most common inflammatory disorder of the central nervous system, characterized by focal immune cell infiltration, activation of microglia and astrocytes, along with progressive damage to myelin sheaths, axons, and neurons. The current review begins with the identification of memory cell types in the previous literature and a recent description of the modulation of these cell types in T, B, and resident memory cells in the presence of different clinically approved multiple sclerosis drugs. Overall, this review paper tries to determine the potential of memory cells to act as a target for the current or newly-developed drugs.

Keywords: multiple sclerosis; MS; central memory T cells; T_{CM}; effector memory T cells; T_{EM}; resident memory T cells; T_{RM}

1. Recent Insights into Inflammatory Neuronal Injury in Multiple Sclerosis

Multiple sclerosis is one of the most prominent demyelinating disorders, and makes a bridge between immune and neuronal systems by degenerating the neuronal myelin sheath through a series of inflammatory mechanisms. Scientists over the decades have attempted to investigate the exact immune mechanisms underlying the degeneration of the myelin sheath.

The classification of multiple sclerosis is as clinically isolated syndrome (CIS), primary progressive multiple sclerosis (PPMS), secondary progressive multiple sclerosis (SPMS), and relapsing-remitting multiple sclerosis (RRMS), depending on the progression and relapses of the disease [1]. The roles of memory T or B cells are prominent in each of the different forms of the disease, but its role is more prominent in the relapsing forms, as explained later in detail. The interesting fact is that the multiple sclerosis drugs prescribed for various forms of multiple sclerosis have a major impact on the functionality and abundance of T and B memory cells. Memory cells by definition are a group of cells which have the experience of antigen recognition in a lifetime of T or B cells. They represent the distinctive features of the adaptive immune functionality, and their mode of action and phenotypic features are distinct depending on the cell types. Human memory T cells, B cells, and resident memory T cells are $CD45RO^+CD45RA^-$, IgD^+CD27^+, and $CD69^+CD103^+$, respectively (cell surface antigens). The origins and functions of the T and B cells are different, but both T and B cells have the same division of labor: plasma cells secreting antibodies in the B cell part does the job of protective memory, and effector memory T cells (T_{EM}s) does the same function by migrating immediately to the inflamed peripheral tissue and displaying necessary effector functions. Memory B cells perform the function of proliferation and stimulation in response to antigenic stimulation, whereas central memory T cells (T_{CM}s) do the same job by homing in the secondary lymphoid organs and readily transform into T_{EM}s while encountering the antigens [2]. To support this function, T_{CM}s express chemokine receptor CCR7 and the adhesion molecule L selectin (CD62L), allowing them to access the lymph nodes from blood, and T_{EMS} express low levels of these two chemokine receptors, permitting them to approach peripheral tissue such as skin [3,4]. T_{CMS} home to the lymph nodes and have a limited capacity to have effector functions until they are stimulated by the secondary responses, whereas T_{EMS} home to peripheral tissue and rapidly produce effector cytokines upon antigenic stimulation. The effector cell type can give rise to the long-lasting tissue-resident memory T cells (T_{RMS}) which might protect against multiple encounters of the similar group of pathogens, and which might help to develop vaccines or drugs in future [5].

2. Role of Memory T Cells in the Pathogenesis of Multiple Sclerosis

An earlier report from the group of Hedlund, G. et al. has shown that the sustained increase of CD4+ memory T cells in the cerebrospinal fluid of multiple sclerosis patients compared to the peripheral blood was a normal phenomenon [6]. A later report in the 1990s by the group of Zaffaroni, M. et al. observed the augmented conversion from naïve to memory cells in chronic-progressive multiple sclerosis [7]. Further, there is a definite trend of increase in memory CD45RO+CD4+ T cells and a decrease in naïve CD45RA+ T cells in the peripheral blood of multiple sclerosis patients. Additionally, there is a significant elevation of CD4/CD8 ratio [8]. In parallel, the role of memory cells in identifying the myelin basic protein (MBP) or myelin antigen-specific T cells was continuously explored in several publications. Most of the myelin-reactive T cells were shown to exist in the memory T cell subset [9]. Memory T cells are activated and proliferated even with the lack of CD28 co-stimulation [10,11]. Thus, this kind of co-stimulation blockade is not an effective strategy to prevent the MS responses. Besides CD28, later study initiated the chance of Inducible COStimulator (ICOS)-co-stimulation as an effective target for the autoimmune demyelinating disorder [12–14]. As mentioned, CD4+CD28− cells have the full potential to proliferate in the central nervous system—a site which is devoid of any professional antigen-presenting cells [15]. During this period, there was also a search to determine if any cytokine has the potential to enhance the effector function of memory cells upon adoptive transfer. It was indeed possible to find that the transforming growth factor-beta has the efficiency to increase the memory phenotype of the cultured cells and effector function of the cells upon adoptive transfer into an experimental autoimmune encephalomyelitis animal model [16]. An enhanced expression of CD45RO+ memory T cells and decreased expression of CD45RA+ naïve T cells while immunophenotyping the peripheral blood from the patients of another form of neurodegeneration (Parkinson's Disease) was also observed around this time [17]. After the establishment of the role of memory T cells as one of the major culprits, there was a continuous trial to determine which memory subset is important in case of the presence of disease or application of the drug. Some of the examples from this investigation are the following: In a transcriptomic study, fingolimod increases the effect of CCR7− T_{EMS} in the peripheral blood of the patients [18–21]. On the other hand, another important drug for this disease, dimethyl fumarate (methyl ester of fumaric acid), was shown to lower the proportion of circulating T_{CM} and T_{EM} in compared to naïve T cells [22]. Further, there was a decreased presence of Th1 CD4+ cells, increase in the abundance of Th2 CD8+ cells, and an unaltered presence of Th17+ cells in the presence of this drug [23]. Interestingly, another clinically approved drug, natalizumab (monoclonal antibody targeting adhesion molecule α4-integrin), increased the IFN-γ and IL17A cytokines secreted by CD4+ memory T cells and reduced the CD49d+ Treg cells more than the Th1 or Th17 cells [24]. In contrast, a later study showed unchanged memory, naïve, or effector T cells with the affected B cell population [25]. In the presence of other two approved drugs *viz* interferon beta (glycoprotein) and glatiramer acetate (immunomodulator), there was a beneficial decline of T_{CMS} and an increase of naïve cells [26]. In a recent paper, there was an attempt to explain the association between MS, viral infection, and MS-drugs (fingolimod and natalizumab). They pointed out Th1/Th17 central memory cells can be targeted to protect from both the MS-induced relapses and virus-induced encephalomyelitis [27]. The investigation also found that memory cells have a favorable phenotype compared to the naïve cells to breach the blood–brain barrier. The reason being was the invadosome-like protrusions in them were 2–3 fold increased compared to the crawling naïve

T cells that helped them to cross long distances (150 μm) on endothelial tight junctions before crossing the blood–brain barrier [28,29]. As the functions and origins of T_{CM} and T_{EM} differ, the modulation of these populations either in the lymph node or periphery in the presence of several MS drugs can also have an aftermath effect on relapses after exposure to the drugs.

3. Role of Memory B Cells in the Pathogenesis of Multiple Sclerosis

The depletion of $CD19^+CD27^+$ B memory cells in the presence of natalizumab and the long-term persistence of this status in the presence of other depleting factors like CD52 and CD20 strengthened the importance of B memory cells in this autoimmune demyelinating disorder [30]. Along with this line, the investigations supported the depletion of memory B cells in presence of other MS drugs, as observed in case of memory T cells. Exploring different kinds of memory cells also resolves the underlying mechanism of action of the drugs. For example, the therapeutic mode of action of dimethyl fumarate (DMF) in treating relapsing-remitting multiple sclerosis is still not properly understood. In a recent paper, memory B cells in circulating mature/differentiated B cell type was significantly diminished while treating with this drug. The DMF-mediated decrease leads to the reduction of the pro-inflammatory signals (GM-CSF, IL-6, TNF-α) compounded with reduced phosphorylation of STAT5/6 and NF-κB in surviving B cells [31]. An earlier report mentioned that this drug increased the amount of B cells with regulatory capacity (IL-10 producing B cells) [32]. Fingolimod used for treating relapsing-remitting multiple sclerosis was shown to have broad effects on the increase/decrease of the cell populations similar to DMF. It increases the naïve to memory cell phenotype, modulates the circulatory B cells with an abundance of regulatory capacity and an increase of anti-inflammatory cytokines [33]. Another first-line disease-modifying drug, interferon-beta (IFN-β), has both anti-inflammatory properties and can effectively target the memory B cells [34]. To determine whether a single dose of the drug is sufficient to eradicate the disease-causing cell subsets, it is elucidated that a single dose of rituximab did not eliminate the IgG memory B cells and might facilitate the presence of auto-reactive immune cells [35]. Along with memory B cells, the exploration of CD40 co-stimulation helped in identifying the mechanistic pathway of the currently existing drugs. To support that, CD40-mediated elevation in pP65 (NF-κB) level was observed in the naïve and memory cells from the relapsing-remitting and progressive multiple sclerosis patients compared to the control subjects [36]. Further, the combination therapy of IFN-β-1a (Avonex) and mycophenolate mofetil (Cellcept) and glatiramer acetate leads to the modulation of hyperphosphorylation of P65 in B cells [36]. There was an intention to search for the signaling molecule responsible for the propagation of granulocyte macrophage colony-stimulating factor (GM-CSF) memory B cells, and it was found that the signal transducer and activator of transcription 5/6 (STAT5/6)-regulated mechanistic pathway is upregulated in untreated MS patients, and this also reciprocally regulates the IL-10 secretion [37]. It is also interesting to observe how different external factors (e.g., Epstein-Barr virus, EBV) modulate the self- and poly-reactivity of memory B cells. In the case of EBV infection, memory cells have evolved to have less self-reactivity which gives the virus an opportunity to propagate more in B-memory cell type in contrast to others [38].

Table 1 explains a brief overview of the relationship between clinically-approved MS drugs and modulation of memory cell types.

Table 1. The effect of multiple sclerosis (MS) drugs on different kinds of memory cells.

MS Drugs	Memory T Cells	Memory B Cells
Fingolimod	Increase T_{CM}s [39]	Decreased [40]
Dimethyl fumarate	Decrease of T_{EM} and T_{CM}s [41]	Decreased [32]
Natalizumab	Unchanged [25]	Increased [25]
Interferon-β	Decrease of T_{CM}s [26]	Decreased [34]
Glatiramer acetate	Decrease of T_{CM}s [26]	Decreased [42]
Teriflunomide	Not known	Decreased [43]
Dalfampridine	Not known	Not known

4. Role of Resident Memory Cells in Mediating Demyelinating Disorders

T_{RM}s are the new bunch of memory cells having different transcriptional programs than effector and central memory T cells. They are mostly present in the barrier tissues like the skin and gut. Among the two populations of memory T cells (recirculating and resident) present in the peripheral tissue, residency of the cells in the case of both CD4$^+$ and CD8$^+$ are determined by CD103 expression [44,45]. Other than CD103, the prominent activation marker used to identify the T_{RM}s is CD69 [44,46]. In certain tissues (e.g., skin and intestinal epithelium), there is no requirement of antigen presentation for the CD103$^+$CD8$^+$ T_{RM}s formation as a consequence of TGF-β signaling [46,47]. Along with the barrier tissues, there are recent reports of the presence of T_{RM}s in other non-barrier tissues like kidney and joint inflammation. T_{RM}s protect the barrier tissue against environmental pathogens, but a recent report observed that T_{RM} is generated in response to a topically-applied allergen. In a recent publication, T_{CM} was shown to match T_{RM} in terms of their functions, *viz* being stimulated by the secondary responses [48]. The T_{RM}s were shown to present in the brain, evading the blood–brain barrier [49]. In this kind of CD103+CD8+ expression, the local antigen stimulation for CD103$^+$ is necessary. T_{RM}s were present in the brain tissue after the in vitro infection with vesicular stomatitis virus, and the effector population here was CD8$^+$CD103$^+$ type, but the factor required for the continuous stimulation of T cells is still unclear. At the transcriptional level, brain T_{RM} resembles well with the skin, gut, and lungs but they are transcriptionally distinct from central and effector memory population [49,50]. There is still a lack of evidence as to whether these kinds of memory cells are indeed present in the brain. There is a recent report that supports the presence of CD8$^+$ T_{RM}s in MS patients. In this report, relapsing-remitting and chronic forms of the disease were mediated by the tissue-resident CD8$^+$ lymphocytes, and the acute form of the disease was regulated by the effector memory population residing in the meninges and perivascular space [51].

5. Novel Therapies Targeting Memory Cells with a Future in Clinical Development

The most important knowledge that the modulation of memory cells brings to us is the modification of the MS patients' immune profiles while taking the clinically-approved drugs. The immune-modulating mechanism in the case of both T or B cells is the elevation of naïve immune cells compared to the memory cells and the shift towards the anti-inflammatory paradigm, both of which ensure the elimination of auto-aggressive immune cells. With the increasing knowledge, the final goal will be to use different immunomodulators which may prevent the relapsing of MS. One such example of the new class of modulator is VitD3, the application of which in vitro in the peripheral blood mononuclear cells can abrogate the proportion of effector memory T cells and enhance the abundance of naïve cells [52]. Further investigations in this direction may yield innovative treatment either with the existing approved drugs, or in combination with other new classes of immunomodulators.

Acknowledgments: I would like to acknowledge Fred Lühder, Institut für Multiple Sklerose Forschung, Göttingen for his critical comments on the manuscript.

Conflicts of Interest: The author declares no conflict of interest.

References

1. Hurwitz, B.J. The diagnosis of multiple sclerosis and the clinical subtypes. *Ann. Indian Acad. Neurol.* **2009**, *12*, 226–230. [CrossRef] [PubMed]
2. Sallusto, F.; Geginat, J.; Lanzavecchia, A. Central memory and effector memory T cell subsets: Function, generation, and maintenance. *Annu. Rev. Immunol.* **2004**, *22*, 745–763. [CrossRef] [PubMed]
3. Fuhlbrigge, R.C.; Kieffer, J.D.; Armerding, D.; Kupper, T.S. Cutaneous lymphocyte antigen is a specialized form of PSGL-1 expressed on skin-homing T cells. *Nature* **1997**, *389*, 978–981. [PubMed]
4. Mackay, C.R.; Marston, W.L.; Dudler, L.; Spertini, O.; Tedder, T.F.; Hein, W.R. Tissue-specific migration pathways by phenotypically distinct subpopulations of memory T cells. *Eur. J. Immunol.* **1992**, *22*, 887–895. [CrossRef] [PubMed]

5. Park, C.O.; Kupper, T.S. The emerging role of resident memory T cells in protective immunity and inflammatory disease. *Nat. Med.* **2015**, *21*, 688–697. [CrossRef] [PubMed]

6. Hedlund, G.; Sandberg-Wollheim, M.; Sjogren, H.O. Increased proportion of CD4+CDw29+CD45R-UCHL-1+ lymphocytes in the cerebrospinal fluid of both multiple sclerosis patients and healthy individuals. *Cell. Immunol.* **1989**, *118*, 406–412. [CrossRef]

7. Zaffaroni, M.; Rossini, S.; Ghezzi, A.; Parma, R.; Cazzullo, C.L. Decrease of CD4+CD45+ T-cells in chronic-progressive multiple sclerosis. *J. Neurol.* **1990**, *237*, 1–4. [CrossRef] [PubMed]

8. Muraro, P.A.; Pette, M.; Bielekova, B.; McFarland, H.F.; Martin, R. Human autoreactive CD4+ T cells from naive CD45RA+ and memory CD45RO+ subsets differ with respect to epitope specificity and functional antigen avidity. *J. Immunol.* **2000**, *164*, 5474–5481. [CrossRef] [PubMed]

9. Burns, J.; Bartholomew, B.; Lobo, S. Isolation of myelin basic protein-specific T cells predominantly from the memory T-cell compartment in multiple sclerosis. *Ann. Neurol.* **1999**, *45*, 33–39. [CrossRef]

10. Lovett-Racke, A.E.; Trotter, J.L.; Lauber, J.; Perrin, P.J.; June, C.H.; Racke, M.K. Decreased dependence of myelin basic protein-reactive T cells on CD28-mediated costimulation in multiple sclerosis patients. A marker of activated/memory T cells. *J. Clin. Investig.* **1998**, *101*, 725–730. [CrossRef] [PubMed]

11. Perrin, P.J.; Lovett-Racke, A.; Phillips, S.M.; Racke, M.K. Differential requirements of naive and memory T cells for CD28 costimulation in autoimmune pathogenesis. *Histol. Histopathol.* **1999**, *14*, 1269–1276. [PubMed]

12. Sporici, R.A.; Beswick, R.L.; von Allmen, C.; Rumbley, C.A.; Hayden-Ledbetter, M.; Ledbetter, J.A.; Perrin, P.J. ICOS ligand costimulation is required for T-cell encephalitogenicity. *Clin. Immunol.* **2001**, *100*, 277–288. [CrossRef] [PubMed]

13. Sporici, R.A.; Perrin, P.J. Costimulation of memory T-cells by ICOS: A potential therapeutic target for autoimmunity? *Clin. Immunol.* **2001**, *100*, 263–269. [CrossRef] [PubMed]

14. Fan, X.; Jin, T.; Zhao, S.; Liu, C.; Han, J.; Jiang, X.; Jiang, Y. Circulating CCR7+ICOS+ Memory T Follicular Helper Cells in Patients with Multiple Sclerosis. *PLoS ONE* **2015**, *10*, e0134523. [CrossRef] [PubMed]

15. Markovic-Plese, S.; Cortese, I.; Wandinger, K.P.; McFarland, H.F.; Martin, R. CD4$^+$CD28$^-$ costimulation-independent T cells in multiple sclerosis. *J. Clin. Investig.* **2001**, *108*, 1185–1194. [CrossRef] [PubMed]

16. Weinberg, A.D.; Whitham, R.; Swain, S.L.; Morrison, W.J.; Wyrick, G.; Hoy, C.; Vandenbark, A.A.; Offner, H. Transforming growth factor-beta enhances the in vivo effector function and memory phenotype of antigen-specific T helper cells in experimental autoimmune encephalomyelitis. *J. Immunol.* **1992**, *148*, 2109–2117. [PubMed]

17. Fiszer, U.; Mix, E.; Fredrikson, S.; Kostulas, V.; Link, H. Parkinson's disease and immunological abnormalities: increase of HLA-DR expression on monocytes in cerebrospinal fluid and of CD45RO+ T cells in peripheral blood. *Acta Neurol. Scand.* **1994**, *90*, 160–166. [CrossRef] [PubMed]

18. Fujii, C.; Kondo, T.; Ochi, H.; Okada, Y.; Hashi, Y.; Adachi, T.; Shin-Ya, M.; Matsumoto, S.; Takahashi, R.; Nakagawa, M.; et al. Altered T cell phenotypes associated with clinical relapse of multiple sclerosis patients receiving fingolimod therapy. *Sci. Rep.* **2016**, *6*, 35314. [CrossRef] [PubMed]

19. Hunter, S.F.; Bowen, J.D.; Reder, A.T. The Direct Effects of Fingolimod in the Central Nervous System: Implications for Relapsing Multiple Sclerosis. *CNS Drugs* **2016**, *30*, 135–147. [CrossRef] [PubMed]

20. Roch, L.; Hecker, M.; Friess, J.; Angerer, I.C.; Koczan, D.; Fitzner, B.; Schroder, I.; Flechtner, K.; Thiesen, H.J.; Meister, S.; et al. High-Resolution Expression Profiling of Peripheral Blood CD8+ Cells in Patients with Multiple Sclerosis Displays Fingolimod-Induced Immune Cell Redistribution. *Mol. Neurobiol.* **2017**, *54*, 5511–5525. [CrossRef] [PubMed]

21. Teniente-Serra, A.; Hervas, J.V.; Quirant-Sanchez, B.; Mansilla, M.J.; Grau-Lopez, L.; Ramo-Tello, C.; Martinez-Caceres, E.M. Baseline Differences in Minor Lymphocyte Subpopulations may Predict Response to Fingolimod in Relapsing-Remitting Multiple Sclerosis Patients. *CNS Neurosci. Ther.* **2016**, *22*, 584–592. [CrossRef] [PubMed]

22. Longbrake, E.E.; Ramsbottom, M.J.; Cantoni, C.; Ghezzi, L.; Cross, A.H.; Piccio, L. Dimethyl fumarate selectively reduces memory T cells in multiple sclerosis patients. *Mult. Scler.* **2016**, *22*, 1061–1070. [CrossRef] [PubMed]

23. Gross, C.C.; Schulte-Mecklenbeck, A.; Klinsing, S.; Posevitz-Fejfar, A.; Wiendl, H.; Klotz, L. Dimethyl fumarate treatment alters circulating T helper cell subsets in multiple sclerosis. *Neurol. Neuroimmunol. Neuroinflamm.* **2016**, *3*, e183. [CrossRef] [PubMed]

24. Kimura, K.; Nakamura, M.; Sato, W.; Okamoto, T.; Araki, M.; Lin, Y.; Murata, M.; Takahashi, R.; Yamamura, T. Disrupted balance of T cells under natalizumab treatment in multiple sclerosis. *Neurol. Neuroimmunol. Neuroinflamm.* **2016**, *3*, e210. [CrossRef] [PubMed]

25. Planas, R.; Jelcic, I.; Schippling, S.; Martin, R.; Sospedra, M. Natalizumab treatment perturbs memory- and marginal zone-like B-cell homing in secondary lymphoid organs in multiple sclerosis. *Eur. J. Immunol.* **2012**, *42*, 790–798. [CrossRef] [PubMed]

26. Praksova, P.; Stourac, P.; Bednarik, J.; Vlckova, E.; Mikulkova, Z.; Michalek, J. Immunoregulatory T cells in multiple sclerosis and the effect of interferon beta and glatiramer acetate treatment on T cell subpopulations. *J. Neurol. Sci.* **2012**, *319*, 18–23. [CrossRef] [PubMed]

27. Paroni, M.; Maltese, V.; De Simone, M.; Ranzani, V.; Larghi, P.; Fenoglio, C.; Pietroboni, A.M.; De Riz, M.A.; Crosti, M.C.; Maglie, S.; et al. Recognition of viral and self-antigens by TH1 and TH1/TH17 central memory cells in patients with multiple sclerosis reveals distinct roles in immune surveillance and relapses. *J. Allergy Clin. Immunol.* **2017**, *140*, 797–808. [CrossRef] [PubMed]

28. Kawakami, N.; Bartholomaus, I.; Pesic, M.; Mues, M. An autoimmunity odyssey: How autoreactive T cells infiltrate into the CNS. *Immunol. Rev.* **2012**, *248*, 140–155. [CrossRef] [PubMed]

29. Lyck, R.; Engelhardt, B. Going against the tide—How encephalitogenic T cells breach the blood-brain barrier. *J. Vasc. Res.* **2012**, *49*, 497–509. [CrossRef] [PubMed]

30. Baker, D.; Marta, M.; Pryce, G.; Giovannoni, G.; Schmierer, K. Memory B Cells are Major Targets for Effective Immunotherapy in Relapsing Multiple Sclerosis. *EBioMedicine* **2017**, *16*, 41–50. [CrossRef] [PubMed]

31. Li, R.; Rezk, A.; Ghadiri, M.; Luessi, F.; Zipp, F.; Li, H.; Giacomini, P.S.; Antel, J.; Bar-Or, A. Dimethyl Fumarate Treatment Mediates an Anti-Inflammatory Shift in B Cell Subsets of Patients with Multiple Sclerosis. *J. Immunol.* **2017**, *198*, 691–698. [CrossRef] [PubMed]

32. Lundy, S.K.; Wu, Q.; Wang, Q.; Dowling, C.A.; Taitano, S.H.; Mao, G.; Mao-Draayer, Y. Dimethyl fumarate treatment of relapsing-remitting multiple sclerosis influences B-cell subsets. *Neurol. Neuroimmunol. Neuroinflamm.* **2016**, *3*, e211. [CrossRef] [PubMed]

33. Blumenfeld, S.; Staun-Ram, E.; Miller, A. Fingolimod therapy modulates circulating B cell composition, increases B regulatory subsets and production of IL-10 and TGFbeta in patients with Multiple Sclerosis. *J. Autoimmun.* **2016**, *70*, 40–51. [CrossRef] [PubMed]

34. Rizzo, F.; Giacomini, E.; Mechelli, R.; Buscarinu, M.C.; Salvetti, M.; Severa, M.; Coccia, E.M. Interferon-beta therapy specifically reduces pathogenic memory B cells in multiple sclerosis patients by inducing a FAS-mediated apoptosis. *Immunol. Cell Biol.* **2016**, *94*, 886–894. [CrossRef] [PubMed]

35. Maurer, M.A.; Tuller, F.; Gredler, V.; Berger, T.; Lutterotti, A.; Lunemann, J.D.; Reindl, M. Rituximab induces clonal expansion of IgG memory B-cells in patients with inflammatory central nervous system demyelination. *J. Neuroimmunol.* **2016**, *290*, 49–53. [CrossRef] [PubMed]

36. Chen, D.; Ireland, S.J.; Remington, G.; Alvarez, E.; Racke, M.K.; Greenberg, B.; Frohman, E.M.; Monson, N.L. CD40-Mediated NF-kappaB Activation in B Cells Is Increased in Multiple Sclerosis and Modulated by Therapeutics. *J. Immunol.* **2016**, *197*, 4257–4265. [CrossRef] [PubMed]

37. Li, R.; Rezk, A.; Miyazaki, Y.; Hilgenberg, E.; Touil, H.; Shen, P.; Moore, C.S.; Michel, L.; Althekair, F.; Rajasekharan, S.; et al. Proinflammatory GM-CSF-producing B cells in multiple sclerosis and B cell depletion therapy. *Sci. Transl. Med.* **2015**, *7*, 310ra166. [CrossRef] [PubMed]

38. Tracy, S.I.; Kakalacheva, K.; Lunemann, J.D.; Luzuriaga, K.; Middeldorp, J.; Thorley-Lawson, D.A. Persistence of Epstein-Barr virus in self-reactive memory B cells. *J. Virol.* **2012**, *86*, 12330–12340. [CrossRef] [PubMed]

39. Song, Z.Y.; Yamasaki, R.; Kawano, Y.; Sato, S.; Masaki, K.; Yoshimura, S.; Matsuse, D.; Murai, H.; Matsushita, T.; Kira, J. Peripheral blood T cell dynamics predict relapse in multiple sclerosis patients on fingolimod. *PLoS ONE* **2015**, *10*, e0124923. [CrossRef] [PubMed]

40. Grützke, B.; Hucke, S.; Gross, C.C.; Herold, M.V.; Posevitz-Fejfar, A.; Wildemann, B.T.; Kieseier, B.C.; Dehmel, T.; Wiendl, H.; Klotz, L. Fingolimod treatment promotes regulatory phenotype and function of B cells. *Ann. Clin. Transl. Neurol.* **2015**, *2*, 119–130. [CrossRef] [PubMed]

41. Wu, Q.; Wang, Q.; Mao, G.; Dowling, C.A.; Lundy, S.K.; Mao-Draayer, Y. Dimethyl Fumarate Selectively Reduces Memory T Cells and Shifts the Balance between Th1/Th17 and Th2 in Multiple Sclerosis Patients. *J. Immunol.* **2017**, *198*, 3069–3080. [CrossRef] [PubMed]

42. Ireland, S.J.; Guzman, A.A.; O'Brien, D.E.; Hughes, S.; Greenberg, B.; Flores, A.; Graves, D.; Remington, G.; Frohman, E.M.; Davis, L.S.; et al. The effect of glatiramer acetate therapy on functional properties of B cells from relapsing-remitting multiple sclerosis. *JAMA Neurol.* **2014**, *71*, 1421–1438. [CrossRef] [PubMed]

43. Gandoglia, I.; Ivaldi, F.; Laroni, A.; Benvenuto, F.; Solaro, C.; Mancardi, G.; Kerlero de Rosbo, N.; Uccelli, A.M. Teriflunomide treatment reduces B cells in patients with MS. *Neuroimmunol. Neuroinflamm.* **2017**, *4*, e403. [CrossRef] [PubMed]

44. Gebhardt, T.; Wakim, L.M.; Eidsmo, L.; Reading, P.C.; Heath, W.R.; Carbone, F.R. Memory T cells in nonlymphoid tissue that provide enhanced local immunity during infection with herpes simplex virus. *Nat. Immunol.* **2009**, *10*, 524–530. [CrossRef] [PubMed]

45. Bromley, S.K.; Yan, S.; Tomura, M.; Kanagawa, O.; Luster, A.D. Recirculating memory T cells are a unique subset of CD4+ T cells with a distinct phenotype and migratory pattern. *J. Immunol.* **2013**, *190*, 970–976. [CrossRef] [PubMed]

46. Casey, K.A.; Fraser, K.A.; Schenkel, J.M.; Moran, A.; Abt, M.C.; Beura, L.K.; Lucas, P.J.; Artis, D.; Wherry, E.J.; Hogquist, K.; et al. Antigen-independent differentiation and maintenance of effector-like resident memory T cells in tissues. *J. Immunol.* **2012**, *188*, 4866–4875. [CrossRef] [PubMed]

47. Mackay, L.K.; Rahimpour, A.; Ma, J.Z.; Collins, N.; Stock, A.T.; Hafon, M.L.; Vega-Ramos, J.; Lauzurica, P.; Mueller, S.N.; Stefanovic, T.; et al. The developmental pathway for CD103(+)CD8+ tissue-resident memory T cells of skin. *Nat. Immunol.* **2013**, *14*, 1294–1301. [CrossRef] [PubMed]

48. Gaide, O.; Emerson, R.O.; Jiang, X.; Gulati, N.; Nizza, S.; Desmarais, C.; Robins, H.; Krueger, J.G.; Clark, R.A.; Kupper, T.S. Common clonal origin of central and resident memory T cells following skin immunization. *Nat. Med.* **2015**, *21*, 647–653. [CrossRef] [PubMed]

49. Wakim, L.M.; Woodward-Davis, A.; Liu, R.; Hu, Y.; Villadangos, J.; Smyth, G.; Bevan, M.J. The molecular signature of tissue resident memory CD8 T cells isolated from the brain. *J. Immunol.* **2012**, *189*, 3462–3471. [CrossRef] [PubMed]

50. Wakim, L.M.; Woodward-Davis, A.; Bevan, M.J. Memory T cells persisting within the brain after local infection show functional adaptation to the tissue of residence. *Proc. Natl. Acad. Sci. USA* **2010**, *107*, 17872–17879. [CrossRef] [PubMed]

51. Hussain, R.Z.; Hayardeny, L.; Cravens, P.C.; Yarovinsky, F.; Eagar, T.N.; Arellano, B.; Deason, K.; Castro-Rojas, C.; Stüve, O. Immune surveillance of the central nervous system in multiple sclerosis–relevance for therapy and experimental models. *J. Neuroimmunol.* **2014**, *276*, 9–17. [CrossRef] [PubMed]

52. Bhargava, P.; Gocke, A.; Calabresi, P.A. 1,25-Dihydroxyvitamin D3 impairs the differentiation of effector memory T cells in vitro in multiple sclerosis patients and healthy controls. *J. Neuroimmunol.* **2015**, *279*, 20–24. [CrossRef] [PubMed]

MDPI
St. Alban-Anlage 66
4052 Basel
Switzerland
Tel. +41 61 683 77 34
Fax +41 61 302 89 18
www.mdpi.com

Brain Sciences Editorial Office
E-mail: brainsciences@mdpi.com
www.mdpi.com/journal/brainsciences

www.ingramcontent.com/pod-product-compliance
Lightning Source LLC
Chambersburg PA
CBHW051858210326
41597CB00033B/5943